OPTIMIZATION AND COMPUTATIONAL LOGIC

WILEY-INTERSCIENCE
SERIES IN DISCRETE MATHEMATICS AND OPTIMIZATION

ADVISORY EDITORS

RONALD L. GRAHAM
AT & T Bell Laboratories, Murray Hill, New Jersey, U.S.A.

JAN KAREL LENSTRA
Centre for Mathematics and Computer Science, Amsterdam, The Netherlands
Erasmus University, Rotterdam, The Netherlands

ROBERT E. TARJAN
Princeton University, New Jersey, and
NEC Research Institute, Princeton, New Jersey, U.S.A.

A complete list of titles in this series appears at the end of this volume

Optimization and Computational Logic

KEN McALOON
CAROL TRETKOFF

A WILEY-INTERSCIENCE PUBLICATION
JOHN WILEY & SONS, INC.
New York • Chichester • Brisbane • Toronto • Singapore

This text is printed on acid-free paper.

Copyright © 1996 by John Wiley & Sons, Inc.

All rights reserved. Published simultaneously in Canada.

Reproduction or translation of any part of this work beyond that permitted by Section 107 or 108 of the 1976 United States Copyright Act without the permission of the copyright owner is unlawful. Requests for permission or further information should be addressed to the Permissions Department, John Wiley & Sons, Inc., 605 Third Avenue, New York, NY 10158-0012

Library of Congress Cataloging in Publication Data:

McAloon, Ken.
    Optimization and computational logic / Ken McAloon and Carol Tretkoff.
      p.   cm. — (Wiley-Interscience series in discrete mathematics and optimization)
    "A Wiley-Interscience publication."
    Includes bibliographical references and index.
    ISBN 0-471-11533-9 (cloth : alk. paper)
    1. Linear programming.   2. Mathematical optimization.   3. Logic, symbolic and mathematical.   4. Computer science.   5. Artificial intelligence.   6. Programming languages.   I. Tretkoff, Carol.   II. Title.
III. Series.
T57.74.M39   1996
519.7'2—dc20   96-23734

Printed in the United States of America

10 9 8 7 6 5 4 3 2 1

Contents

Preface   ix

1 Constraints and Optimization   1

    1.1 Linear constraints   1
    1.2 Feasible regions and the witness point   6
    1.3 Formulating an LP model   13
    1.4 Data types   17
    1.5 Affine expressions   24
    1.6 Symbolic constants   32

2 Loop Constructs   39

    2.1 and loops, sigma loops   39
    2.2 Basic roundoff techniques   45
    2.3 Bounding constraints   53

3 Structured Linear Programming   63

    3.1 Program structure and procedures   63
    3.2 Parameters   67
    3.3 Sparse data   79
    3.4 Stochastic programming   89

4 Conjunction and Implication   97

    4.1 Conjunction   97
    4.2 Implication   103
    4.3 Network models   112
    4.4 Smoothing out extreme solutions   128

5 Conditional Disjunction — 139

 5.1 A programming interpretation of disjunction   139
 5.2 Goal programming   150
 5.3 Pattern recognition   155

6 Negation — 165

 6.1 Negation-as-failure   165
 6.2 Evaluation functions and lookaheads   170
 6.3 Randomization and local search   175
 6.4 Genetic algorithms   190

7 Sensitivity Analysis — 199

 7.1 Shadow prices and reduced costs   200
 7.2 Parametric analysis   211

8 Backtracking — 227

 8.1 Persistent disjunction   227
 8.2 Generate-and-test   233
 8.3 Constrain-and-generate   243

9 Classical Disjunction and Combinatorially Hard Problems — 253

 9.1 Classical disjunction   253
 9.2 Disjunctive programs   260
 9.3 P, NP, and co-NP   276

10 Soundness and Completeness — 283

 10.1 Loop invariants   283
 10.2 Discrete models   301
 10.3 Recursion and nesting   312
 10.4 Unit resolution   320

11 Depth-First Branch-and-Bound Search — 331

 11.1 Finding all solutions   331
 11.2 Discrete branch-and-bound search   337
 11.3 Linear relaxations and branch-and-bound search   346
 11.4 State space optimization   363

12 The Injury Method — 371

 12.1 The eureka effect   371
 12.2 Gaps   387
 12.3 Branching and duality   391

13 Tightening the Linear Relaxation — **401**

    13.1 Fuzzy booleans   401
    13.2 Valid cuts   412
    13.3 Branch-and-cut   425
    13.4 Penalties   433

14 Further Search Methods — **443**

    14.1 Iterative deepening   444
    14.2 Breadth-first search   455
    14.3 A* search and IDA* search   458
    14.4 Breadth-first branch-and-bound search   466

15 Mathematical Underpinnings — **475**

    15.1 Dual multipliers   475
    15.2 Geometry and algebra   483
    15.3 The revised simplex method   500

Getting Started — **513**

References — **515**

List of Models — **525**

Name Index — **527**

Subject Index — **531**

Preface

Modern decision making often involves several disciplines including operations research, artificial intelligence and mathematical logic. This book brings them together using the methodology of computer science. The applications discussed include resource allocation, scheduling, pattern recognition, classical and probabilistic logic, financial planning, expert systems, and many others. The text presents concepts, programming templates, and tools for the task of attacking thorny problems that are encountered again and again. For the programming examples, we employ a small language 2LP, which stands for "linear programming and logic programming." A Windows version of the software accompanies this book; other versions and additional materials are available on the World Wide Web: http://www.brooklyn.cuny.edu/lbslab.

This text is written for practitioners and students of artificial intelligence, operations research or decision support software. It is designed to be read broadly or selectively. The prerequisite is a certain amount of computational maturity.

The opening chapters of the book deal with linear programming which has been a mainstay of decision support systems for many years now. Although linear programming was one of the first success stories of digital computing, the word programming in this context means planning or scheduling and not computer programming in the contemporary sense of the term. The 2LP language is designed to integrate linear programming into the modern programming scheme of things, using the technology of constraint logic programming. As a result the modeling power of constraints and logic as well as basic software engineering techniques can be used for linear programming and extensions such as goal programming and stochastic programming.

Then the emphasis shifts to search techniques and heuristics. It turns out that the 2LP embedding of linear programming practice into a programming language environment, in fact, embeds the larger disciplines of mixed integer programming and disjunctive programming. As a result linear programming methods can be integrated into artificial intelligence techniques, and artificial intelligence methods can be readily applied to a broader range of applications than is usually the case. To solve hard problems, we use variations on old and new search strategies such as randomized local search, genetic algorithms, constrain-and-generate, marking-and-trailing, branch-and-bound, fuzzy logic, branch-and-cut, and iterative deepening. An important theme throughout is the back-and-forth between discrete and continuous techniques.

The closing chapter is devoted to the mathematical underpinnings of linear programming. The basic theorems are proved and the revised simplex method is presented, which brings together the topics of search, discrete methods and continuous mathematics.

We would like to thank the research communities of artificial intelligence, mathematics, theoretical computer science, operations research, management science, and constraint-based programming. Members of these communities have been uniformly helpful and encouraging. We would also like to thank our colleagues and our students at Brooklyn College and the CUNY Graduate Center. A special note of thanks must go to Jean-Louis Lassez, a tireless supporter of the Cartesian approach to computational science. It is also a pleasure to thank the Office of Naval Research and the National Science Foundation; their research support was invaluable.

Naturally the text builds on previous material as it proceeds. The *Leitfaden* on the facing page outlines the logical dependencies among sections of the book.

Ken McAloon
Carol Tretkoff

Preface

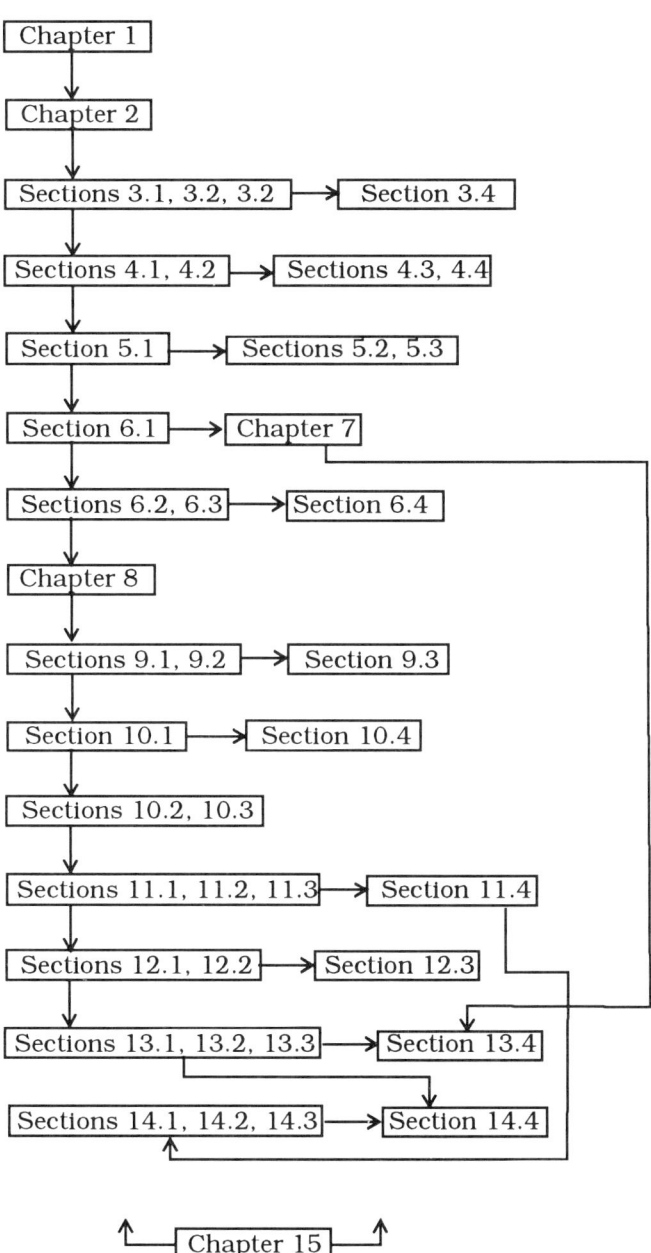

OPTIMIZATION AND COMPUTATIONAL LOGIC

1

Constraints and Optimization

The subject matter of this text is constructed from logic, search, combinatorial algorithms, linear programming, and computer science. In this chapter we begin the presentation of a unified view of the common kernel of these areas using the programming language 2LP, which stands for "linear programming and logic programming."

1.1 Linear constraints

Our starting point is the linear constraint. In algebra, a student encounters linear constraints in the form of simultaneous linear equations. A typical application goes something like this: An oil company wants to produce 18,000 gallons of oil by blending refined oil and crude oil. The amount of refined oil in the blend must be equal to 2/3 the amount of crude oil. How many gallons of each kind of oil are needed? Letting x denote the refined oil and y the crude oil leads to the equations

$$x + y = 18000$$
$$3x = 2y$$

These equations are *declarative* in the sense that once they are postulated they remain in force throughout the process of solving the problem. Also when the equations are formulated, the variables x and y do not have fixed values but range over the real numbers. Solving the equations requires a procedure or algorithm that prescribes a sequence of actions to take. Typically, in the procedural part of the process to solve a system of linear equations, one of the equations is rewritten to express one variable in terms of the rest. Then this variable is eliminated in the remaining equations using the expression found for it. This process is repeated until all the variables are eliminated or no equations remain. At this point one has the information to determine if there are no solutions to the system of equations, exactly one solution, or an infinite number of solutions. In our example the equations have one solution. The method of solving linear equations by eliminating variables is called *Gaussian elimination*.

If the linear constraints are all equations, the subject is known as *linear algebra*. When inequalities involving ≤ or ≥ are added, the subject becomes *linear programming*. For example, suppose that the above blending problem is changed: The amount of refined oil in the blend still must be 2/3 that of the crude, but now the amount of oil to be produced must be a quantity between 18,000 and 30,000 gallons, and the blend must have at least 8500 gallons of refined oil but no more than 16,500 gallons of crude. The conditions on the oils in the blend now translate to

$$x + y \geq 18000$$
$$x + y \leq 30000$$
$$x \geq 8500$$
$$y \leq 16500$$
$$3x = 2y$$

Determining values for x and y that satisfy these conditions requires more steps than solving one set of linear equations. For solving complex systems of constraints, powerful procedural algorithms must be used. The oldest go back to the work of Fourier in the early 1800s who introduced two techniques. One is now known as *Fourier-Motzkin elimination* and the other as the *simplex method*. The work of Dantzig in the 1940s and 1950s made the simplex method a powerful computer-based algorithmic tool and helped create the modern science of operations research. An important theoretical breakthrough was made by Khachian (1979) using the *ellipsoid method*. A great deal of work is currently being done on *interior point methods*, propelled by the theoretical and algorithmic contribution of Karmarkar (1984). Fourier-Motzkin elimination and the simplex method are discussed in detail in Chapter 15; theoretical properties of the other methods are discussed in Section 9.3.

In formal mathematical notation, a *linear constraint* is defined to be an expression of one of the following forms:

$$a_1 x_1 + \ldots + a_n x_n \leq b$$

$$a_1 x_1 + \ldots + a_n x_n = b$$

$$a_1 x_1 + \ldots + a_n x_n \geq b$$

In these expressions the a_i and b are fixed values and the x_i are variables. The first goal in a linear programming problem is to determine if the constraints have a solution and, if so, typically to find a numerical solution that maximizes or minimizes the value of one of the variables subject to the constraints.

1.1 Linear constraints

The variables in linear constraints have a continuous range and so are called *continuous variables*. An important thing about continuous variables and constraints is that the inequality constraints as well as the equality constraints are declarative, and once stated, they all stay in force throughout the application. The notion of continuous variable is supported in 2LP by the data type continuous, and constraints on these variables are maintained in a declarative way by the programming language. Returning to the blending example, the variables x and y are continuous variables, and we can formulate this problem as a 2LP program. In programming one uses more meaningful identifiers and so we will replace x and y by the names Refined and Crude. In 2LP we would set up the constraints by means of the following code:

```
2lp_main()
{
continuous Refined, Crude;

    Refined + Crude >= 18000;
    Refined + Crude <= 30000;
    Refined >= 8500;
    Crude <= 16500;
    3*Refined == 2*Crude;
}
```

We have used initial capitals for the identifiers of continuous variables declared in the program, a practice we will follow for readability. Following C and C++ conventions, 2LP uses == as the relational operator for equality and reserves = for the assignment operator.

The workings of this 2LP program can be interpreted geometrically. Each constraint limits the variables Refined and Crude to a more restricted region of two-dimensional space. In Figure 1.1 the successive restrictions are sketched; for reasons of space, the axes are labeled in thousands. Thus as the program progresses, the constraints combine to determine a sequence of geometric objects which are regions in a Cartesian space. Communication with these geometric objects is an important part of 2LP programming. In this example, at termination the region is a line segment. Internally the 2LP system's constraint solver uses the simplex method. This method determines if a system of linear constraints has a solution and yields important information about the solution set.

A region defined by an inequality constraint is called a *half space*; a region defined by an equality constraint is called a *hyperplane*. In three dimensions, hyperplanes are simply called *planes* and in two dimensions they are called *lines*.

The results of the procedural analysis of the constraints carried out by the internal 2LP constraint solver can be accessed by means of print statements and other built-in procedures and functions. In particular,

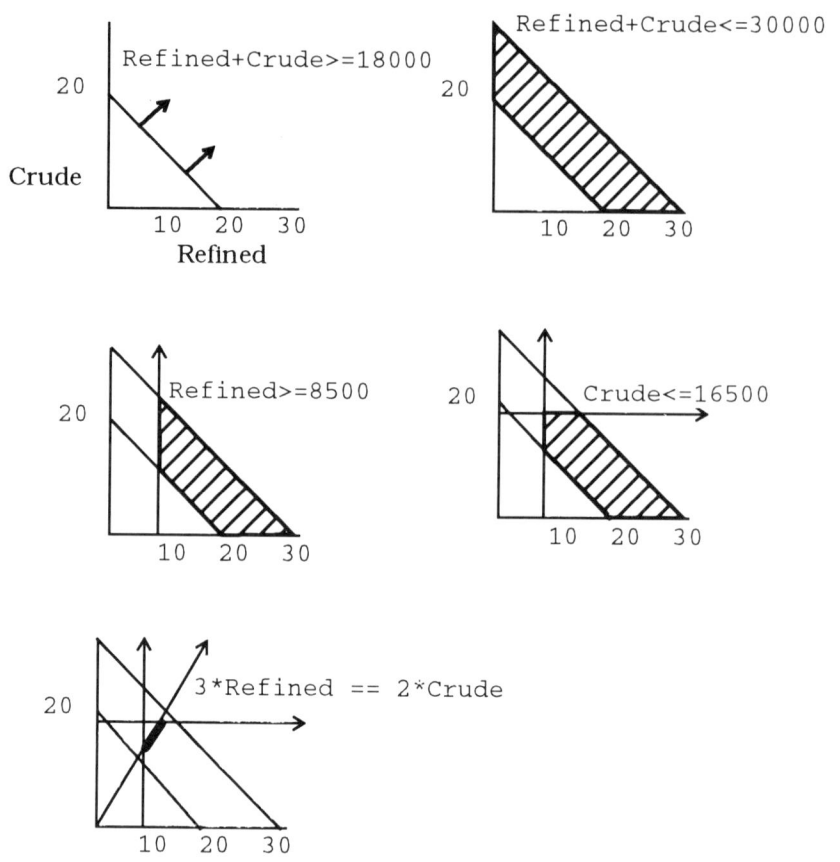

FIGURE 1.1 Adding constraints.

to extract a solution to the blending problem, we can add a print statement that will output numerical values for Refined and Crude and satisfy the constraints of the program:

```
2lp_main()
{
continuous Refined, Crude;

    Refined + Crude >= 18000;
    Refined + Crude <= 30000;
    Refined >= 8500;
    Crude <= 16500;
    3*Refined == 2*Crude;
```

1.1 Linear constraints

```
        printf("Refined is %.2f and Crude is %.2f\n",
                                            Refined,Crude);
}
```

The printout to the screen is

Parsing ...
Parse succeeded

Refined is 8500.00 and Crude is 12750.00

... Program Completed

Here, the message Parsing... means that the interpreter is carrying out syntactic analysis of the model and Parse succeeded means that the code is syntactically correct. The third line of output is produced by the `printf` statement; the syntax and semantics of this statement are the same as in C. The last line of output, the completion message, is self-explanatory. We will usually omit the parsing messages and the completion message in listing program output.

There are three interwoven themes here: the declarative meaning of constraint, the geometric object defined by constraints, and the procedural programming apparatus that supports constraints. These themes will remain important throughout this book. Other themes will be added to this list, such as programming with logic and interweaving declarative and procedural methods.

By convention in a 2LP program the continuous variables are constrained to be nonnegative and so vary over the interval $[0,+\infty)$. Thus it is not necessary to specify X >= 0 for a continuous variable X. Geometrically this means that the regions associated with a program's execution will be restricted to the first quadrant in the two-dimensional case and to the first orthant in higher dimensions.

Exercises

1.1.1. Sketch the sequence of feasible regions generated by the following code:

```
2lp_main()
{
continuous X,Y;

        X + 3*Y <= 15;
        5*X - 3*Y >= 0;
        X <= Y;
}
```

1.2 Feasible regions and the witness point

As we saw in Figure 1.1, as constraints accumulate during the run of a 2LP program, they combine to define a geometric region. The following 2LP program leads to a simple polygon in two-dimensional space.

```
2lp_main()
{
continuous X,Y;

    -X + Y <= 4;
    X + 4*Y <= 36;
    2*X + Y <= 23;
    3*X - 2*Y <= 24;
    X + Y >= 4;
}
```

Graphing the polygon resulting from these five constraints gives Figure 1.2. Any point inside the polygon labeled "Feasible Region" satisfies the constraints.

When each of the constraints is added in the 2LP program above, the constraint *succeeds*. When a constraint succeeds, we say that the set of constraints accumulated thus far is *consistent* or *feasible* and the set of points satisfying them is called the *feasible region*. Another name for a feasible region defined by linear constraints is *polyhedral set* or, more simply, *polyhedron*. However, if a new constraint is added and no points lying in the current feasible region satisfy the new constraint, then the new constraint *fails*. Thus, when the following 2LP program is run,

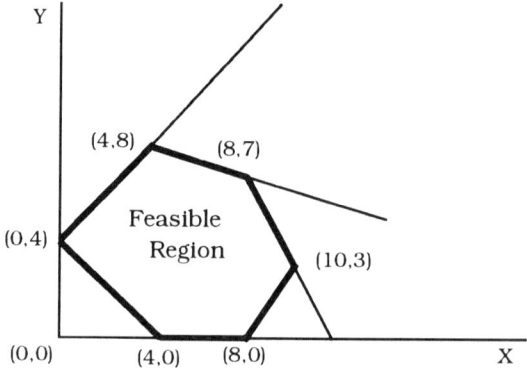

FIGURE 1.2 A constrained region in two-dimensional space.

1.2 Feasible regions and the witness point

```
2lp_main()
{
continuous X,Y;

    /* The first five constraints form the feasible region
         of FIGURE 1.2 */

    -X +  Y <=  4;
     X + 4*Y <= 36;
    2*X +  Y <= 23;
    3*X - 2*Y <= 24;
     X +  Y >=  4;
     Y >= X + 10;    // Constraint outside the feasible
                     // region of the other constraints
}
```

we obtain the output

No feasible solution exists

The reason there is no solution is illustrated in Figure 1.3. However if the last constraint in the above program is changed to Y <= X + 10, then the feasible region remains the feasible region of Figure 1.2.

Note that there are two types of comments in the above program. Comments can be written anywhere in the code of a 2LP program. A double slash // means that the rest of the line it appears on is a comment; a /* opens a comment that must be closed by a */.

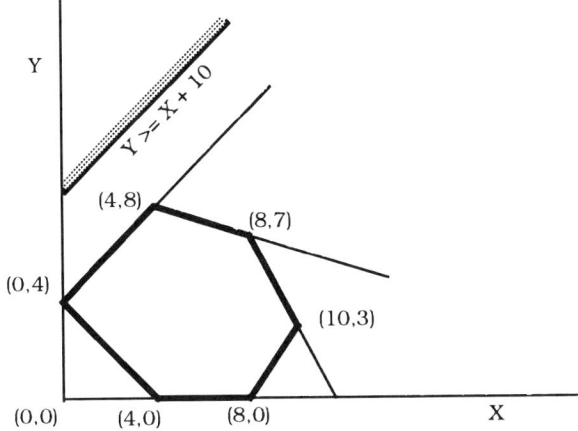

FIGURE 1.3 An infeasible region.

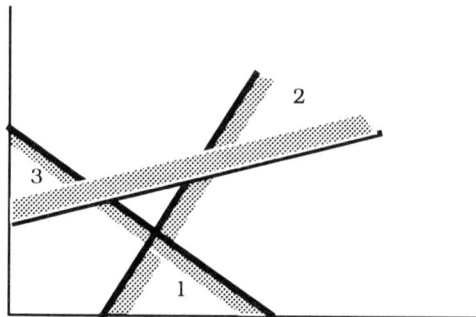

FIGURE 1.4 Interrelationships among constraints. The shading indicates the region where each constraint is satisfied.

In the example of Figure 1.3, the constraint that caused the infeasibility is in fact inconsistent with the first constraint of the program. However, it is possible that the cause of the infeasibility is not so simple and that it is due to the interrelationship among several of the constraints. Figure 1.4 illustrates the interrelationship between three constraints. Any two constraints define one of the feasible numbered regions. However, taken together, the three constraints are not feasible.

In a 2LP program the constraints on the continuous variables define a polyhedron in n-dimensional space where n is the number of continuous variables. A corner point on a polyhedron is called a *vertex* or an *extreme point*. (Algebraic definitions are given in Chapter 15.) The polyhedron defined by the constraints in a 2LP model will always have a vertex because continuous variables are restricted to the first orthant; thus the initial polyhedron has a vertex and adding constraints preserves this property. The simplex-based constraint solver of 2LP maintains the current polyhedron together with a distinguished vertex on the polyhedron; this vertex is called the *witness point*.

Since a witness point is always maintained during run time, the 2LP system has a concrete floating point value associated with each continuous variable. These witness point coordinates are made available to the program through built-in procedures and functions. For example, when the `printf` statement is called with a continuous variable as a parameter, the value printed to the screen will be the value determined by the witness point.

In computer science terminology, the polyhedron or feasible region defined by the constraints and maintained by the linear programming solver is an *object*. The notion of a variable of type `continuous` is an example of an *abstract data type*. The program communicates with this object by sending it "messages" and receiving "replies." Basically a constraint on continuous variables is a message that asks the polyhedron

1.2 Feasible regions and the witness point

to alter its shape. A reply of success is made if the constraint is feasible; otherwise, failure is the reply.

In the next example, we illustrate how the witness point can move about as the feasible region defined by the constraints evolves. Consider the following 2LP code:

```
2lp_main()
{
continuous X,Y,Z;
    printf("The coordinates of the witness point are\n");
    printf("( %.1f, %.1f, %.1f )\n",X,Y,Z);

    X + Y + Z <= 1;
    printf("The coordinates of the witness point are\n");
    printf("( %.1f, %.1f, %.1f )\n",X,Y,Z);

    X + Y == 1;
    printf("The coordinates of the witness point are\n");
    printf("( %.1f, %.1f, %.1f )\n",X,Y,Z);

    X == Y;
    printf("The coordinates of the witness point are\n");
    printf("( %.1f, %.1f, %.1f )\n",X,Y,Z);
}
```

The output of the above program will be

```
The coordinates of the witness point are
( 0.0, 0.0, 0.0 )
The coordinates of the witness point are
( 0.0, 0.0, 0.0 )
The coordinates of the witness point are
( 1.0, 0.0, 0.0 )
The coordinates of the witness point are
( 0.5, 0.5, 0.0 )
```

The declarations

```
continuous X,Y,Z;
```

initialize the polyhedron to be the positive orthant of three-dimensional space as in Figure 1.5 (a). Here the witness point must be (0, 0, 0) since this is the only vertex. When the constraint X + Y + Z <= 1 is generated, the polyhedron becomes the four-sided solid of Figure 5.1 (b); the witness point can remain at (0, 0, 0) since it is still a vertex of the feasible region. Upon adding the constraint X + Y == 1, the witness point must change to an end point of the line segment joining the points (1, 0, 0) and (0, 1, 0); as illustrated in Figure 1.5 (c), the point chosen by the sys-

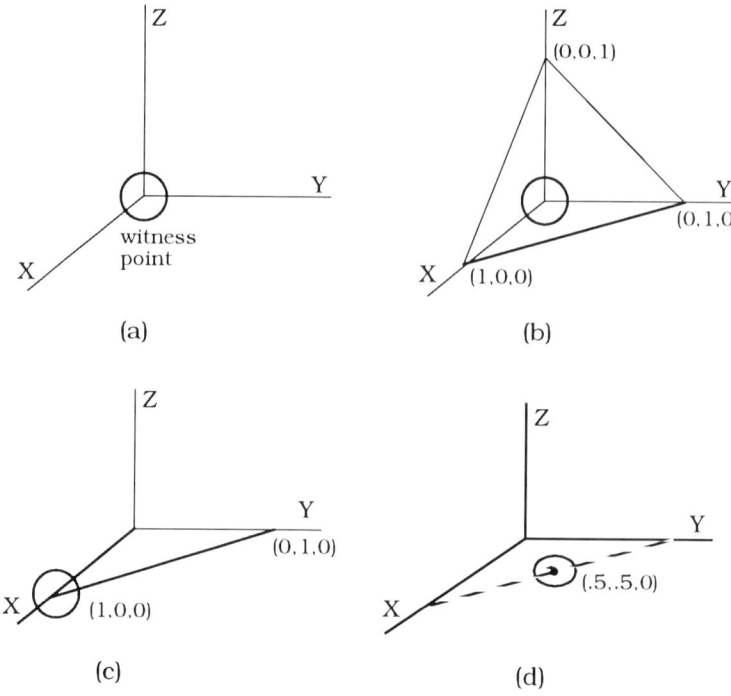

FIGURE 1.5 (a) The positive orthant, (b) the triangular solid defined by the constraints, (c) the line joining (1, 0, 0) and (0, 1, 0), and (d) the point (.5,.5, 0).

tem is (1, 0, 0). Adding the constraint X == Y means that the constraints now reduce the polyhedron to the single point (.5, .5, 0), and hence the witness point must be this point as in Figure 1.5 (d).

In this last example the constraints combined to determine a feasible region consisting of a single point; this point therefore provides the unique solution to the constraints. In other situations the feasible region will contain an infinite number of points, but we will often want a witness point that is optimal relative to various criteria.

An application that requires the "best possible" solution is called an *optimization problem*. Part of the power of the simplex method for solving a system of linear constraints is the ability to determine a solution to the constraints that yields the largest or smallest possible value for one of the continuous variables of the system. The mathematics of the situation is such that when a variable has an optimal value, this value is assumed at a vertex of the feasible region. The simplex method can thus determine a witness point that is optimal for the application and allow the program to access the coordinates of this point. In 2LP the built-in operator

1.2 Feasible regions and the witness point

```
max: X;
```

takes a variable of type `continuous` as its parameter; its effect is to move the witness point to a vertex of the feasible region that has the maximum possible X-coordinate. Similarly the built-in operator

```
min: X;
```

moves the witness point to a vertex of the feasible region that has the minimum possible X-coordinate. The variable X to be optimized is called the *objective function*.

If in the blending example of Section 1.1, we wish to find a blend that uses as much crude oil as possible, we would write the program as follows:

```
2lp_main()
{
continuous Refined,Crude;

    Refined + Crude >= 18000;
    Refined + Crude <= 30000;
    3*Refined == 2*Crude;
    Refined >= 8500;
    Crude <= 16500;
    max: Crude;

    printf("Refined is %.2f and Crude is %.2f",
                                    Refined,Crude);
}
```

The output is

Refined is 11000.00 and Crude is 16500.00

which is different from that obtained previously. Note, however, that this point (11000, 16500) is still an end point of the line segment defined by the constraints of Figure 1.1.

The operators `min: X` and `max: X` leave the current polyhedron unchanged and send the message to move the witness point to a vertex at which the value of the X coordinate is optimized. It is possible that no such optimal value exists, and linear programming systems will detect this. Thus the code

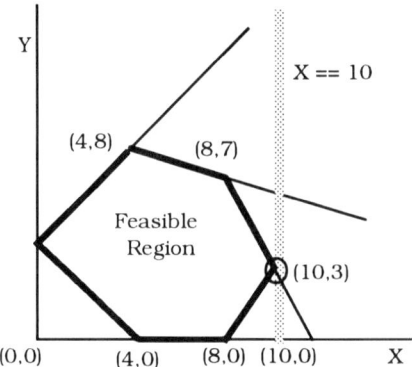

FIGURE 1.6 The vertex that maximizes X is shown as (10, 3).

```
2lp_main()
{
continuous X,Y,Z;
    max: Z;
}
```

will produce a message to the effect that Z has no possible maximum value and fail. This happens because the variable Z can take arbitrarily large values in the feasible region, which in this case is the entire first orthant. The operation `min: Z` would have succeeded.

To illustrate further the point that the optimum value of a continuous variable has to occur at a vertex of the feasible region, let us look again at the polygon created in Figure 1.2. Suppose that we add the call `max: X` to the constraints that define this region. Then we want the greatest value of c such that the line `X == c` intersects the polygon. The power of linear programming is that it enables us to find the optimal value of c such that `X == c` intersects the polygon and also to find all the coordinates of a vertex at which this happens. In this example this best value for c is 10. In Figure 1.6 the graph of `X == 10` has been drawn with a gray line. The maximum occurs at the vertex (10, 3).

Exercises

1.2.1. Graph the following constraints in the positive quadrant of the Cartesian plane, and then use the graph to find the maximum value of the objective function x and the objective function y. Code the problem in 2LP, and check the answer.

   (a) $x \geq 0,\ y \geq 1$,
           $3x - 2y \geq 0,\ x + 2y \leq 4$

1.3 Formulating an LP model

(b) $\quad x \geq 0, y \geq 0,$
$\quad x + y \leq 4, 2x + y \leq 5, 4y \geq x + 2$

1.2.2. Put the following constraints in a 2LP program. Check that changing their order will change the sequence of witness points and produce different final witness points.

```
2*Y  >=  .5 - X;
Y    >=  1  - X;
```

1.3 Formulating an LP model

The goal of this section is to discuss how to set up applications in a linear programming format and then use the linear programming solver built into 2LP to obtain a solution. When trying to solve a problem, a model or program is formulated to reflect the "real-world" situation. In this text we will consider many different situations, starting here with applications that can be done using simple linear programming models. Later we will work with more complex situations that require much more elaborate techniques. Formulating a correct model requires attention to the nature and structure of the problem. A classic example where linear programming is effective is in determining a food regimen that meets given nutritional requirements at an acceptable cost, a *diet problem*. This type of application was formulated by Stigler (1945) and predates the modern development of the simplex method.

Model 1.1 The Wildlife Conservation Park

A specialist at a wildlife conservation park has the task of providing a nutritious snack for the park's omnivorous animals. The snack will be composed of bonemeal, corn, and soybean, and the food will be shipped to the park in 125-pound sacks. Each meal must include certain percentages of three different nutritional elements. The nutritional requirements are that each meal must be composed of

1. Not more than 1.2% calcium.
2. But at least 0.8% calcium.
3. At least 22% protein.
4. At most 5% fiber.

Table 1.1 gives the nutritional content of the ingredients that will go into the feed mixture. The nutritional content of each of these foods measures how much of that foodstuff is composed of the nutrient; thus a pound of bonemeal is 38% calcium, while corn is .1% calcium.

In addition, the cost of a 125-pound sack must be kept as low as possible. The costs in dollars per pound of the three ingredients are

TABLE 1.1 Nutrition Information per Pound

	Bonemeal	Corn	Soybean
Calcium	0.38	0.001	0.002
Protein	0.0	0.09	0.50
Fiber	0.0	0.02	0.08

TABLE 1.2 Cost Information

Bonemeal	Corn	Soybean
.164	.463	.125

given in Table 1.2. The challenge is to mix quantities of the ingredients so as to meet the nutritional constraints on each 125-pound sack and then to minimize its cost. Simply by looking at the data, one might find a solution that satisfies the four nutritional requirements above. For example, we see that the mix cannot contain more than 3.95 pounds of bonemeal because then the animal would receive too much calcium. Continuing along these lines, one might try a mix of 3 pounds of bonemeal, 75 pounds of corn, and 47 pounds of soybeans. This mixture satisfies the nutritional constraints but costs $41.09. One would hope to do somewhat better, but quite a bit of manual analysis of this kind would be required to bring this figure close to the true minimum.

To solve this problem with linear programming in 2LP, we let the variables Bonemeal, Corn, and Soybean denote the number of pounds of each ingredient to be included in the 125-pound sack and let Cost denote the cost of producing the feed sack. The variables are declared to be of type continuous. This type is required when all that is known *a priori* about the variables is that the desired values lie in some continuous range. In this case the range is $[0,+\infty)$.

We must have the sum of the three variables equal to 125 pounds, that is,

```
Bonemeal + Corn + Soybean == 125;
```

Note again that when variables of type continuous are used, equality is written with ==. The dietary requirements above lead to four constraints. Since the snack can be composed of at most 1.2% calcium and since the sacks are each 125 pounds, there must be less than .012*125 pounds of calcium in the sack. The calcium contribution of each foodstuff is in the Calcium row of the table. This leads to a constraint that describes how the limit on calcium content affects the contents of the feed sack:

```
.38*Bonemeal + .001*Corn + .002*Soybean <= .012*125;
```

1.3 Formulating an LP model

The code corresponding to the fact that the animal needs at least 0.8% calcium is

```
.38*Bonemeal + .001*Corn + .002*Soybean >= .008*125;
```

The remaining two requirements are

```
0.0*Bonemeal + .09*Corn + .50*Soybean >= .22*125;
```

and

```
0.0*Bonemeal + .02*Corn + .08*Soybean <= .05*125;
```

The 2LP program for this problem consists basically of writing the constraints given above and the equation for the cost:

```
Cost == .164*Bonemeal + .463*Corn + .125*Soybean;
```

It is also part of the model that the ingredients cannot take negative values:

```
    Bonemeal >= 0.0;
    Corn >= 0.0;
    Soybean >= 0.0;
```

However, these constraints are assumed by default and do not need to be written explicitly in the program. Here is the complete 2LP code for the model:

```
2lp_main() // Wildlife Conservation Park
{
continuous Cost,Bonemeal,Corn,Soybean;

    Bonemeal + Corn + Soybean == 125;
    .38*Bonemeal + .001*Corn + .002*Soybean <=  .012*125;
    .38*Bonemeal + .001*Corn + .002*Soybean >=  .008*125;
    0.0*Bonemeal +  .09*Corn +  .50*Soybean >=  .22*125;
    0.0*Bonemeal +  .02*Corn +  .08*Soybean <=  .05*125;
    Cost == .164*Bonemeal + .463*Corn + .125*Soybean;

    min: Cost;

    // Output
    printf("Use %.2f pounds Bonemeal \n",Bonemeal);
    printf("Use %.2f pounds Corn \n",Corn);
```

```
      printf("Use %.2f pounds Soybean \n",Soybean);
      printf("The Cost will be $%.2f\n",Cost);
}
```

If this code is executed, the following messages will come to the screen:

Use 3.46 pounds Bonemeal
Use 57.89 pounds Corn
Use 63.65 pounds Soybean
The Cost will be $35.33

The model determines the best possible mix to keep the cost down. This result is more than 10% better than the one we found by hand.

<u>*End of Model 1.1*</u>

Model 1.1 is also an example of a *blending problem*, one that requires an optimal mix of ingredients. There are many examples like this where linear programming is a very effective tool.

Differential calculus is also a powerful tool for solving optimization problems; however, it provides but little help in the context of linear programming applications. Let us see why. If we were doing Model 1.1 as a calculus problem, we would consider the function

```
      Cost == .164*Bonemeal + .463*Corn + .125*Soybean;
```

where `Bonemeal`, `Corn` and `Soybean` vary over the region defined by the nutritional constraints:

```
      Bonemeal + Corn + Soybean == 125;
      .38*Bonemeal +.001*Corn + .002*Soybean <= .012*125;
      .38*Bonemeal +.001*Corn + .002*Soybean >= .008*125;
             .09*Corn + .5*Soybean >= .22*125;
             .02*Corn + .08*Soybean <= .05*125;
```

We would compute the partial derivatives of `Cost` with respect to the other variables. Then we would set the partial derivatives to zero in order to find a point where the partial derivatives all vanish. But, since `Cost` is a linear function of `Bonemeal`, `Corn`, and `Soybean`, the partial derivatives are constant; in fact, here they are equal to .164, .463, and .125 and do not vanish anywhere. From this, we can conclude that the optimal values of the objective function must occur at some point on the boundary of the region defined by the nutritional constraints on `Bonemeal`, `Corn` and `Soybean`. This is as far as calculus will take us. With further analysis, it can be proved that the optimum occurs at a vertex of the feasible region, and then an algorithm to identify such a vertex must be developed. The simplex method is based on an efficient and intelli-

1.4 Data types

gent enumeration of vertices together with a test for determining that the optimal vertex has been reached.

Exercises

1.3.1. How does the food mixture of Model 1.1 change when the price of soybean goes up to .425 per pound? To .525 per pound?

1.3.2. A micro mill produces a non-oxidizing steel for engine valves. Two materials *A* and *B* are blended to make this product; these materials cost 320 and 210 dollars a ton respectively. The product must have a manganese content of at least 4.6% and a chromium content of at least 1.9%. Material *A* is 6.2% manganese and 2.9% chromium; material *B* is 2.0% manganese and 1.1% chromium. Up to 12 tons of material *A* are available right now and up to 10 tons of *B* are available. How much of each should the mill use in order to make 15 tons of product, while minimizing the cost of the materials used.

1.4 Data types

The 2LP system supports the three data types int, double, and continuous. The types double and int are the same types as in C and C++. Thus int is the integer type and double is the floating point type. Arrays of each of the three basic types can be declared and used in a program. For readability, we use initial capitals for the names of variables of type continuous but lowercase only names for identifiers of type int and double. Variables of type double may contain nonintegral values, while variables of type int must contain integers. Some sample declarations are

```
double d[3];            // 3 doubles have been created:
                        // d[0], d[1], d[2]
int j,i_array[44];      // 45 integers have been created:
                        // j,i_array[0],...,i_array[43]
continuous X[2],Cost;   // 3 continuous variables have
                        // been created: X[0], X[1], Cost
```

The numerical constants and variables of type int and double can serve as constants and as coefficients in constraints. The fundamental operation on variables of type double and int is assignment. For example, if the array d is declared as above, the statement

```
d[0] = 3.3;
```

assigns the value 3.3 to the 0 th element of d. The right-hand side of an assignment statement may be an expression involving *arithmetic operators*. In 2LP the arithmetic operators are *, /, +, and -.

Let us note that numerical constants have a type according to the way they are entered. If a decimal point or an exponentiation symbol is used in a constant, the constant is of type double; otherwise, it is of type int. Thus, 6.0 and 1e3 are of type double, while 6 and 1000 are of type int. The types of constants and other identifiers determine how a computing system performs arithmetic on them. Thus, as in C, C++, and most other programming languages, the result of the operation 1/3 is 0 but that of 1.0/3 is .333333.

The variables of type continuous are constrained by linear equalities and inequalities built with the relational operators ==, <=, and >=. These are called the *closed relational operators*. There are also the *strict relational operators* !=, < and >, where != is read "not equal to." We note once again that following the usage in C and C++, we retain = for the assignment statement and use == for equality.

A constraint basically has the form

$$t_1 \text{ <closed relational operator> } t_2;$$

where t_1 and t_2 are combinations of the form

$$c_1 * X_1 + \ldots + c_n * X_n + b_1 + \ldots + b_m$$

Here $c_1, \ldots, c_n$ are coefficients, $b_1, \ldots, b_m$ are constants, and $X_1, \ldots, X_n$ are continuous variables. These combinations are called *affine combinations*. When there are no constant terms $b_1, \ldots, b_m$, the expression is called a *linear combination*. Numerical constants and identifiers of type int and double can serve as constants and as coefficients in constraints but variables of type continuous can not serve as coefficients.

In the next model we will use the data types double and continuous. It is an application where constraints link several periods of activity; an application of this kind is called a *multiperiod problem*.

Model 1.2 The Library Fund

The library committee in a small city has raised a fund of $100,000 which it plans to use for improvements in one year's time. The treasurer has been asked to invest this money for the coming year in certificates of deposits (CDs) with the local bank. The bank has 3-month, 6-month, and 12-month CDs. The committee has been known to change its mind and so has asked the treasurer to invest the capital so that at least $75,000 will be available at each 3 month breakpoint. In this way, if conditions require, this much money could be made available for other purposes. The 3-month CD returns 2% for the 3-month period, the 6-month CD returns 4.25% at the end of the 6-months, and the 12-month CD has a yield of 8.75%. The general assumption is that interest rates

1.4 Data types

will not change during the next year. The committee has also let it be known that an annual return of at least 8.25% would keep the fund from being eroded by inflation. The treasurer hopes to be able to present the committee with an investment strategy that will meet these goals. Can it be done?

The numerical data we are given consist of the amount of the initial capital, the liquidity requirement that $75,000 be available at each three month breakpoint, the annual return that the committee expects and the rate of return on each of the CDs. We can encapsulate this information by means of variables of type `double`. We make the declarations

```
double infd, lq, annual_ret;
double ret3, ret6, ret12;
```

In the program the first three identifiers will be initialized by assignment statements as follows:

```
    infd = 100000;
    lq = 75000;
    annual_ret = .0825;
```

The total return on any CD is the amount placed in that CD times (1 + the rate of return). So we will initialize the identifiers `ret3`, `ret6`, and `ret12` to 1 plus the given percentage. This way we can multiply the total return by the current investment to obtain the investment's value at the end of the investment period.

```
    ret3 = 1.02;
    ret6 = 1.0425;
    ret12 = 1.0875;
```

There are quite a few decisions to be made here. The treasurer's plan must determine the amounts to invest now, 3 months from now, 6 months from now, and 9 months from now. More precisely, the plan must determine the amounts to invest in 3-, 6- and 12-month CDs now, then the amount to invest in 3-month and 6-month CDs in both 3 month's time and in 6 month's time, and finally the amount to invest in a 3-month CD in 9 month's time. So let us start by introducing three arrays of continuous variables to represent these decisions. We have the declarations

```
continuous Three[4], Six[3], Twelve[1];
```

where `Three[0]` will represent the amount placed in a 3-month CD at the beginning; `Three[1]` will represent the funds placed in a 3-month CD at month 3, and so on. These investments must be made so that the liquid capital at each breakpoint will be greater than or equal to the liquidity requirement `lq`. At the starting point the liquid capital is simply the $100,000 to which we initialize the variable `infd`. Then there are three breakpoints at which the liquid capital must meet the liquidity requirement. At the end of the year, the fund should be worth at least `(1 + annual_ret)*infd`.

Let us introduce a continuous variable `FundValue` to represent what the fund will be worth in one year. Clearly we want

```
FundValue >= (1 + annual_ret)*infd;
```

Working backward, at the end of the 12-month period, the capital will be that coming from the 3-month CD that was purchased at month 9, the 6-month CD that was purchased at month 6, and the 12-month CD that was purchased at the beginning. So we have the equality constraint

```
FundValue ==
    ret3*Three[3] + ret6*Six[2] + ret12*Twelve[0];
```

It also must be the case that at the 9 month breakpoint, there be sufficient liquid capital. The liquid capital at this point is `ret3*Three[2] + ret6*Six[1]`. The liquidity requirement can be stated as

```
ret3*Three[2] + ret6*Six[1] >= lq;
```

Moreover the liquid capital at this point will be now be reinvested; at month 9, the only choice is the 3-month CD, so we can express this as

```
ret3*Three[2] + ret6*Six[1] == Three[3];
```

Similarly we have the liquidity and reinvestment constraints at the 6-month and 3-month breakpoints

```
ret3*Three[1] + ret6*Six[0] >= lq;
ret3*Three[1] + ret6*Six[0] == Three[2] + Six[2];
ret3*Three[0] >= lq;
ret3*Three[0] == Three[1] + Six[1];
```

Finally, at the outset we have

```
Three[0] + Six[0] + Twelve[0] == infd;
```

We can encode this analysis in a 2LP program:

```
2lp_main()
{
double infd,lq;
```

1.4 Data types

```
double ret3,ret6,ret12,annual_ret;

continuous Three[4],Six[3],Twelve[1];
continuous FundValue;

    infd = 100000;
    ret3 = 1.02;
    ret6 = 1.0425;
    ret12 = 1.0875;
    lq = 75000;
    annual_ret = .0825;

    FundValue >= (1 + annual_ret)*infd;

    FundValue ==
        ret3*Three[3] + ret6*Six[2] + ret12*Twelve[0];
    ret3*Three[2] + ret6*Six[1] >= lq;
    ret3*Three[2] + ret6*Six[1] == Three[3];
    ret3*Three[1] + ret6*Six[0] >= lq;
    ret3*Three[1] + ret6*Six[0] == Three[2] + Six[2];
    ret3*Three[0] >= lq;
    ret3*Three[0] == Three[1] + Six[1];
    Three[0] + Six[0] + Twelve[0] == infd;

    printf("A feasible plan is:\n\n");
    printf("Invest %.2f in a 3 month CD now\n",Three[0]);
                            .
                            .
                            .
    printf("Invest %.2f in a 3 month CD in 9 months\n",
                            Three[3]);
    printf("The fund will be worth %.2f\n",FundValue);
}
```

The output will read

A feasible plan is:

Invest 73529.41 in a 3 month CD now
Invest 0.00 in a 6 month CD now
Invest 26470.59 in a 12 month CD now
Invest 75000.00 in a 3 month CD in 3 months
Invest 0.00 in a 6 month CD in 3 months
Invest 76500.00 in a 3 month CD in 6 months
Invest 0.00 in a 6 month CD in 6 months
Invest 78030.00 in a 3 month CD in 9 months
The fund will be worth 108377.36

The plan determined has a certain structure to it. At first invest just enough in the 3-month CD each time to ensure liquidity at the next breakpoint, and invest all other money in the 12-month CD; from then on, only the 3-month instrument is used. This plan yields a return of 8.387%.

To eliminate the suspense and determine an optimal investment strategy, we can insert the call

```
max: FundValue;
```

before the print statements in the program and now the output will be

A feasible plan is:

Invest 73529.41 in a 3 month CD now
Invest 26470.59 in a 6 month CD now
Invest 0.00 in a 12 month CD now
Invest 46474.91 in a 3 month CD in 3 months
Invest 28525.09 in a 6 month CD in 3 months
Invest 44375.10 in a 3 month CD in 6 months
Invest 30624.90 in a 6 month CD in 6 months
Invest 75000.00 in a 3 month CD in 9 months
The fund will be worth 108426.46

This result is somewhat better than the one previously found and yields an annual return of 8.426%. Note that it does not use the 12-month CD but instead makes an investment in the 6-month instrument at the outset.

In Section 3.4 we revisit this model and consider the situation where it is not assumed that interest rates will remain the same during the coming year.

End of Model 1.2

The print statements in this last model can be organized in a more compact way using loops and identifiers of type `string`, as we will see in Chapters 2 and 3.

This last model contains a structure that will arise in many applications, that of a network. For we can think of a breakpoint as a node and the flow of an investment from node to node as an arc. This flow is described by the equality constraints of the model. This gives rise to the diagram of Figure 1.7 Models with network structure are the topic of Section 4.3.

Exercises

1.4.1. Rewrite Model 1.1 using arrays and individual identifiers for the data.

1.4 Data types

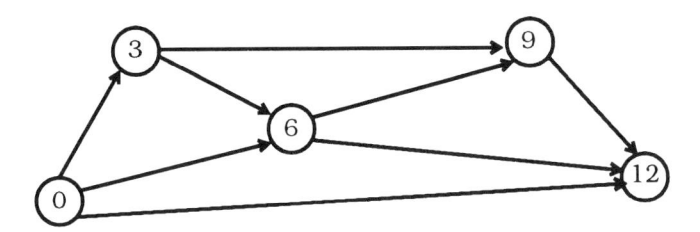

FIGURE 1.7 The flow of investment.

1.4.2. How much can the fund of Model 1.2 be worth if only $65,000 is needed to meet the liquidity requirement at each breakpoint, but with the proviso that no more than $20,000 can be invested in a 6 month CD?

1.4.3. A financial house has positions in a currency in three overseas locations with code names A, B, and C. Table 1.3 gives the commissions that must be paid to transfer funds from one location to another. Thus if 1.0 million are transferred from A to B, only 999,000 are credited to B, since there is a shrinkage of .1%. The house is planning its strategy for the next 3 days. Right now, its holdings (in millions) are 1.5 at A, 2.3 at B and .9 at C. However, funds must be in different locations on different days to meet equity requirements. The equity requirements for each of the next 3 days are given in Table1.4. The house must prepare an exchange strategy for each of the 3 days. Develop a plan without using linear programming. Then find the strategy with the least amount of transaction costs.

TABLE 1.3 Transaction Rates in Percent

From/To	A	B	C
A	0.000	0.10	0.15
B	0.20	0.000	0.13
C	0.19	0.25	0.000

TABLE 1.4 Equity Requirements

	A	B	C
Wednesday	1.0	2.8	.8
Thursday	1.6	2.1	.95
Friday	1.2	2.2	1.0

1.5 Affine expressions

The 2LP system supports different methods for communicating with the geometric object defined by the constraints. Constraints themselves alter the shape of the feasible region. The `max:` and `min:` operators move the witness point to an optimal vertex. The system also supports several functions that query the geometric object and return values of type `int` or `double`.

The function `wp(X)` takes a continuous argument X and returns the X coordinate of the current witness point. With the declarations

```
continuous X;
double a;
```

the assignment `a = wp(X)` will record this value by assigning it to a variable of type `double`. The signature of this function is

```
double wp(continuous)
```

Some built-in functions that can be applied to variables of type `double` can also be applied to continuous variables. For example, the function `floor(X)` will return the floor of `wp(X)` and similarly `ceil(X)` will return the ceiling of `wp(X)`. The function `nint(X)` returns the integer nearest to `wp(X)`. Thus, if we run the code that generates the constraints of Figure 1.5 and print the values of these functions applied to the variable X, we have the following program:

```
2lp_main()
{
continuous X,Y,Z;

    X + Y + Z <= 1;
    X + Y == 1;
    X == Y;

    printf("The X coordinate of the witness point:\n\n");
    printf("Its value is %.1f\n",wp(X));
    printf("Its floor is %.1f\n",floor(X));
    printf("Its ceiling is %.1f\n",ceil(X));
    printf("Its nearest integer is %d\n",nint(X));
}
```

The resulting printout to the screen is

1.5 Affine expressions

The X coordinate of the witness point:

Its value is 0.5
Its floor is 0.0
Its ceiling is 1.0
Its nearest integer is 1

Note that the values returned by `wp`, `floor`, `ceil` and `fabs` are of type `double`, while the `nint` function returns a value of type `int`.

The basic building blocks for constraints are affine combinations of continuous variables that take the form

$$c_1*X_1 + \ldots + c_n*X_n + b_1 + \ldots + b_m$$

Working with constraints is made easier by the fact that affine combinations of affine combinations are equivalent to affine combinations, as is easily seen using the associative and distributive laws of arithmetic. Also, parentheses and the minus sign can be used and division can be applied to constants and coefficients. So for example,

```
Net == (5.0/8.0)*(GrossIncome - Deductions);
```

is a bonafide constraint.

To ensure that complex expressions are equivalent to affine combinations of continuous variables, 2LP follows a rule borrowed from informal mathematical usage. In linear algebra, one writes

$$a(x-3y) + b(4-z) = cz$$

but instinctively avoids writing zc. In a similar way, in working with vectors, one writes

$$a(\mathbf{i}-2\mathbf{j}) + b(\mathbf{j}-\mathbf{k})$$

but one eschews $\mathbf{j}2$ and $(\mathbf{j}-\mathbf{k})b$.

In order to formalize this practice, we first need a definition. We define an *affine expression* and its *type* by induction. The most basic affine expressions are identifiers and constants; both have their given types. This is the base case of the induction. An affine expression is defined inductively as an identifier, a constant, or a compound expression built up from simpler affine expressions using arithmetic operators or built-in functions. A compound expression inherits its type from its constituents. Table 1.5 gives some examples.

For compound terms formed by applying a function, the resulting affine expression has the type that is returned by the function. Putting parentheses around a term produces a new term of the same type; similarly, prefixing an expression with a unary plus or minus yields an expression of the same type.

In combining expressions by means of binary operators, the type `continuous` dominates `double`, and the types `continuous` and

TABLE 1.5 Expressions and Types

Expression	Type
a	int or double
b	int or double
X	continuous
X + a	continuous
(X + a)	continuous
c/d*b*(X + a)	continuous
floor(c/d*b*(X+a))	double
Y - floor(c/d*b*(X + a))	continuous

double both dominate int. There are two restrictions. In forming a compound affine expression, if the binary operator is *, then the left-hand argument cannot be of type continuous; if the operator is /, neither argument can be of type continuous. The formal definition of affine expression as a context-free language is left to the exercises. For the semantics of these expressions, the usual precedence rules are used.

To reiterate, the basic rule is that an affine expression of type continuous cannot be used as the left-hand argument of a multiplication or as either argument to a division. The 2LP parser checks that complex expressions yield linear constraints by verifying that this rule is observed. So, for example, with X declared as continuous, the line of code

```
b <= 1.5*(X + 1)*(Y - 12*X);
```

cannot be parsed successfully because the expression (X + 1) is being used as a coefficient. This discussion has been informal and bottom-up. A formal and top-down definition of affine expression is left to the exercises.

A constraint is a declarative statement that enforces relations on continuous variables. Formally, a *constraint* is a statement of the form

<affine expression> <closed relational operator> <affine expression> ;

where at least one of the affine expressions is of type continuous.

A constraint is different from a test. A test compares two expressions that are of type double or int. Formally, a *test* is a statement of the form

<affine expression> <relational operator> <affine expression> ;

1.5 Affine expressions

where the affine expressions are *not* of type `continuous`. As with constraints, tests either succeed or fail. However, in contrast to constraints, tests will not affect the current feasible region or witness point. Tests are critically important in working with loops, conditional statements, and other logical constructs, as we will see in the succeeding chapters.

The strict operators `!=`, `>`, and `<` can only be used in tests and not in constraints. This restriction is due to the fact that mathematically these operators define open sets. The relations `<=`, `==`, and `>=` all define closed sets in n-dimensional space; continuous functions, in particular linear ones, attain maximum and minimum values on bounded closed sets. This is not true of course for linear functions and open sets.

The fundamental operation on variables of types `int` and `double` is the assignment operation. This takes the form

<variable identifier> = <affine expression> ;

where the variable and the affine expression are both of type `int` or `double`. Unlike constraints and tests, an assignment statement always succeeds.

It is often convenient not to have to introduce a continuous variable for the objective function but simply to use an affine expression instead. The optimum value of an objective function given as an affine expression will also be found at a vertex, as in the case when the objective function is a single variable.

If the programmer wants to move the witness point to a vertex that optimizes the value of an objective function given as an affine expression of type `continuous`, the commands are

min: <affine expression> ;

and

max: <affine expression> ;

As a simple illustration of this notation, the introduction of the variable `Cost` in Model 1.1 can be eliminated and the model rewritten as

```
2lp_main()
{
continuous Bonemeal,Corn,Soybean;

    Bonemeal + Corn + Soybean == 125;
    .38*Bonemeal + .001*Corn + .002*Soybean   <=   .012*125;
    .38*Bonemeal + .001*Corn + .002*Soybean   >=   .008*125;
                   .09*Corn + .5*Soybean      >=   .22*125;
                   .02*Corn + .08*Soybean     <=   .05*125;
    min: .164*Bonemeal + .463*Corn + .125*Soybean;
```

```
                // Output
                printf("Use %.2f pounds Bonemeal \n",Bonemeal);
                printf("Use %.2f pounds Corn \n",Corn);
                printf("Use %.2f pounds Soybean \n",Soybean);
                printf("The Cost will be $%.2f\n",
                        .164*Bonemeal + .463*Corn + .125*Soybean);
}
```

The next model presents elements of work of Boole on probabilistic logic. For a comprehensive exposition of these ideas, we refer the reader to Hailperin (1976). The story takes place in the mid 19th century and is recounted in the historical present.

Model 1.3 The Fortune-teller

An Irish country lass enters the tent of the fortune-teller at the fair and asks about the future. She wants to know the answers to three questions: Will she be rich, will she happy, and will she be long-lived. Most especially, she wants to know if she will be happy and long-lived. Of course the fortune-teller doesn't come straight out with answers to these questions. Rather she skirts around them and points out likelihoods of various combinations. Thus she says that the likelihood of the lass' being poor and unhappy is no more than 5%, and so on. Naturally the country lass has an excellent memory and when she leaves the fortune-teller's tent, she quickly writes down what she was told. This has been transcribed into Table 1.6.

The country lass still wants to have a clearer answer to the question whether she will be both happy and long-lived. So she journeys to the city of Cork and consults Professor George Boole there who has a method for extracting the prediction she is interested in.

The professor formulates the information he is given. Each of the three categories, Rich, Happy, and LongLived is considered a propositional variable that can be either true or false. Therefore there are eight possible outcomes for her life. In the truth table given in Table 1.7 each outcome is associated with the corresponding row of 0-1 values for the propositional variables.

TABLE 1.6 What the Fortune-teller Said

Predictions	Probabilities
NotRich and NotHappy	At most .05
Rich and NotLongLived	.10
NotRich Happy and LongLived	.38
Rich and Happy	At least .33

1.5 Affine expressions

TABLE 1.7 The Truth Values

Outcome	Rich	Happy	LongLived
0	0	0	0
1	0	0	1
2	0	1	0
3	0	1	1
4	1	0	0
5	1	0	1
6	1	1	0
7	1	1	1

With each of these eight outcomes, the professor associates a continuous variable. This leads to the declaration

```
continuous X[8];
```

The analysis proceeds. One of these eight outcomes must occur. This leads to the constraint

```
X[0] + X[1] + X[2] + X[3] + X[4] + X[5] + X[6] + X[7] == 1.0;
```

The set of outcomes where the country lass will be neither rich nor happy has likelihood at most .05; these are outcomes 0 and 1:

```
    X[0] + X[1] <= .05;
```

Continuing this way, a constraint is written down for each entry in Table 1.6. Then the task is to find the minimum and maximum likelihoods that she will be happy and long lived. This outcome is given by X[3] + X[7] and so this expression must be minimized and maximized. This all leads to the following program:

```
2lp_main()
{
continuous X[8];

    X[0] + X[1] + X[2] + X[3] + X[4]
    + X[5] + X[6] + X[7] == 1.0;
    X[0] + X[1] <= .05;     // Not Rich and Not Happy
    X[4] + X[6] == .10;     // Rich and Not LongLived
    X[3] == .38;            // Not Rich, Happy and LongLived
    X[6] + X[7] >= .33;     // Rich and Happy

    printf("Happiness and Long Life lies between ");
```

```
    min: X[3] + X[7];        // Minimize Happy and LongLived
    printf("%.2f and ",X[3] + X[7]);
    max: X[3] + X[7];        // Maximize Happy and LongLived
    printf("%.2f",X[3] + X[7]);
}
```

The results of this consultation are

Happiness and Long Life lies between 0.61 and 0.90

The country lass goes away with hope in her heart that her dreams will come true.

<u>End of Model 1.3</u>

In this last model there were 3 propositional variables and 2^3 continuous variables. Clearly, if the number of propositional variables is much greater, the resulting linear programs will be intractable. These issues are addressed in Jaumard, Hansen and Poggi de Aragaö (1991) where a technique known as *column generation* is used to deal with oversized linear programs.

Probabilistic logic is a very active field. For the relationship of Boole's probabilistic logic with those of Nilsson, Dempster-Shafer, and others, see Andersen and Hooker (1996).

To end this section, let us make some remarks on the structure of constraints in order to broach the important concept of *duality*.

A constraint can always be put in the form

<linear expression> <closed relational operator> <constant>

or, in conventional mathematical notation, one of the forms

$$a_1 x_1 + ... + a_n x_n \leq b$$
$$a_1 x_1 + ... + a_n x_n = b$$
$$a_1 x_1 + ... + a_n x_n \geq b$$

When a constraint is written this way, the linear expression is called the *left-hand side* of the constraint and the constant term is called the *right-hand side*.

In mathematical terminology, when there are two complementary ways of looking at a situation, the two views are said to be *dual* to one another and the link between them is called a *duality relation*. Duality is a powerful concept that will come up often. We will use the term both formally and informally.

A laundry list of dual pairs includes *landlord* and *tenant* in economics, *element* and *set* in set theory, *vertex* and *edge* in a graph, *arc* and *node* in a network, *or* and *and* in logic, *there exists* and *for all* in logic, *program* and *data* in programming, *rows* and *columns* in a matrix or

1.5 Affine expressions

table, *linear functionals* and *points* in Euclidean space and Hilbert space, *position* and *momentum* in Hamiltonian mechanics and quantum mechanics.

In linear programming there is always an important duality relation between the continuous variables and the left-hand sides of the constraints. Thus, in Model 1.3, The Fortune-teller, the continuous variables represent atomic events and the left-hand sides represent sets of atomic events; the constraints themselves assert conditions on the probabilities of different sets of atomic events. In Model 1.2, The Library Fund, there is a network structure; the continuous variables represent cash flows and the left-hand sides represent breakpoints, the variables are arcs and the left-hand sides are nodes. The equality constraints in this model assert equilibrium conditions on the cash flows, to wit, the money going out is equal to the money coming in; the $\geq$ constraints assert that liquidity must be maintained at each breakpoint. In Model 1.1, The Wildlife Conservation Park, the continuous variables Bonemeal, Corn, and Soybean represent ingredients and the left-hand sides of the $\leq$ and $\geq$ constraints represent nutritional content; these continuous variables correspond to the columns of Table 1.1 and the left-hand sides of these constraints correspond to rows. As is shown in Chapter 15, duality is always present in a linear programming model in a very dramatic way. Till then, we will use the notion informally, both for the linear programming aspect of a model and for other aspects such as logical structure.

Exercises

1.5.1. Redo Exercise 1.2.2 with the objective functions $x + y$ and $x - y$.

1.5.2. Suppose the country lass of Model 1.3 had also asked the fortune teller if she would emigrate to America and that this brought the additional responses that the likelihood of her being happy and going to America was less than 20% and that the likelihood of her going to America and not being rich was more than 44%. How does this change the outcome of the professor's analysis?

1.5.3. An executive wants to follow industry trends in allocating the advertising budget. The executive has gathered the following information about the competition. For the coming season, all these firms will use TV, radio, or print media. Between 30% and 66% will use both TV and print media; over 60% will use both TV and radio, at least 50% will not use both radio and print media, and no more than 10% will not use TV at all. What can be said about the likelihood of firms' using radio advertising?

1.5.4. (The elements of formal language theory are needed for this and the following exercise.) Give a formal definition of *affine expression* by means of a context-free grammar. For this assume that there are only a finite number of identifiers and constants of each type. Treat unary plus and minus as symbols separate from + and −.

1.5.5. Give a formal definition of *linear expression* modeled on that of affine expression. Your definition should account for expressions such as `-1.0/3*floor(X+Y+4)*(U - 2*a*V)`.

1.6 Symbolic constants

Arrays of type `double` and `int` are often used to store numerical data that are given in tables or lists. To facilitate initializing arrays with external data, 2LP supports an array assignment feature that is modeled on the initialization syntax used in C and C++. For example, to enter the sequence 0, 1, 2, 3, 4, 5, 6, 7, 8, 9 into the array `a[10]` of type `int`, we can write the array assignment

```
a = { 0, 1, 2, 3, 4, 5, 6, 7, 8, 9 } ;
```

The assignment

```
a = { 0, 2, 4, 6, 8 } ;
```

will enter the even digits into `a[0]`, `a[1]`, `a[2]`, `a[3]`, and `a[4]` successively. Multidimensional arrays are stored in row-major order; that is, row by row for two dimensional arrays. So to initialize the array `nutrition[3][3]` of type `double` to the values of the nutritional information in Table 1.1, we can write

```
nutrition = {
    0.38 ,  0.001,  0.002,
    0.00,   0.09,   0.50,
    0.00,   0.02,   0.08
} ;
```

In general, for multidimensional arrays it is the rightmost subscript that varies fastest as values are assigned to elements of the array.

Code can be made more readable if numerical constants that occur in models are replaced by more meaningful names. This is done in 2LP by defining identifiers to be *symbolic constants*. As an example of this syntax, placing the following definition at the beginning of a model

```
#define SIZE 20
```

1.6 Symbolic constants

makes the identifier SIZE a synonym for the integer 20 throughout the model. These definitions normally appear at the very top of the program, although the only requirement is that they be made before they are used. Symbolic constants make code more abstract, more readable, and easier to change. As a practice we will use all capitals for names of symbolic constants. This is illustrated in the next model which deals with the issue of *outsourcing*.

Model 1.4 Inside vs. Outside

A small petroleum refining company is doing well and has run into the pleasant but thorny problem of having more orders than it can fill. That is, the company's production capacity is limited by various constraints and the orders exceed the company's production capability. The company wants to fill the orders on time and ship promptly at the end of the week, so it must turn to outside suppliers to provide the needed extra production. The petroleum products that the company has committed to shipping are known as regular, super, and detergent gasoline. For these products, at the end of the week, the company must ship the quantities given in Table 1.8

Each of these products is produced by two basic chemical processes, cracking and blending. These processes can only be run for a limited number of hours during the week. In fact, the cracking process can be run for up to 100 hours and the blending process for up to 85 hours, information given in tabular form in. Table 1.9. Table 1.10 shows how much time in hours of each process is needed in order to yield a gallon of product.

TABLE 1.8 Product Demand

	Regular	Super	Detergent
Gallons	20,000	35,000	18,000

TABLE 1.9 Polymerization Processes

	Cracking	Blending
Hours	100	85

TABLE 1.10 Hours per Gallon

	Regular	Super	Detergent
Cracking	.005	.004	.003
Blending	.002	.004	.006

TABLE 1.11 Cost per Gallon

	In-House	Outside
Regular	.75	.85
Super	.78	.90
Detergent	.90	.98

Purchasing finished product from the outside is of course more expensive than in-house production. The costs of purchasing from the outside as against the costs of internal production are summarized in Table 1.11. Thus there are two kinds of information on in-house production. First, for each product, in-house production is limited by the need to employ the blending and cracking processes. Second, we know the total cost per gallon of producing a product in-house when the production constraints allow it. Because of the availability of outsourcing, the company has two ways of obtaining each product for shipping. The challenge is to commit the company's own production apparatus to working on the products in such a way as to minimize the total cost to the company of meeting its shipping commitment.

Data given in lists or tables lend themselves naturally to array storage. Let us set up one-dimensional and two-dimensional arrays of type `double` to store the data given by the tables above. We will use symbolic constants to represent the dimensions of the data in the tables and to represent the indices that run over these dimensions.

The shipping commitment data can be stored in a one-dimensional array whose size will be the number of products and whose entries will be indexed by the products themselves. So we set

```
#define PRODS 3
#define REG 0
#define SUP 1
#define DET 2
```

The shipping array can be declared as

```
double shpp[PRODS];
```

Then we will have the initialization

```
    shpp = { 20000, 35000, 18000 };
```

The data for the time that the chemical processes are available is given in a one-dimensional array whose size is the number of processes and so we define

1.6 Symbolic constants

```
#define PROCS 2
```

The entries in this array are indexed by the processes themselves; we set

```
#define CRCKNG 0
#define BLNDNG 1
```

The array can now be declared as `double limit[PROCS]` and initialized as

```
limit[CRCKNG] = 100;
limit[BLNDNG] = 85;
```

or as

```
limit = { 100, 85 };
```

The table that links processes and products can be stored in a two-dimensional array declared as `double prcprd[PROCS][PRODS]` and initialized as follows:

```
prcprd = {
        .005,.004,.003,
        .002,.004,.006
};
```

This form of initialization emphasizes the relation between the structure of the table and that of the array. More compactly we can write

```
prcprd = { .005, .004, .003, .002, .004, .006 };
```

Similarly the table giving the cost of in-house production and of outsourcing is a two-dimensional array, and we declare an array `double cost[PRODS][2]` to record these data. The columns in the table represent the in-house cost and the outsourcing cost of obtaining a gallon of each product. For completeness we introduce symbolic constants for these columns

```
#define IN 0
#define OUT 1
```

The initialization of the `cost` array will be

```
cost = { .75, .85, .78, .90, .90, .98 };
```

The quantities that must be determined are the amount of each product to be produced in-house and the amount of each product to be purchased on the outside. Let us introduce an array of continuous variables to represent the amount of in-house production and another array for the outsourced production

```
continuous InHouse[PRODS];
continuous OutSource[PRODS];
```

These quantities must be combined to meet the company's shipping commitment. So we have the constraints

```
// Shipping constraints
InHouse[REG] + OutSource[REG] == shpp[REG];
InHouse[SUP] + OutSource[SUP] == shpp[SUP];
InHouse[DET] + OutSource[DET] == shpp[DET];
```

The constraints on the amount of in-house production are

```
// In-house production constraints
prcprd[CRCKNG][REG]*InHouse[REG] +
prcprd[CRCKNG][SUP]*InHouse[SUP] +
prcprd[CRCKNG][DET]*InHouse[DET]
    <= limit[CRCKNG];

prcprd[BLNDNG][REG]*InHouse[REG] +
prcprd[BLNDNG][SUP]*InHouse[SUP] +
prcprd[BLNDNG][DET]*InHouse[DET]
    <= limit[BLNDNG];
```

Finally, we introduce a variable `FinalCost` of type `continuous` to represent the total cost to the company of meeting its goal

```
// Objective function
FinalCost ==
    cost[REG][IN]*InHouse[REG] +
    cost[REG][OUT]*OutSource[REG] +
    cost[SUP][IN]*InHouse[SUP] +
    cost[SUP][OUT]*OutSource[SUP] +
    cost[DET][IN]*InHouse[DET] +
    cost[DET][OUT]*OutSource[DET];
```

The goal of the model is to find a solution to these constraints that minimizes the variable `FinalCost`. Putting things together, we have the program

```
#define REG 0
#define SUP 1
#define DET 2
```

1.6 Symbolic constants

```
#define CRCKNG 0
#define BLNDNG 1
#define PRODS 3
#define PROCS 2
#define IN 0
#define OUT 1

21p_main()
{
double shpp[PRODS], limit[PROCS];
double prcprd[PROCS][PRODS],cost[PRODS][2];

continuous InHouse[PRODS];
continuous OutSource[PRODS];
continuous FinalCost;

    shpp = { 20000, 35000, 18000 };
    limit = { 100, 85 };
    prcprd = { .005, .004, .003, .002, .004, .006 };
    cost = { .75, .85, .78, .90, .90, .98 };

    // Shipping constraints
    ...
    // In-house production constraints
    ...
    // Objective function
    ...
    min: FinalCost;

    printf("Purchase %.2f of regular\n",OutSource[REG]);
                    .
                    .
                    .
    printf("The final cost will be %.2f\n",FinalCost);
}
```

The printout is

```
Purchase 15000.00 of regular
Purchase 16250.00 of super
Purchase 18000.00 of detergent
Produce 5000.00 of regular
Produce 18750.00 of super
Produce 0.00 of detergent
The final cost will be 63390.00
```

The solution found suggests that the company's way of producing detergent grade gasoline might not be economically competitive.

In this model the continuous variables `InHouse[PRODS]` are used to represent products and the left-hand sides of the ≤ constraints represent resources. The ≤ constraints themselves assert limits on the use of the resources. The variables `OutSource[PRODS]` simply measure the gap between in-house production and the shipping goals.

<u>End of Model 1.4</u>

Exercises

1.6.1. Keeping all other costs fixed in Model 1.4, determine the point (within one cent) at which the cost of outsourcing production of regular gasoline will mean that the company will purchase less regular gasoline to meet its delivery commitment. Do the same for the super and detergent grades.

1.6.2. A company manufactures shafts for motors and toolposts for lathes. To make a shaft 3, 2 and 1 hours are required on a drill press, lathe and milling machine respectively. To make a toolpost, 3 and 4 hours are required on a lathe and milling machine, respectively. No machine can be used more than 60 hours in a week. The company has an order for 8 shafts that it must fill at the end of the week. Otherwise a shaft can be sold for $100 and a toolpost for $180. How should the company plan production for this week to maximize its revenue?

2

Loop Constructs

In this chapter we introduce some basic loop constructs that give us greater flexibility and enable us to write models more simply. We also will encounter the important topic of applications that require integer-valued solutions to linear constraints.

2.1 and loops, sigma loops

Loops are useful when it is necessary to generate a collection of constraints of a very similar form or to execute a sequence of statements in a repetitive way. In this section and the next, we look at two types of loop constructs: the logical and loop and the arithmetic sigma loop. In subsequent chapters we will see disjunctive loop constructs, the or loop and c_or loop.

In 2LP, logical and loops may be used in a way that is analogous to the for loop of classical programming languages. Thus, to generate the 100 constraints

```
X[1]   >= X[0]  + a[0];
X[2]   >= X[1]  + a[1];
         . . .
X[100] >= X[99] + a[99];
```

we write the loop

```
and(int i=0;i<100;i=i+1)
    X[i+1] >= X[i] + a[i];
```

The keyword and begins the loop and is followed by a *loop header*. The general form of a loop header is

(*<initial assignment>*; *<loop condition>*; *<step assignment>*)

The initial assignment creates and initializes a new variable, called the *loop control variable*. It is important to note that the loop control variable is only available inside the scope of the loop; after the loop is finished, the value of the loop control variable is no longer available. The loop control variable is introduced with the keyword `int`. In the and loop above, `i` is the loop control variable, and it is initialized to 0.

The loop condition is a test. When this test fails, the loop is exited successfully. The step assignment assigns a value to the loop control variable. This allows for a wide range of ways of updating the loop variable. When we do use unit increments for updating the loop variable, we may substitute `i++` for the statement `i=i+1`. Similarly, for unit decrements, we may use `i=i-1` or `i--`.

The syntax of the and loop header is flexible. We will see examples where one or more of the elements of the loop header are omitted. Let us also note that the test that serves as the loop condition of an and loop does not need to involve the loop control variable itself.

The and loop header is followed by a statement, which is called the *body* of the loop. So the formal syntax of the and loop is

and <loop header> <statement>

The loop control variable in the and loop can only be assigned in the loop header and not within the body of the loop, nor can its value be accessed after the loop is exited. These restrictions are imposed in order to provide support in 2LP for important disjunctive programming constructs.

Model 2.1 The Fixed-Rate Mortgage

Let us compute the monthly payment on a 15-year fixed-rate mortgage. The mortgage is for $100,000, and the annual rate of interest is 9%. Let us use the symbolic constants MORTGAGE to represent the size of the mortgage, RATE to denote the annual interest rate, and MONTHS for the number of months in 15 years:

```
#define MORTGAGE 100000.0    // Initial balance
#define RATE 0.09            // 9% annual rate
#define MONTHS 180           // Number of months
```

In a mortgage the balance due changes from month to month. The initial balance is the amount of the mortgage, and then each payment from 1 through 180 results in a new balance. The balance after the final payment must be 0.0.

To set up this model, we introduce a `continuous` variable Pymt to denote the monthly payment and an array Bal[MONTHS+1] of type `continuous` where Bal[k] is the balance left after k payments. Thus the initial balance Bal[0] is the amount of the mortgage; this relation-

2.1 and loops, sigma loops

ship is expressed by the constraint `Bal[0] == MORTGAGE`. The final balance must be zero, so we need the constraint `Bal[MONTHS] == 0.0`. The relation between each balance `Bal[k]` and the next balance `Bal[k+1]` after a monthly payment is

```
Bal[k+1] == (1 + RATE/12)*Bal[k] - Pymt
```

Recapitulating this in coded form, we have

```
2lp_main()
{
continuous Pymt, Bal[MONTHS+1];

    Bal[0] == MORTGAGE;
    Bal[MONTHS] == 0.0;
    and(int k=0;k<MONTHS;k++)
        Bal[k+1] == (1 + RATE/12)*Bal[k] - Pymt;

    printf("The monthly payment will be $%.2f\n",Pymt);
}
```

For the above code, the output will be

The monthly payment will be $1014.27

In this mortgage program, all the constraints are equality constraints. Applying linear algebra and variable elimination to these constraints would yield the well-known formula relating `Pymt` and the initial balance `Bal[0]`. Here it has been sufficient to set up the constraints declaratively. Moreover, using constraints, we can turn things around and compute the initial balance given the monthly payment. To pursue this idea, let us start with a given monthly payment and determine the size of the mortgage. This is the case where the prospective home-buyers know what size payment they can afford and want to determine how expensive a house they can consider buying.

```
#define RATE .09        // 9% annual rate
#define MONTHS 180      // Number of months

2lp_main()
{
continuous Pymt,Bal[MONTHS+1];

    Pymt == 1014.27;    // Should allow $100,000 mortgage
    Bal[MONTHS] == 0.0; // Final balance must be 0.0
```

```
   and(int k=1;k<=MONTHS;k++)
      Bal[k] == (1 + RATE/12)*Bal[k-1] - Pymt;

   printf("The mortgage is $%.0f\n",Bal[0]);
}
```

The above code will print

The mortgage is $100000

End of Model 2.1

The underlying mathematical programming apparatus that supports the continuous variables in 2LP is done by means of floating-point arithmetic; this may lead to roundoff errors and slight perturbations in the results generated by a program. To illustrate this, let us take another look at the reverse mortgage situation. In computing the monthly payment given the initial balance, we found that for a $100,000 mortgage, the monthly payment would be $1014.27; if we had printed out six decimal places, we would have obtained $1014.266584 as the answer. If in the reverse mortgage code, we use this number as the monthly payment

```
   Pymt == 1014.266584;
```

and determine the initial balance and print it out to six decimal places by changing the print statement to

```
   printf("The mortgage is $%f\n",Bal[0]);
```

the output would be

The mortgage is $99999.999984

So the reverse computation finds the original input with accuracy to nine significant digits. In any programming system, floating-point arithmetic will cause such perturbations. Thus in 2LP the values of the witness point's coordinates are subject to this kind of variation and so are determined only "within epsilon."

Statements in 2LP can be grouped into a single statement by enclosing them in braces. The statements within the block are invoked as a unit. By way of example, the following code segment will execute both constraints each time through the loop:

2.1 and loops, sigma loops

```
and(int i=j;i!=k;i=i+2) {
    X[i] >= Y[j];
    X[i] <= 100.0;
}
```

This construct will prove useful in the next model.

To simplify the expression of constraints, 2LP uses the keyword `sigma` to write a constraint in a notation similar to the ordinary mathematical usage. For example, for the equation

$$\sum_{i=0}^{99} a_i x_i = b$$

the 2LP expression is

```
sigma(int i=0;i<100;i=i+1) a[i]*X[i] == b;
```

In a `sigma` loop, the loop header has the same form as the `and` loop, namely

(<*initial assignment*>; <*loop condition*>; <*step assignment*>)

Unlike the and loop, however, the loop condition *must* be a test on the loop control variable. The test is of the form

<*loop variable*> <*relational operator*> <*affine expression*>

where <*affine expression*> is of type `int` or `double`. The relational operator can be <=, >=, ==, <, > or !=.

The value of the loop variable is changed after each pass through the loop by the step assignment, which is a statement that assigns a value to the loop control variable. Note that this form allows change to the loop variable in increments other than unit increments. For example, to sum the even-indexed elements of a 10-element array called b, we would write

```
a = sigma(int i=0;i<10;i=i+2) b[i];
```

Again, when we do use unit increments, we may substitute i++ for the statement i=i+1. Similarly, for unit decrements, we use i=i-1 or i--.

Syntactically, the `sigma` is a unary operator and does not change the type of the term it is applied to. It has higher precedence than +; it has higher precedence than * and / when it occurs to their right and lower precedence when it occurs to their left. Thus

```
... + a/b*sigma(int i=0;i<N;i++) c/d*X[i] + ...
```

is interpreted as

```
... + a/b*(sigma(int i=0;i<N;i++) c/d*X[i]) + ...
```

The expression

```
sigma(int i=0;i<N;i++) a[i]*(X[i] - b[i])
```

is of type `continuous` and

```
sigma(int i=0;i<N;i=i+2) i
```

is of type `int`. One more point: If the test in the loop never succeeds, then the loop returns 0. For example, the expression `sigma(int i=0;i<0;i=i+2) i` evaluates to 0 because the test 0 < 0 fails.

Exercises

2.1.1. Rewrite Model 1.4, Inside vs. Outside, using loops.

2.1.2. Redo Exercise 1.4.3, using loops.

2.1.3. The situation in Model 2.1 has changed. The homeowners have opted for a balloon mortgage that leaves a balance of $50,000 at the end of the 15 year period and that has an annual interest rate of 10%. What will the monthly payment be?

2.1.4. Solve the following linear program

$$3.0x_0 + 2.0x_1 + 1.2x_2 + 4.4x_3 + 2.2x_4 + 9.0x_5 \leq 33$$
$$5.0x_0 + 0.0x_1 + 2.2x_2 + 6.4x_3 + 4.9x_4 + 4.8x_5 \leq 45$$
$$0.0x_0 + -7.0x_1 + -1.9x_2 + 0.0x_3 + 3.8x_4 + -8.2x_5 \leq -18$$
$$1.0x_0 + 5.0x_1 + 4.3x_2 + 5.0x_3 + 3.2x_4 + 0.0x_5 \leq 21$$
$$3.0x_0 + 8.0x_1 + 5.3x_2 + 3.7x_3 + 5.4x_4 + 7.1x_5 \leq 36$$
$$8.0x_0 + 2.0x_1 + 6.6x_2 + 7.0x_3 + 5.6x_4 + 9.5x_5 \leq 28$$
$$0.0x_0 + -5.0x_1 + 0.2x_2 + -4.9x_3 + -2.7x_4 + -6.0x_5 \leq -15$$
$$x_0 \geq 0, \, x_1 \geq 0, \, x_2 \geq 0, \, x_3 \geq 0, \, x_4 \geq 0, \, x_5 \geq 0$$
$$x_0 \leq 50, \, x_1 \leq 50, \, x_2 \leq 50, \, x_3 \leq 50, \, x_4 \leq 50, \, x_5 \leq 50$$
$$\max: x_1 + x_3 + x_5$$

Enter the coefficients of the ≤ constraints in an array; set up these constraints by means of an `and/sigma` combination.

2.1.5. Extend the formal definition of affine expression of Exercise 1.2.4 to include `sigma`. For this you do not have to account for issues of array indices and scoping.

2.2 Basic roundoff techniques

The next model is a classic that is attributed to Kemeny by Dantzig (1963). In it we use both `sigma` and `and` loops. We also address the issue of rounding off fractional solutions so as to avoid plans that have a hen and a half lay an egg and a half in a day and a half.

Model 2.2 The Chicken or the Egg

A farmer has 100 hens and is in the business of selling fresh eggs and baby chicks. The eggs laid by the hens during the week can either be sold on a daily basis or kept in an incubator room for hatching the following week. In a week a hen can either hatch 4 eggs or lay 7 new eggs. The eggs that are sold fetch 10 cents, while the chicks bring in 60 cents. The farmer wants a 10-week plan to assign hens to hatching and to laying each week so as to maximize the income earned from selling eggs and chicks. The farmer has just returned from a vacation in the city, and there are no eggs currently in the incubator room. At the end of the 10 weeks, the farmer plans again to go to the city on vacation and so wants all eggs and chicks sold off by the end of the last week of the plan.

First, let us formulate a solution to this planning problem that does not take into account the fact that chicks and eggs must be counted in integers. We begin by defining symbolic constants for the basic data:

```
#define WKS 10          // Number of weeks to plan for
#define LAY_RATE 7      // Eggs laid per week
#define HATCH_RATE 4    // Eggs hatched per week
#define CHK_PRICE .60   // Amount earned by a chick
#define EGG_PRICE .10   // Amount earned by an egg
#define HENS 100        // Number of hens available
```

The key continuous variables of the model are `Hatchers[i]` and `Layers[i]`, which represent the hatching hens and laying hens in each week `i`. These variables are related by the equation

```
Hatchers[i] + Layers[i] == HENS;
```

We also introduce variables `Incub[i]` to represent the eggs placed in the incubator room during week `i` and variables `EggsSold[i]` to represent the number of eggs laid and sold during week `i`. The eggs sold plus the eggs placed in incubation must sum to the number of eggs laid during the week, so

```
EggsSold[i] + Incub[i] == LAY_RATE*Layers[i];
```

No hens can be hatching during the first week so we have

```
    Hatchers[0] == 0.0;
```

The number of hens that may hatch each week depends on the number of eggs placed in the incubator room the previous week. Since each hen can hatch HATCH_RATE many eggs, we have

```
    HATCH_RATE*Hatchers[i] == Incub[i-1];
```

Finally, no eggs should be placed in the incubator room during the last week, so

```
    Incub[WKS-1] == 0.0;
```

The objective function to be maximized is the affine expression

```
    sigma(int i=0;i<WKS;i++)
        (CHK_PRICE*Incub[i] + EGGS_PRICE*EggsSold[i])
```

Putting all of this together yields the model

```
#define WKS 10     // Number of weeks to plan for
...

2lp_main()
{
continuous Incub[WKS],EggsSold[WKS],
                    Layers[WKS],Hatchers[WKS];

    and(int i=0;i<WKS;i++) {
        Hatchers[i] + Layers[i] == HENS;
        EggsSold[i] + Incub[i] == LAY_RATE*Layers[i];
    }
    Hatchers[0] == 0.0;
    and(int i=1;i<WKS;i++)
        HATCH_RATE*Hatchers[i] == Incub[i-1];
    Incub[WKS-1] == 0.0;

    max: sigma(int i=0;i<WKS;i++)
        (CHK_PRICE*Incub[i] + EGG_PRICE*EggsSold[i]);
    and(int i=0;i<WKS;i++) {
        printf("In week %d: ",i);
        printf("Plan %.2f Hatchers ",Hatchers[i]);
        printf("and %.2f Layers \n",Layers[i]);
    }
}
```

2.2 Basic roundoff techniques

The output is

```
In week 0: Plan 0.00 Hatchers and 100.00 Layers
In week 1: Plan 64.05 Hatchers and 35.95 Layers
In week 2: Plan 62.91 Hatchers and 37.09 Layers
In week 3: Plan 64.90 Hatchers and 35.10 Layers
In week 4: Plan 61.42 Hatchers and 38.58 Layers
In week 5: Plan 67.51 Hatchers and 32.49 Layers
In week 6: Plan 56.85 Hatchers and 43.15 Layers
In week 7: Plan 75.51 Hatchers and 24.49 Layers
In week 8: Plan 42.86 Hatchers and 57.14 Layers
In week 9: Plan 100.00 Hatchers and 0.00 Layers
```

The solution found prescribes assigning fractional hens to hatching and laying. However, in this model we can adjust the output of the fractional solution to produce a plan that prescribes integer numbers of hatchers and layers each month by means of a roundoff technique.

When rounding floating point solutions to constraint problems, one must take care to ensure that the rounded values continue to satisfy the constraints of the system. In the present case we can use the `ceil` and `floor` functions to round the number of layers up and the number of hatchers down; this will preserve the constraint

```
Hatchers[i] + Layers[i] == HENS;
```

It will also preserve the other constraints since raising the number of layers in week `i-1` and lowering the number of hatchers in week `i` means that there will be sufficient eggs laid in week `i-1` for hatching in week `i`. What happens is that more whole eggs will be sold. This would not be the case if we raised the number of hatchers and lowered the number of layers. So we can output a more practical plan by changing the output code to read

```
and(int i=0;i<WKS;i++) {
    printf("In week %d: ",i);
    printf("Plan %.0f Hatchers ",
                    floor(Hatchers[i]));
    printf("and %.0f Layers \n",ceil(Layers[i]));
}
```

Another important point about moving from fractional to integer-valued solutions is that finding an integer solution by roundoff does not guarantee that the *optimal integer-valued* solution has been found. That is to say, there may be other integer-valued solutions to the constraints that provide a better value for the objective function. In fact, the roundoff solution above gives the farmer a plan that will earn $1468.30, while there are many other integral solutions that are somewhat better. Using techniques that we will discuss in Chapter 11, it can be determined that the very best plan would earn the farmer $1472.20.

However, without having to know the exact value of the optimal integer-valued solution, we can estimate how close to such a solution the roundoff process does bring us. The optimal integer-valued solution cannot yield a better value for the objective function than the linear optimum. The linear optimum for this model is 1474.82; this is an upper bound on the return that the best integer-valued solution could provide. By considering the ratio 1468.30/1474.82, we see that the roundoff solution must be within 0.5% of the best possible integer-valued solution, even without knowing this value. In other situations, as we will see, roundoff solutions will not be so satisfactory (or even feasible), and other methods must be used.

End of Model 2.2

A strategy that aims for an expedient solution to a difficult problem but not necessarily the optimal solution is called a *heuristic strategy* or simply a *heuristic*. The term heuristic is derived from the Greek word meaning *to find* and is used frequently in situations where a search of some sort must be carried out. Solutions to relaxed versions of a problem can be a helpful heuristic for finding a solution to the full problem. When the relaxed problem is a linear program, it is called a *linear relaxation*. In the previous model we solved the linear relaxation of a problem that required an integer-valued solution and used the solution to the relaxed problem to move to a solution to the more difficult problem.

In the next model we will use linear programming to help solve another planning problem. Here we will again need an integer-valued solution, but the naive roundoff techniques we have used will be inadequate. However, a second roundoff technique will yield much better results. Rounding up will be applicable in this next model because all the constraints will have the form

```
sigma(int i=0;i<=N;i++) a[i]*X[i] >= b ;
```

with all coefficients nonnegative. Constraints of this type are called *demand constraints*; these constraints are not violated when a solution is rounded upwards.

The following model is based on Anderson and Patny (1994). It is an example of an important kind of planning problem.

Model 2.3 Call 911

The service provided by the 911 system is a critically important part of the city's emergency response effort. The phones must be manned 24 hours a day, 7 days a week; the people answering the

2.2 Basic roundoff techniques

TABLE 2.1 Staffing Needs

	Number of Personal Communication Technicians Needed for Each of 168 Hours											
Mon	30	24	18	15	14	14	15	25	34	36	38	40
	41	43	46	57	57	59	61	59	55	50	45	38
Tues	32	25	20	17	15	13	17	25	32	35	38	40
	42	43	47	58	57	57	59	57	55	52	47	41
Wed	33	25	20	17	15	13	15	25	32	33	37	39
	42	43	47	57	56	57	57	56	53	50	47	41
Thurs	34	27	22	19	16	15	16	25	31	35	37	40
	44	45	48	57	57	56	58	56	53	53	46	41
Fri	34	28	23	19	16	15	17	25	33	37	39	42
	45	47	51	59	58	60	61	61	57	56	57	55
Sat	48	41	35	30	26	20	18	22	26	32	42	46
	49	53	54	56	56	56	59	59	57	57	56	56
Sun	52	46	41	34	29	23	18	19	25	31	36	41
	46	50	52	53	52	53	54	53	50	49	45	40

phones are known as Personal Communication Technicians or PCTs, a job title that describes the combination of skills required. After a careful analysis of the patterns of calls and demographic information for the coming year, analysts have drawn up a schedule of the number of operators who should be on hand at each of the 168 hours during the week in order to assure the necessary response rate. These figures are given in Table 2.1.

Both in order to economize on labor costs and to avoid the problems of overstaffing, what is needed is a way to meet this demand with a minimal or near-minimal number of PCTs. Because there is a union structure, a collective contract is negotiated, and a single schedule for the entire work force can be used. On the other hand, the union has insisted that no one be asked to arrive at work between the hours of 1 AM and 4 AM. Otherwise, all workers arrive on the hour, spend 8 hours at their post except for a one hour break after 4 hours. Each PCT has a work week of 5 consecutive days and then has 2 days off.

The task is to determine how many workers should start their shifts at each hour of the day so as to keep the total number of PCTs small. The model requires, of course, a solution that determines an integral number of people to start at each hour. Unavoidably there will be periods where the number of people available exceeds the number required for that period.

To model this application, we overlook the law of nature that people come in whole numbers and develop the constraints that correspond to a world with fractional workers. First, we can introduce an array for the number of PCTs needed during each hour that will be initialized with the data in Table 2.1.

```
int wneeded[168];
```

Since the constants of this model such as 168 and 24 are self-explanatory, we will not need to introduce symbolic constants. To represent the number of PCTs who come to work at each hour h during the week, we will have a continuous variable Pct[h]:

```
continuous Pct[168];
continuous Z; // For the objective function
```

The goal is to meet staffing needs while minimizing the total number of workers involved. The objective function is

```
Z == sigma(int h=0;h<168;h++) Pct[h];
```

The next thing, and the most difficult thing about this particular model, is to express the number of PCTs working at hour h. Certainly those who start the work week at h are there, as are those who started their work week an hour earlier and so on going back to hour h - 8, with the exception of those who started at h - 4 and who are now on break. This set of workers can be accounted for by the term

```
sigma(int j=0;j<8;j++) Pct[h-j] - Pct[h-4];
```

However, we must guard against h-j becoming negative. What we need here is arithmetic modulo 168, the number of hours in the week. If h-j is negative, we want to replace it by (h-j) + 168. On the other hand, if h-j is nonnegative, we still want h-j. In both cases the value we want is given by (h - j + 168)%168, where % is the operator for modular integer arithmetic.

In addition to the workers who have started their work week in the last 8 hours are those who started at the same times yesterday, the day before, and so on, for 4 previous days in all. The following constraint takes all these workers into account:

```
wneeded[h] <=
    sigma(int k=0; k<5; k++) ( // For 5 days
        sigma(int j=k*24; j<k*24+8; j++)// For 8 hours
            Pct[ (h+168-j)%168 ]
        -
        Pct[ (h+168-(k*24+4))%168 ]    // Less lunch
    )
```

2.2 Basic roundoff techniques 51

Notice that we have written this constraint with <= rather than ==. This will afford us a great deal of extra flexibility.

Again we have a situation where the linear optimum does not directly provide a usable solution, since we cannot engage a fractional number of workers. Here the naive roundoff strategy of rounding up the number of workers that the linear optimum prescribes for each hour will not violate the constraints. However, as we will see, this can lead to hiring significantly more PCTs than are really needed. A much better strategy is *progressive roundoff*: First, one rounds up the number of workers Pct[0] that the linear optimum prescribes for hour 0 to ceil(Pct[0]), and fixes Pct[0] at this integer value. It is here that we are using the fact that the constraints were not entered as equality constraints. Then, one resolves the linear programming problem to determine a new linear optimum, fix Pct[1] at ceil(Pct[1]), and repeats this process for each of the remaining hours of the week. This strategy can be coded by means of the and loop

```
and(int i=0;i<168;i++) {
    min: Z;
    Pct[i] == ceil(Pct[i]);
}
```

The union requirement that no one start work between 1 AM and 4 AM inclusive can be accounted for by fixing the number of workers who start at these hours at 0.0. So the main procedure can take the form

```
2lp_main(){

    wneeded = { ... } ; // Data from Table 2.1

    Z == sigma(int h=0;h<168;h++) Pct[h]; // Objective

    and(int k=0;k<7;k++)    // Union requirements
        and(int j=1;j<=4;j++)
            Pct[24*k+j] == 0.0;

    and(int h=0;h<168;h++) // Staffing requirements
        wneeded[h] <=
            sigma(int k=0; k<5; k++) (
                sigma(int j=k*24; j<k*24+8; j++)
                    Pct[ (h+168-j)%168 ]
                    -
                    Pct[ (h+168-(k*24+4))%168 ]
            ) ;

    min: Z;
    printf("The linear minimum is %.2f\n",Z);
```

```
    printf("Naive roundoff yields %.0f PCTs\n",
              sigma(int h=0;h<168;h++)ceil(Pct[h]));

    and(int i=0;i<168;i++) {    // Progressive roundoff
        min: Z;
        Pct[i] == ceil(Pct[i]);
    }

    printf("Progressive roundoff yields %.0f PCTs\n",Z);
}
```

Here is the output:

The linear minimum is 204.67
Naive roundoff yields 259 PCTs
Progressive roundoff yields 209 PCTs

Simply rounding the numbers up preserves the constraints but gives a solution that is very far from optimal; the progressive roundup gives a much better solution. In fact, from the linear minimum we know that the very best integer-valued solution to the constraints can be no better than 205 PCTs. So the result obtained is very good. One reason the heuristic used here is so good is that if we simplify the problem by eliminating the lunch break, we have a situation where the heuristic is guaranteed to find the optimal solution; see Bartholdi, Orlin, and Ratliff (1980). In other words, progressive roundoff provides the optimal solution to a closely related problem. Using techniques for simpler problems as a heuristic in working on more difficult problems is a time-honored technique.

An arithmetic note about solutions found by roundoff techniques is in order. The roundoff results depend on the specific sequence of optimal vertices obtained. If a linear programming problem has more than one optimal vertex, different vertices can be found by different implementations of the simplex algorithm and by the "same" simplex code on different platforms. This means that the roundoff results can be somewhat different.

End of Model 2.3

In the two models in this section, we have used continuous variables to solve problems that require discrete solutions. There are many situations where continuous methods can serve this way, as we will see.

The progressive roundoff technique of the last model is an example of an important family of problem-solving methods known as *local search* algorithms: The idea is to start with an initial solution or partial solution to the application at hand and then to improve this state iteratively by means of small changes. We will encounter many examples of this kind of search method.

Exercises

2.2.1. Compare the result of Model 2.3 with that you obtain when the restriction on not starting work between 1 AM and 4 AM is lifted. Would it be cost efficient for the city to pay $30 for car service for those employees arriving during these hours? Assume that the annual salary for a PCT is $30,000. Comment on the time required to solve the linear relaxation.

2.2.2. Consider the possibility in Model 2.3 of having workers put in a six hour day but with no lunch break. Find a schedule. Can you prove that the schedule found is the best possible in terms of minimizing the number of workers needed under this six hour day plan?

2.3 Bounding constraints

A very common kind of constraint is one that bounds a single continuous variable. For example,

```
Expenses <= 1000;
```

constrains the model to seek a solution in which expenses do not exceed one thousand. Constraints on a single continuous variable are called *bounding constraints*. Given the three continuous variables X, Y, Z and the bounding constraints

```
X <= 500; Y == 250.0; Z >= 100;
```

we say that X has an *upper bound* of 500 and Z a *lower bound* of 100. As for Y, we say that Y is *fixed* at 250.0. Binding a continuous variable to a fixed value is different, of course, from an assignment statement in that the continuous variable cannot now be fixed successfully at a different value.

Let's consider once again the geometry of constraints. A critically important fact about polyhedra is that the projection of a polyhedron onto a coordinate subspace is again a polyhedron. This is proved in Chapter 15. In particular, this means that the projection of a polyhedron onto a given coordinate axis is a one-dimensional polyhedron. In one dimension, a polyhedron is defined by bounding constraints on a single variable and reduces to an interval; this interval will be a finite interval of the form $[a, b]$ or an unbounded interval of the form $[a, +\infty)$.

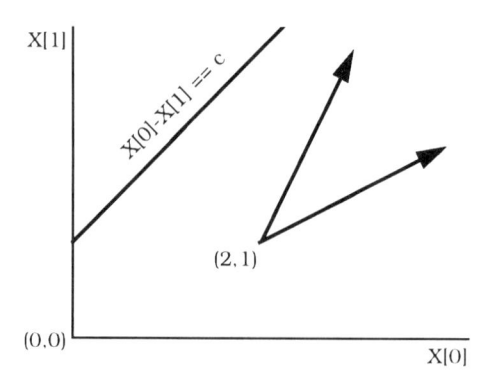

FIGURE 2.1 An unbounded region.

If a feasible region is such that its projection on every coordinate axis is a finite interval $[a, b]$, the region is said to be *bounded* and is called a *polytope*. If all the continuous variables of a program have upper bounds, then the constraints will always define a polytope; moreover every objective function will have a finite minimum and a finite maximum. However, it is possible for a feasible region to be unbounded. In this case the constraints will be consistent, but it might not be possible to determine an optimal value for a given objective function. Figure 2.1 illustrates the unbounded region resulting from the constraints

```
X[1] >= 0.5*X[0];
X[1] <= 2*X[0] - 3;
```

Here the objective `max: X[0]` would be unrealizable but the objective `min: X[0]` would be realizable. Moreover both `max: X[0] - X[1]` and `min: X[0] - X[1]` are unrealizable. In the case where an optimization call is not realizable, the 2LP system will print a message to the screen and the call will fail. Thus it is important to use bounding constraints in working with optimization applications. In fact, stating bounding constraints in linear programming is very analogous to proper initialization of variables in ordinary programming.

Constraints on more than one continuous variable are called *multi-variable constraints*. From the computational point of view, in maintaining the polyhedron defined by a collection of constraints, the multivariable constraints generate the bulk of the work for the simplex method. The reason for this is that with the simplex method, if M is the number of multivariable constraints, then the most compute-intensive tasks are to (cleverly) invert an M by M nonsingular matrix and to take $M + N$ scalar products of pairs of M-dimensional vectors, where N is the number of continuous variables declared in the program. Bounding information, on the other hand, can speed up the process by which the

2.3 Bounding constraints

simplex algorithm makes decisions on its way to determining an optimal vertex or in determining if a new constraint is feasible.

We can apply the above remarks to a situation that arises quite often. A pair of constraints of the form

```
a <= sigma(int i=0;i<N;i++) c[i]*X[i] + M ;
sigma(int i=0;i<N;i++) c[i]*X[i] + M <= b ;
```

constitute a *range constraint*. Since a and M can be adjusted, we can assume, without loss of generality, that a >= 0. Computation with a pair of this kind can then be simplified by introducing a new continuous variable Y, say, and rewriting the range constraint pair as

```
Y == sigma(int i=0;i<N;i++) c[i]*X[i] + M;
a <= Y;
Y <= b;
```

We have thus replaced two multivariable constraints by a single multivariable constraint and two bounding constraints. Linear programming packages typically have a facility for supporting range constraints in this or similar way.

Let us look at an example of range constraints. In Model 2.3, Call 911, we found a solution with only 209 workers. However, there are times in this schedule when there are as many as 20 extra workers for a given hour. It would make more sense to distribute the excess workers more evenly, both because of the confusion that can result when too many unnecessary people are milling around and because it would be better to have extra hands at other periods just in case. Let us try to bring down the spike in the number of excess workers to 15. The number of workers who are there at hour h is given by the affine expression

```
sigma(int k=0; k<5; k++) (
    sigma(int j=k*24; j<k*24+8; j++)Pct[(h+168-j)%168]
    -
    Pct[ (h+168-(k*24+4))%168 ]
)
```

We can introduce continuous variables Y[168] to represent these numbers

```
and(int h=0;h<168;h++)
    sigma(int k=0; k<5; k++) (
    sigma(int j=k*24; j<k*24+8; j++)Pct[(h+168-j)%168]
    -
    Pct[ (h+168-(k*24+4))%168 ] )
    == Y[h];
```

Then the constraints become

```
and(int i=0;i<168;i++) {
   wneeded[h] <= Y[h];
   Y[h] <= wneeded + 15;
}
```

At this point the progressive roundoff can proceed as before, though it is not guaranteed to succeed. With the current data it does succeed and finds a solution again with 209 total workers, but now one where the work force is more smoothly distributed.

The simplest nontrivial linear program has but one multivariable constraint. This is called the *knapsack problem*, and it is the subject of the next model.

Model 2.4 The Linear Knapsack

In this model we wish to fill a knapsack with quantities of several different items so as to optimize the value of its contents. Let us describe the situation and introduce the notation at the same time.

There are N variables X[i], with i ranging from 0 to N-1, which represent the quantity of item i to be placed in the knapsack. A unit of item i has weight w[i] and value p[i]. The knapsack cannot be loaded beyond its maximum weight M. The amount available of item i is given by hi[i], and the amount of item i that must be included is lo[i]. The objective is to maximize the value of what can be packed in the knapsack, that is,

```
max: sigma(int i=0;i<N;i++) p[i]*X[i];
```

The only multivariable constraint on the X[i] that must be satisfied is

```
sigma(int i=0;i<N;i++) w[i]*X[i] <= M;
```

Knapsack models occur frequently, often as a subproblem in a larger context. In typical situations, the X[i] will represent products to be manufactured or investments to be made, w[i] will be the amount of resource or capital required for each unit of X[i], and p[i] will be the profit or return on each unit of X[i]. The quantity M is the upper bound on the resources available and hi[i] is a limit on X[i].

We can easily formulate the code for solving this linear programming problem; sample data are included in the program.

```
#define M 85
#define N 8
double p[N],w[N],lo[N],hi[N];
continuous X[N];
```

2.3 Bounding constraints

```
21p_main()
{
    p  = {  9.5,  12.25, 10.75, 7.0,  9.0,  12.50, 4.75, 5.0  };
    w  = {  5.00,  7.00,  6.0,  6.50, 9.0,  13.75, 6.25, 9.50 };
    lo = {  0.2,   0.1,   0.2,  0.0,  2.1,  1.4,   0.4,  1.2  };
    hi = {  1.0,   0.4,   1.2,  0.2,  5.1,  3.4,   2.4,  4.2  };

    and(int i=0;i<N;i++) {
        lo[i] <= X[i];
        X[i]  <= hi[i];
    }

    sigma(int i=0;i<N;i++) w[i]*X[i] <= M;

    max: sigma(int i=0;i<N;i++) p[i]*X[i];

    and(int i=0;i<N;i++)
        printf("Pack %.2f of item %d in knapsack\n",X[i],i);
}
```

The output is

```
Pack 1.00 of item 0 in knapsack
Pack 0.40 of item 1 in knapsack
Pack 1.20 of item 2 in knapsack
Pack 0.20 of item 3 in knapsack
Pack 3.95 of item 4 in knapsack
Pack 1.40 of item 5 in knapsack
Pack 0.40 of item 6 in knapsack
Pack 1.20 of item 7 in knapsack
```

Notice that the variables are ordered so that the ratios `p[i]/w[i]` go in order from largest to smallest as `i` goes from 0 to N-1; that is, these ratios form a nonincreasing sequence. In other words, the items `X[i]` are ordered with the "most profitable" items first. Note also that the solution found follows a simple pattern: As much as possible of item `X[0]` is used, then as much as possible of `X[1]`, `X[2]`, and `X[3]`. At this point the maximum possible amount of item `X[4]` that still allows for the minimum amounts of items `X[5]`, `X[6]`, and `X[7]` is added. The remaining items are only used at the minimum required. So the solution follows the strategy of using as much as possible of the most profitable item still available.

A strategy that always chooses the most gainful next move is called a *greedy strategy*, and an algorithm that uses this strategy is a *greedy algorithm*. It can be shown directly that for the linear knapsack problem with nonnegative `p[i]` and `w[i]`, the greedy strategy always yields the optimal solution; the proof is left as an exercise. Thus this is a situation where a simple discrete method can be used to solve a problem that is naturally formulated in terms of continuous variables. For an introduction to greedy algorithms in other settings, see Kozen (1992).

Let us also note that we have declared the variables of the model as global variables, rather than as local variables in `2lp_main`. This distinction will stay somewhat moot until the next chapter, where subprocedures and parameters are discussed. In Section 4.2 we return to this model and at that point the use of global variables will serve to simplify the discussion.

End of Model 2.4

In 2LP models, it is not necessary to state a nonnegativity constraint such as

```
X[i] >= 0;
```

since nonnegativity is a property of the type `continuous`. Relaxing this constraint and freeing a variable to take on negative values is sometimes required in an application. There is a classic mechanism for doing this. We break the variable X into positive and negative parts, Xp and Xm, and substitute the difference Xp - Xm for every occurrence of X. This device can be very useful indeed. For example, if we introduce a variable Net to represent cash flow and set

```
Net == Income - Expenses;
```

then, unless we are sure that Income will always exceed Expenses, we have to allow for negative values. So it's more prudent to express this as

```
Netp - Netm == Income - Expenses;
```

This technique will be used in the next model, which is another example of a multiperiod problem.

Model 2.5 The Paris Bourse

A day trader at the Paris Bourse works for an important pension plan. Her job is to trade foreign currencies during the day; her expertise is in trading francs, marks, florins, and eurodollars. Prices in these currencies fluctuate during the day, and the trader earns money for the pension fund by "buying low and selling high." Typically, she begins the trading day with 10,000,000 francs in capital and buys and sells francs and the other currencies from 9 AM until closing at 3 PM. At close, she consolidates her position completely in francs.

2.3 Bounding constraints

TABLE 2.2 Yesterday's Currency Exchange Prices in Francs

Time	Mark	Florin	Dollar	Time	Mark	Florin	Dollar
9:00	2.50	3.50	5.50	12:00	2.50	3.495	5.50
9:15	2.495	3.495	5.50	12:15	2.505	3.50	5.505
9:30	2.495	3.49	5.50	12:30	2.505	3.50	5.505
9:45	2.50	3.485	5.50	12:45	2.50	3.50	5.50
10:00	2.50	3.49	5.50	1:00	2.50	3.50	5.50
10:15	2.495	3.495	5.505	1:15	2.495	3.495	5.498
10:30	2.495	3.485	5.505	1:30	2.495	3.495	5.498
10:45	2.50	3.49	5.50	1:45	2.50	3.50	5.50
11:00	2.50	3.49	5.50	2:00	2.50	3.50	5.50
11:15	2.495	3.485	5.50	2:15	2.505	3.505	5.501
11:30	2.495	3.49	5.50	2:30	2.50	3.505	5.50
11:45	2.50	3.495	5.50	2:45	2.50	3.50	5.50

It will often be advantageous to buy significant amounts of another currency when its price is attractive. The trader is authorized to run a deficit of up to 2,500,000 francs; however, in the firm's tradition, it is considered *gauche* to use more than 25% of this allowance on the average. The exchange tacks on a .1% charge for every trade. The posted prices for the various currencies are updated approximately every 15 minutes. Yesterday's postings in francs are given in Table 2.2.

At the end of this day, the trader had a position of 10,050,321 francs. She would like to be able to evaluate her performance by comparing it to the best possible results a trader could have had that day.

The task is to use hindsight to determine what would have been an optimal trading strategy for the day. Since we have knowledge of the exchange rates only at each 15 minute breakpoint, we will assume trading is done at these prices during these intervals. Let us define symbolic constants for the basic dimensions of the problem

```
#define CURRS 3    // Kinds of currency other than Francs
#define BRKPTS 24  // 15-minute breakpoints
```

and an array `pr[BRKPTS][CURRS]` of type `double` for the data. What the program must determine are the values of the trader's holdings in francs and the other currencies at each breakpoint during the day. Since the position in francs can become negative, let us set up the two continuous arrays `Fp[BRKPTS+1]`, `Fm[BRKPTS+1]` for the position in francs at each breakpoint and at closing. We will also need the continuous array `Holdings[BRKPTS+1][CURRS]` to represent the positions in the other currencies.

The initial and final positions are captured by the code

```
// Initial and final positions
Fp[0] - Fm[0] == 10000000;
and(int j=0;j<CURRS;j++)
    Holdings[0][j] == 0;
and(int j=0;j<CURRS;j++)
    Holdings[BRKPTS][j] == 0;
```

Since the final position is completely in francs, we want to maximize `Fp[BRKPTS] - Fm[BRKPTS]`. For each of the currencies we will need to determine the amount of that currency to buy or to sell during the period from breakpoint `i-1` to breakpoint `i`. For this purpose let us use the arrays `Buy[BRKPTS+1][CURRS]` and `Sell[BRKPTS+1][CURRS]` of type `continuous`. The relations that link the positions in the different currencies from breakpoint `i-1` to `i` are therefore summed up by

```
// Links
and(int i=1;i<BRKPTS+1;i++){
  and(int j=0;j<CURRS;j++)
    Holdings[i][j] ==
            Holdings[i-1][j] + Buy[i][j] - Sell[i][j];
  Fp[i] - Fm[i] == Fp[i-1] - Fm[i-1] +
       sigma(int j=0;j<CURRS;j++)
       pr[i-1][j]*(.999*Sell[i][j] - 1.001*Buy[i][j]);
}
```

Note that the position in francs, the local currency, reflects the commission of .1% that the exchange exacts on all transactions.

To allow the debit in the account to go no lower than 2.5 million francs and to control the average negative balance, we require the following code:

```
// Allow negative positions
and(int i=1;i<BRKPTS+1;i++)
    Fp[i] - Fm[i] >= -2500000; //Absolute bound on debit
(1.0/BRKPTS)*sigma(int i=0;i<BRKPTS;i++)Fm[i]
<= .25*2500000;       // Bound on average debit
```

The trader would also like to know what the best possible outcome would have been had she not been able to take a debit position in francs. This can be easily modeled by adding the constraints `Fp[i] - Fm[i] >= 0` and re-optimizing of the objective function. So the header and the main routine can be written as

2.3 Bounding constraints

```
2lp_main()
{
double pr[BRKPTS][CURRS];

continuous Fp[BRKPTS+1],Fm[BRKPTS+1];
continuous Holdings[BRKPTS+1][CURRS];
continuous Buy[BRKPTS+1][CURRS],Sell[BRKPTS+1][CURRS];

    pr = {  ...  };            // Data from Table 2.2

    // Initial and final positions
    ...
    // Links
    ...
    // Allow negative positions
    ...
    // First output
    max: Fp[BRKPTS] - Fm[BRKPTS];
    printf("With debit, the optimal result was %.2f\n",
                        Fp[BRKPTS] - Fm[BRKPTS]);
    // Eliminate debit
    and(int i=0;i<BRKPTS+1;i++)
        Fp[i] - Fm[i] >= 0.0;          // Reset bounds at 0.0
    // Second output
    max: Fp[BRKPTS] - Fm[BRKPTS];
    printf("Without debit, it was %.2f\n",
                        Fp[BRKPTS] - Fm[BRKPTS]);
}
```

The output is

With debit, the optimal result was 10061796.26
Without debit, it was 10051776.21

The trader's performance was certainly creditable, given the results of the program. But with hindsight it means that significantly more money could have been made. The trader can console herself with the thought that had she been trading with the optimal strategy provided by hindsight, her behavior itself would eventually have influenced the market and the prices. Therefore this strategy would have been unusable after awhile.

<u>End of Model 2.5</u>

Exercises

2.3.1. Prove that the greedy algorithm finds the optimal solution to the linear knapsack problem. Hint: Use induction on the number of variables.

2.3.2. Using range constraints, run the problem of Model 2.3, Call 911, with limits between 10 and 19 on the spike in the number of excess workers at any one time. Report on the results.

2.3.3. Solve the following system of linear equations using 2LP:

$$
\begin{aligned}
x_0 + x_1 + 4x_2 + 2x_3 + -2x_4 + -4x_5 &= -3.3 \\
2x_0 + 3x_1 + x_2 + -4x_3 + 5x_4 + x_5 &= -2.2 \\
x_0 + 2x_1 + 5x_2 + x_3 + -1x_4 + -2x_5 &= 4.0 \\
3x_0 + 5x_1 + x_2 + 2x_3 + -3x_4 + -6x_5 &= 0.0 \\
3x_0 + -9x_1 + 6x_2 + 3x_3 + -3x_4 + -6x_5 &= 0.0 \\
x_0 + 4x_1 + -2x_2 + 3x_3 + 2x_4 + -2x_5 &= 0.0
\end{aligned}
$$

3
Structured Linear Programming

In this chapter we introduce procedures and other programming mechanisms that enable us to write abstract and flexible models. Program structure can be an important aid in the process of modeling applications.

3.1 Program structure and procedures

A 2LP program consists of definitions of symbolic constants, declarations of global variables, a main procedure and other procedures. The symbolic constants can go anywhere in the program but must be defined before they are used. The global declarations begin the program proper. Then come the procedures including the `2lp_main()` procedure. This procedure has the form

```
2lp_main()
{
    <local declarations>
    <statements>
}
```

This procedure can be placed anywhere in the program; the usual practice is to make it either the first or the last procedure. Other procedures may take parameters and so have the form

```
proc_name(<parameters>)
{
    <local declarations>
    <statements>
}
```

In addition to the types `continuous`, `double`, and `int`, the 2LP language also supports the type `string` which can be used for simple output; the valid operations on this type are assignment and printing with the format `%s`.

Variables can be either local or global. Continuous variables are of static storage class; typically all variables of type `continuous` are either global variables or are declared in `2lp_main()`.

Procedures give structure to the logic of a program and enable data and output to be separated more easily from the rest of the program. There are three ways to make variables accessible to procedures in a program. They can be received as parameters, they can be referenced as global variables, or they can be accessed directly as local storage in the subroutine. In this section we consider global variables and procedures without parameters. Parameter passing is discussed in Section 3.2, and local variables in procedures other than `2lp_main` are discussed in Section 3.3.

A *statement* is a constraint, an assignment, a procedure call, a block built up from other statements using braces, or a compound statement constructed with one of the other programming constructs of 2LP. During run time, procedures add constraints and execute assignment statements and tests, as well as call other procedures. As with constraints and tests, procedure calls either succeed or fail. The result of a successful procedure call is the generation of new constraints on the continuous variables and the assignment of new values to the other variables. If the constraints generated by the procedure call prove infeasible, or if a test fails, or if a subroutine fails, then the procedure itself fails. If a procedure fails, it has the same effect as when a constraint or test fails. With the programming constructs discussed so far, failure means that the program will exit without finding a feasible solution; starting with Chapter 4, we will introduce methods for proceeding after failure is encountered.

The next model illustrates how the basic components of a program can be separated by the use of procedure calls. In this model all the constraints will have the form

```
sigma(int i=0;i<=N;i++) a[i]*X[i] <= b ;
```

with all coefficients nonnegative. Constraints of this kind are called *resource constraints* and a model with this structure is called a *resource allocation problem* or a *product mix problem*. The terminology is used because this situation typically arises when there are products competing for limited resources, and the goal is to optimize revenue or some other criterion given the available resources. As an example already encountered, we have Model 2.4, The Linear Knapsack.

3.1 Program structure and procedures

TABLE 3.1 Chip Requirements

	Fuji	Hanla	Huang	Kosciusko	Available
Chip 0	1	1	0	3	6200
Chip 1	2	1	2	1	5500
Chip 2	1	1	1	3	8800
Revenue	70	78.50	42.50	230	

Model 3.1 The Pacific Rim

A new computer company on the Pacific Rim wants to maximize its revenue on its four models of scientific workstations. The models are named for famous mountains in different countries in the region. Each workstation requires various processor chips and memory chips. Supplies of these components are limited. Table 3.1 summarizes the number of components that are required for manufacturing each of the four models and lists the number of pieces of each category that are available. The table also gives the revenue expected from each type of workstation in Australian dollars. How many of each of the models should be made in order to maximize revenue?

Approaching this application in a top-down manner, the program can be separated into four parts: (1) entering the data on the available chips and expected revenue for each type of workstation, (2) setting up the production constraints on the type and number of chips required for each workstation, (3) setting up the objective function and optimizing, and finally (4) producing usable output. These steps are encapsulated in the procedure calls made from `21p_main` in the code that follows:

```
#define FU 0
#define HA 1
#define HU 2
#define KO 3

#define CHIPS 3
#define STATIONS 4

int chips[CHIPS];
double revenue[STATIONS];
int rqd_chps[CHIPS][STATIONS];
string models[STATIONS];
continuous Workstations[STATIONS],Revenue;

data()
{
    chips = { 6200, 5500, 8800 };
```

```
        revenue = { 70, 78.50, 42.50, 230 };
        rqd_chps = {
                1, 1, 0, 3,
                2, 1, 2, 1,
                1, 1, 1, 3
        };
}

constraints()
{
        and(int i=0;i<CHIPS;i++)
                sigma(int j=0; j<STATIONS; j++)
                        rqd_chps[i][j]*Workstations[j]
                <= chips[i];
}

objective()
{
        Revenue ==
                sigma(int i=0;i<STATIONS;i++)
                        revenue[i]*Workstations[i];

        max: Revenue;
}

output()
{
        models =
                {"Fuji","Hanla","Huang","Kosciusko"};

        printf("The plan calls for production of\n");
        and(int m=0;m<STATIONS;m++)
                printf("%8.2f\t%s models\n",
                                Workstations[m],models[m]);
        printf("Revenue will be %.2f\n",Revenue;
}

2lp_main()
{
        data();
        constraints();
        objective();
        output();
}
```

Here we have placed the 2lp_main procedure last. The placement of the main routine in a program is a matter of taste, *de gustibus non est*

3.2 Parameters

disputandum. The solution calls for the production of only Huang and Kosciusko workstations:

```
The plan calls for production of
     0.00  Fuji models
     0.00  Hanla models
  1716.67  Huang models
  2066.67  Kosciusko models
Revenue will be 548291.67
```

Note that the solution calls for fractional numbers of workstations of two types to be produced. The simplest thing to do is to round these fractional quantities down to the nearest integer if necessary for reporting purposes. In fact, since rounding down only reduces the amount of each resource to be used, the constraints are not violated. Even without this, the scale of the quantities involved makes it possible to get a usable plan from the fractional solution. In Chapter 7 we will see other advantages to modeling product mix problems such as this one with continuous variables.

End of Model 3.1

In product mix problems there is a simple duality that can be noted. Continuous variables are used to represent products, and the left-hand sides of the ≤ constraints are used to represent resources.

Exercises

3.1.1. Rewrite Model 2.4, The Linear Knapsack, using procedures.

3.1.2. Rewrite Model 2.2, The Chicken or the Egg, using procedures.

3.2 Parameters

Parameters are used to allow variables to be shared with procedures. For procedures that take parameters, the parameters can be passed either *by value* or *by reference*. The method by which a parameter is received is determined by the signature of the subroutine and not by the calling routine. An individual identifier is called a *scalar*. For scalar parameters of type `int`, `double`, or `string` call by reference is signaled by the use of the ampersand in the parameter list, following C++ conventions. Thus the procedure

```
proc(int i, double & d)
{
    d = i*d + 2.0;
    i = 5;
}
```

receives `i` by value and `d` by reference. The first parameter `i` of `proc` is received by value; hence the procedure receives the value of the argument in the calling statement but not its address. In contrast, when a parameter is passed by reference, any change will simultaneously affect the variable used as the argument in the calling statement. For example, if the above procedure is called by

```
        proc(a,b);
```

the value of `b` will be modified by the procedure, but the value of `a` will remain unchanged.

As noted above, loop control variables can only be assigned in the loop header. Because of this and the fact that 2LP supports logical control structures to be introduced in later chapters, loop variables can only be passed by value.

Since continuous variables do not have values as such, they can only be received by reference. Thus no ampersand is used in the header of a procedure to indicate that an individual continuous variable is received by reference. For example, the procedure

```
proc(continuous X,Y, double & d)
{
    X + Y <= d;
}
```

receives the continuous parameters `X` and `Y`, and the parameter `d`, all by reference.

When a procedure receives an array as a parameter, it is a call by reference. The ampersand is not used when declaring an array parameter. To receive an array, a procedure's signature should include one pair of brackets for each dimension of the array; the size of the first dimension may be omitted. For example, a procedure with the signature

```
data(double price[][KINDS], hardness[])
```

is used in the next model to fill two arrays with initial data. The `price` array is handled within the procedure as a two-dimensional array with rows of length `KINDS`, while the `hardness` array is handled as a one-dimensional array.

To illustrate these points concretely, let us rewrite Model 2.5, The Paris Bourse, passing parameters to routines:

3.2 Parameters

```
#define CURRS 3
#define BRKPTS 24

2lp_main()
{
double prices[BRKPTS][CURRS];

continuous Fp[BRKPTS+1],Fm[BRKPTS+1];
continuous Holdings[BRKPTS+1][CURRS];
continuous Buy[BRKPTS+1][CURRS],Sell[BRKPTS+1][CURRS];

    data(prices);
    initial_and_final_positions(Fp,Fm,Holdings);
    links(Holdings,Buy,Sell,Fp,Fm,prices);
    allow_negative_positions(Fp,Fm);
    first_output(Fp[BRKPTS],Fm[BRKPTS]);
    eliminate_debit(Fp,Fm);
    second_output(Fp[BRKPTS],Fm[BRKPTS]);
}

data(double p[][CURRS])
{
    p = { ... } ;              // Data from Table 2.2
}

initial_and_final_positions(continuous Fp[],Fm[],
                                            H[][CURRS])
{
    Fp[0] - Fm[0] == 10000000;
    and(int j=0;j<CURRS;j++) {
        H[0][j] == 0.0;
        H[BRKPTS][j] == 0.0;
    }
}

links(continuous H[][CURRS],B[][CURRS],S[][CURRS],
                      Fp[],Fm[],double p[][CURRS])
{
    and(int i=1;i<BRKPTS+1;i++) {
        and(int j=0;j<CURRS;j++)
            H[i][j] == H[i-1][j] + B[i][j] - S[i][j];
        Fp[i] - Fm[i] ==
        Fp[i-1] - Fm[i-1] +
        sigma(int j=0;j<CURRS;j++)
            p[i-1][j]*(.999*S[i][j] - 1.001*B[i][j]);
    }
}
```

```
allow_negative_positions(continuous Fp[],Fm[])
{
    and(int i=1;i<BRKPTS+1;i++)
        Fp[i] - Fm[i] >= -2500000;
    (1.0/BRKPTS)*sigma(int i=0;i<BRKPTS;i++)Fm[i]
    <= .25*2500000;
}

....   and so on ...
```

In this example, we have to pass the various arrays and individual variables to the subroutines as parameters because they are all declared within 21p_main. Had they been declared as global storage, they could be passed in the same way to the subroutines without otherwise changing the code.

At times we may wish to pass only a portion of an array to a subroutine. For example, to determine whether all the row sums of the integer array a[M][N] are equal, we can write the procedures

```
first_row(int a_row[N],int & sum)
{
    sum = sigma(int j=0;j<N;j++) a_row[j];
}

other_row(int a_row[N],int sum)
{
    sum == sigma(int j=0;j<N;j++) a_row[j];
}
```

and then call these procedures, using arguments a[0] and a[i] to pass each particular row as follows:

```
    first_row(a[0],row_sum);
    and (int i=1;i<M;i++)
        other_row(a[i],row_sum);
```

More generally, these remarks apply to arrays of any dimension or type. When we pass a one-dimensional portion of an array, we call it a *vector*. A higher-dimensional portion of an array is called a *slice*. As an example, consider an array X of 5 dimensions; we can pass a three-dimensional slice of X using a calling statement such as

```
        proc(X[i][j]);
```

where the signature of proc is

```
proc(continuous Slice[][L][M])
```

3.2 Parameters

In Model 1.1 ingredients for an animal's food were blended in a sack at the minimum cost. The constraints ensured that the animal's diet was nutritious. The next model, which is adapted from Williams (1994), is another example of a blending problem. However, this is a multiperiod application, in that the results of activities in one period influence the resources available for the next period. In this way it is like Model 1.2, Model 2.2, and Model 2.5. In order to encapsulate the activity for a typical period, the code will make use of the ability to pass arrays, slices, and vectors as parameters to subroutines.

Model 3.2 British Cooking

A small company in the UK produces a very fine cooking oil which is used for preparing Fish 'n' Chips and other popular dishes. This oil is made by blending 5 different vegetable oils. The unprocessed oils are purchased from distilleries. Purchase of the raw ingredients and production of the finished product are organized on a monthly basis and management would like to formulate a 6-month plan to optimize revenue. The following table gives the posted price in pounds sterling per 100-liter vat of each raw vegetable oil for the next 6 months. These prices can be locked in now by means of futures contracts, as shown in Table 3.2. Given the product's popularity, all the cooking oil that is produced during a given month can be sold during that month and brings in £150 per vat. The company's process is efficient, and for all practical purposes the amount of cooking oil produced is equal to the sum of the amounts of the vegetable oils used.

Vegetable oil that is in stock but not used during a given month can be kept in storage and used in the following months. In fact, at the start of this 6-month period, there are 500 vats of each kind of vegetable oil in storage, and the planner is asked to provide for the same quantities of oil in storage at the end of the 6 months. The storage capacity for each vegetable oil is 1000 vats; the cost of keeping oil in storage is assessed at the rate of £5 per vat at the beginning of each month.

TABLE 3.2 Raw Oil Prices

	Corn	Peanut	Safflower	Sesame	Sunflower
January	115	110	130	120	110
February	115	90	110	130	130
March	95	100	130	140	110
April	125	120	120	110	120
May	105	110	150	120	100
June	135	80	140	100	90

TABLE 3.3 Hardness

	Corn	Peanut	Safflower	Sesame	Sunflower
Hardness	5.0	4.2	2.0	6.1	8.8

Processing the cooking oil is subject to various requirements. Each month there is a limit of 250 vats on the total amount of corn, peanut, and safflower oil that can be processed and a limit of 200 vats on the total amount of sesame and sunflower oil that can be used. The blended cooking oil must also meet some hardness requirements. Each vegetable oil has a hardness coefficient which is a number between 1.0 and 10.0. The weighted average of hardness of the oils in the blend must be between 3.0 and 6.0. The coefficients are given in Table 3.3.

The object is to formulate a 6-month plan that will specify the amount of each vegetable oil to purchase each month, the amount of cooking oil to produce each month, and the amount of each vegetable oil to hold over in storage for the succeeding month.

Let us begin the design of this model by considering the data and dimensions given above. In the code we will use the symbolic constants MONTHS and KINDS to represent the number of months, 6, and the kinds of oil, 5. Therefore the data of Tables 3.2 and 3.3 will be stored in the arrays price[MONTHS][KINDS] and hardness[KINDS], both of type double.

Based on the above information, the symbolic constants and data arrays used in the program will include

```
#define MONTHS 6         // Number of periods
#define KINDS 5          // Kinds of vegetable oils
#define VATREV 150       // Selling price per vat
#define STORCOST 5       // Storage cost per vat
#define INITSTOR 500     // Initial storage in vats
#define ENDSTOR 500      // Final storage in vats
#define STORCAP 1000     // Monthly storage capacity

double price[MONTHS][KINDS];   // Table 3.2
double hardness[KINDS];        // Table 3.3
```

Now let us turn to the quantities that have to be determined. The revenue can be computed as the sale price for the total amount of oil processed minus the cost of purchasing and storing oil. We will use the continuous variable Revenue to represent this. Naturally we wish to maximize the value of Revenue.

The blended cooking oil must also meet the hardness requirements. Each vegetable oil has a hardness coefficient that is a number between 1.0 and 10.0 stored in the array hardness[KINDS] of type double. The

3.2 Parameters

weighted average of the coefficients of the oils in the blend must be between 3.0 and 6.0, which we can call LBHARD and UBHARD.

Reiterating, the objective is to formulate a 6-month plan that will specify the amount of each raw oil to purchase each month, the amount of cooking oil to produce each month, and the amount of each unprocessed oil to hold over in storage for the succeeding month. We can encapsulate this in an array Plan[MONTHS][MODES][KINDS] of type continuous, where MODES is defined as 3. For improved readability of the code, we will also define PURC, PROC, and STOR to represent the three modes: purchased, processed, and stored:

```
#define MODES 3        // Purchase, Process or Store
#define PURC 0
#define PROC 1
#define STOR 2
```

The declarations of the continuous variables will be

```
continuous Plan[MONTHS][MODES][KINDS];
continuous Revenue;
```

Continuing to approach this application in a top-down manner, we can formulate an outline for the program as follows:

1. Fill the price and hardness arrays.
2. Make constraints on the initial and final storage amounts.
3. Constrain the monthly storage capacities.
4. Constrain the hardness and production limit of the blend for each month.
5. Link the months so that oils stored last month may be used this month.
6. Write an equation for the revenue and maximize the revenue
7. Print the plan.

The procedures to carry out these tasks can now be written as follows:

```
2lp_main()
{
    data(price,hardness);

    first_stor(Plan[0]);
    last_stor(Plan[MONTHS-1][STOR]);

    and(int i=0;i<MONTHS;i++){
        storage_caps(Plan[i][STOR]);
        blend_constraints(Plan[i][PROC]);
    }
```

```
    and(int i=1;i<MONTHS;i++)
        link(Plan[i-1][STOR],Plan[i][PURC],
            Plan[i][PROC],Plan[i][STOR]);

    find_revenue(Plan,price,Revenue);
    max: Revenue;

    output(Revenue,Plan);
}
```

Let us now develop each procedure. First, `data` merely fills its two array parameters (necessarily passed by reference) with the data from the corresponding tables. Note that we have not specified the size of each array's first dimension.

```
data(double price[][KINDS],double hardness[])
{
    price = {
            115, 110, 130, 120, 110,
            115,  90, 110, 130, 130,
             95, 100, 130, 140, 110,
            125, 120, 120, 110, 120,
            105, 110, 150, 120, 100,
            135,  80, 140, 100,  90
    };

    hardness = { 5.0, 4.2, 2.0, 6.1, 8.8 };
}
```

The procedure `first_stor` then needs to set up the storage equality for the end of the first month for each kind of vegetable oil. This will be the initial storage of 500 (or INITSTOR), adjusted by the amounts purchased and used during the first month. This gives us

```
first_stor(continuous FirstMonth[MODES][KINDS])
{
    and(int j=0;j<KINDS;j++)
        FirstMonth[STOR][j] ==
                INITSTOR - FirstMonth[PROC][j] +
                                FirstMonth[PURC][j];
}
```

Since we need to refer only to the first month of the `Plan` array, we pass only the slice `Plan[0]` by reference to `first_stor`, and not the entire array. In essence the procedure treats `FirstMonth` as a two-dimensional array.

3.2 Parameters

The procedure `last_stor` is simpler. The amount of each oil stored at the end of the last month must be bound to 500, or `ENDSTOR`. Only the vector of oils to be stored in the last month is passed to the procedure from the `Plan` array. This procedure can be written as

```
last_stor(continuous LastMonth[KINDS])
{
    and(int j=0;j<KINDS;j++)
        LastMonth[j] == ENDSTOR;
}
```

The `storage_caps` procedure will bound the storage each month by at most 1000 vats, or `STORCAP`. The program calls this procedure for each month and passes only the vector of the `Plan` array that refers to that month's storage, namely `Plan[i][STOR]`. Each of the elements of this vector, called `Stored` in this procedure, should be bound by `STORCAP` as follows:

```
storage_caps(continuous Stored[KINDS])
{
    and(int j=0;j<KINDS;j++)
        Stored[j] <= STORCAP;
}
```

Let us now turn to the production constraints that tell us how oils of different kinds can be processed together. Following our plan, these constraints are to be stated in a procedure `blend_constraints`. This routine requires the vector `Plan[i][PROC]` of the oils to be processed in month `i`. We will use

```
blend_constraints(continuous Oils[KINDS])
```

as the procedure signature, where the identifier `Oils` refers to the vector of this month's processed oils.

Let us develop the `blend_constraints` routine. Processing the cooking oil is subject to various conditions. Each month there is a limit of 250 vats on the combined total amount of corn, peanut, and safflower oil that can be processed and a limit of 200 vats on the combined amount of sesame and sunflower oil that can be used. So let us set

```
#define LIMITA 250      // Limit on group A production
#define LIMITB 200      // Limit on group B production
```

Since we have enumerated the subgroup A of corn, peanut, and safflower oils as 0, 1, and 2 and the subgroup B of sesame and sunflower oils as 3 and 4, we can separate these categories by means of two simple symbolic constants:

```
#define A 2
#define B 3
```

Two blending constraints will enforce these bounds:

```
    sigma(int j=0;j<=A;j++) Oils[j] <= LIMITA;
    sigma(int j=B;j<KINDS;j++) Oils[j] <= LIMITB;
```

There are also constant bounds for the hardness in the blend; so we set

```
#define LBHARD 3.0     // Lower bound on average hardness
#define UBHARD 6.0     // Upper bound on average hardness
```

We now need to write the product's hardness constraint. We may at first be tempted to calculate the weighted average using the equation

$$WAvg = \frac{\sum hardness_i \times Oils_i}{\sum Oils_i}$$

and to bound WAvg by LBHARD below and UBHARD above. Unfortunately, the equation as written involves division by an expression of type continuous, which violates the typing rules of linear programming described in Section 1.4.

However, crossmultiplying allows us to transform the ratio constraint

$$\frac{\sum hardness_i \times Oils_i}{\sum Oils_i} \geq LBHARD$$

into the valid linear constraint,

$$\sum hardness_i \times Oils_i \geq LBHARD \times \sum Oils_i$$

In code, this becomes

```
    sigma(int i=0;i<N;i++) hardness[i]*Oils[i]
            >= LBHARD*sigma(int j=0;j<N;j++) Oils[j];
```

Similarly the constraint for the upper bound of UBHARD can be written as

3.2 Parameters

```
    sigma(int i=0;i<N;i++) hardness[i]*Oils[i]
            <= UBHARD*sigma(int j=0;j<N;j++) Oils[j];
```

We can now join these statements into the procedure

```
blend_constraints(continuous Oils[KINDS])
{
    sigma(int j=0;j<=A;j++) Oils[j] <= LIMITA;
    sigma(int j=B;j<KINDS;j++) Oils[j] <= LIMITB;
    sigma(int i=0;i<KINDS;i++) hardness[i]*Oils[i]
            >= LBHARD*sigma(int j=0;j<KINDS;j++) Oils[j];
    sigma(int i=0;i<KINDS;i++) hardness[i]*Oils[i]
            <= UBHARD*sigma(int j=0;j<KINDS;j++) Oils[j];
}
```

This brings us to the `link` procedure, for which we use the signature

```
link(continuous OldSt[KINDS],Purc[KINDS],
                            Proc[KINDS],NewSt[KINDS])
```

The parameters are vectors that represent the amount of each kind of oil that has been held over in storage from the previous month and the amounts that are to be purchased, processed, and stored this month, respectively. Each kind of oil can be treated separately, so for each kind i, the link routine can call a subroutine `inner_link`. The link procedure will be

```
link(continuous OldSt[KINDS],Purc[KINDS],
                            Proc[KINDS],NewSt[KINDS])
{
    and(int k=0;k<KINDS;k++)
        innerlink(OldSt[k],Purc[k],Proc[k],NewSt[k]);
}
```

In the `innerlink` procedure we need to state that the oil to be processed plus the oil to be held over in storage must be equal to the oil already in storage plus the oil purchased this month. This condition can be written as a constraint:

```
innerlink(continuous OldSt,Purc,Proc,NewSt)
{
    OldSt + Purc == Proc + NewSt;
}
```

Finally, we come to the procedure to find the revenue, which will be an exercise in double summations. Since the production process is conservative, we can compute the amount of oil produced as the sum of the individual vegetable oils processed. Given the entire Plan as a parameter, for each month i and each kind of oil j, the contribution to the revenue will be the amount received for selling processed oil minus the amount spent on buying raw oil and minus the cost of storing oil, which we can write as

```
VATREV*Plan[i][PROC][j] -
    price[i][j]*Plan[i][PURC][j] -
        STORCOST*Plan[i][STOR][j]
```

The net revenue will therefore be the double summation of the terms over i and j, giving us the procedure

```
find_revenue(continuous Plan[MONTHS][MODES][KINDS],
        double price[MONTHS][KINDS], continuous Revenue)
{
    Revenue ==
            sigma(int i=0;i<MONTHS;i++)
                sigma(int j=0;j<KINDS;j++)
                    ( VATREV*Plan[i][PROC][j] -
                        price[i][j]*Plan[i][PURC][j] -
                            STORCOST*Plan[i][STOR][j] );
}
```

There remains the output routine to develop, and this can be done in many ways. Here is one way:

```
output(continuous Revenue, Plan[MONTHS][MODES][KINDS])
{
string months[MONTHS],modes[MODES],kinds[KINDS];

    months = { "Jan","Feb","Mar","Apr","May","June" };
    kinds =
    { "Corn","Peanut","Safflower","Sesame","Sunflower" };
    modes = { "Buy","Process","Store" };

    printf("Maximum revenue is %.2f\n",Revenue);

    and(int m=0;m<MONTHS;m++) {
        printf("\nIn the month of %s\n",months[m]);
        and(int k=0;k<KINDS;k++) {
            printf(" %s oil: ",kinds[k]);
            and(int t=0;t<MODES;t++)
                printf(" %s %.0f vats.\t",
                    modes[t],Plan[m][t][k]);
```

```
            printf("\n");
      }
      if  m == 0; then
      printf("Make a sale of %.0f vats.\n",
            sigma(int k=0;k<KINDS;k++)
              (INITSTOR+Plan[m][PURC][k]-Plan[m][STOR][k]));
      else
      printf("Make a sale of %.0f vats.\n",
            sigma(int k=0;k<KINDS;k++)
              (Plan[m-1][STOR][k]
                 + Plan[m][PURC][k] - Plan[m][STOR][k]));
   }
}
```

With this output routine the maximum revenue and the report for January are

Maximum revenue is 107842.59
In the month of Jan
Corn oil: Buy 0 vats. Process 0 vats. Store 500 vats .
Peanut oil: Buy 0 vats. Process 250 vats. Store 250 vats .
Safflower oil: Buy 0 vats. Process 0 vats. Store 500 vats.
Sesame oil: Buy 0 vats Process 115 vats. Store 385 vats .
Sunflower oil: Buy 0 vats. Process 85 vats. Store 415 vats .
Make a sale of 450 vats.

End of Model 3.2

Exercises

3.2.1. Rewrite Model 2.3, Call 911, using procedures and parameters.

3.2.2. Rewrite Model 1.2, The Library Fund, using procedures and parameters.

3.3 Sparse data

In the last model the data were given in two tables, and all the entries were nonzero values. When all or the greater part of the entries are nonzero, the data are called *dense*. It often happens that the great majority of data elements are zeros; in this case the data are called *sparse*. Representing sparse data is a much discussed topic in the literature.

TABLE 3.4 Sparse Data

Constraint	Number of Nonzeros	Indices	Coefficients	Right Hand Side
0	4	3, 7, 17, 1	1.0, 2.2, 2.4, -.5	8.8
1	3	0, 20, 14	3.3, 5.0, 6.7	4.4
2	3	1, 6, 13	2.7, 8.3, -9.1	3.9

In dense form, a sequence of M constraints on N continuous variables can be written as

```
and(int i=0;i<M;i++)
    sigma(int j=0;j<N;j++) a[i][j]*X[j] <= b[i];
```

However, if most of the a[i][j] are zero, the data for the constraints can be represented more economically. A typical way of representing such a set of constraints in a sparse manner is, for each constraint i < M,

1. List the number of nonzero coefficients in the constraint.
2. List the indices j of the continuous variables X[j] that have nonzero coefficients in the constraint.
3. List the values of these coefficients.
4. List the right hand-side value b[i].

For example, consider the data in Table 3.4. The first constraint is

```
1.0*X[3] + 2.2*X[7] + 2.4*X[17] + -.5*X[1] <= 8.8;
```

To encode this data set, let M be a symbolic constant representing the number of constraints, in this case 3. The total number of nonzero coefficients in the data is 10, and this number can be represented by the symbolic constant, TOTNZ. We let the integer array nnz[M] contain the number of nonzero coefficients in each constraint and the array rhs[M] of type double contain the right-hand side constants. Also we let the integer array indices[TOTNZ] list the indices of the continuous variables with nonzero coefficients, and we let cffs[TOTNZ] be an array of type double that lists the corresponding coefficients. For the data in Table 3.4, the initializations will be

```
nnz = { 4, 3, 3 }; // Number of nonzeros
indices = {         // A one-dimensional array
        3, 7, 17, 1,
        0, 20, 14,
        1, 6, 13
};
```

3.3 Sparse data

```
cffs = {              // Also a one-dimensional array
        1.0, 2.2, 2.4, -.5,
        3.3, 5.0, 6.7,
        2.7, 8.3, -9.1
};
rhs = { 8.8, 4.4, 3.9 };   // Constants in constraints
```

We can generate these constraints using auxiliary identifiers `begin` and `end` of type `int` by means of code such as

```
end = 0;
and(int i=0;i<M;i++) {
    begin = end;     // Start where last constraint ended
    end = end + nnz[i];   // Add number of nonzeros
    sigma(int j=begin;j<end;j++) cffs[j]*X[indices[j]]
       <= rhs[i];
}
```

To encapsulate the process of loading sets of constraints given this way, we can introduce a procedure for sparse constraints with relational operator `<=`:

```
sparslteq(continuous X[],int nnz[],indices[],
                        double cffs[],rhs[],int m)
{
int begin,end;

    end = 0;
    and(int i=0;i<m;i++) {
        begin = end;
        end = end + nnz[i];
        sigma(int j=begin;j<end;j++) cffs[j]*X[indices[j]]
           <= rhs[i];
    }
}
```

and different data collections can be passed to the same routine. In a perfectly similar way we can write procedures `sparsgteq` for collections of sparse `>=` constraints and `sparseq` for `==` constraints.

We make use of this `sparslteq` routine in the next model. It is similar to the Model 3.1, The Pacific Rim, in that it too is a product mix problem.

Model 3.3 Motorcycles Inc.

An enterprise that assembles motorcycles has the ability to produce a line of 12 different models. Weekly production is planned with data

TABLE 3.5 Critical Cycle Parts

Part Number	Number of Cycles Requiring Part	Cycles Requiring Part	Quantity of Part Required by Cycle	Number of Part Available
0	3	0,2,4	2,3,12	1100
1	4	0,1,4,5	10,4,4,4	1200
2	4	1,2,6,11	1,4,1,2	1650
3	3	3,4,5	3,1,2	1822
4	4	4,6,7,8	1,4,16,2	1411
5	3	6,7,8	2,12,12	820
6	5	3,5,6,7,8	1,1,4,3,5	1950
7	5	3,4,9,10,11	12,10,1,1,5	1910
8	6	1,4,6,7,8,9	1,10,2,1,3,3	1130
9	4	3,4,10,11	14,4,1,1	1230
10	4	2,6,9,4	3,4,12,4	950
11	4	3,4,5,9	12,2,10,1	632
12	4	3,4,10,11	4,4,1,1	831
13	4	2,4,6,9	3,1,12,4	920
14	3	5,9,10	2,12,1	950

TABLE 3.6 Cycle Model Data

Cycle	0	1	2	3	4	5	6	7	8	9	10	11
Income	60	100	60	150	50	120	160	150	130	200	120	140
Limit	280	250	300	555	125	125	160	150	333	434	300	400

that are updated right before the start of each week. The data include the availability of certain critical parts both from inventory and from deliveries of ordered parts that will be ready in time for next week's production. As listed in Table 3.5, each model of motorcycle uses a certain number of these different critical components. Also there are the expected income to the company per unit of the different models and an upper bound on how many of each model can be expected to be sold. The job is to decide how many of each model to produce in order to maximize revenue. The relevant data are listed in Tables 3.5 and 3.6.

Notation for this application is easily set up. We can introduce symbolic constants and arrays for the data:

```
#define MODS 12
#define PARTS 15
```

3.3 Sparse data

```
#define TOTNZ 60

double income[MODS],cffs[TOTNZ],hi[MODS],avail[PARTS];
int indices[TOTNZ],nnz[PARTS];
```

Let us introduce continuous variables `Bike[i]` to represent the number of units of each model to make:

```
continuous Bike[MODS];
continuous Z;                    // For objective function
```

The data can be entered in the appropriate arrays by a call to a routine `data`. In partial detail this routine is

```
data()
{
    nnz = { 3,..., 3 };// Number of cycles requiring part

    indices = {
            0, 2, 4,         // Cycles requiring part
               ...
            5, 9, 10
    };

    cffs = {
            2, 3, 12,      // Quantity of part required by cycle
               ...
            2, 12, 1
    };

    avail = { 1100,..., 950 };    // Quantity of part
                                  // available

    income = { 60, ..., 140 };    // Revenue

    hi = { 280,... 400 };         // Limit
}
```

A `setup` routine will be used to specify the upper bounds on the continuous variables, to define the objective function, and to call `sparslteq` to generate the resource constraints:

```
setup()
{
    // Bounds
    and(int i=0;i<MODS;i++)
        Bike[i] <= hi[i];
```

```
    // Objective function
    Z == sigma(int i=0;i<MODS;i++) income[i]*Bike[i];

    // Sparse resource constraints
    sparslteq(Bike,nnz,indices,cffs,avail,PARTS);
}
```

The main routine can now be written as

```
2lp_main()
{
    data();
    setup();

    max: Z;

    printf("Manufacture %.0f motorcycles\n",
        sigma(int i=0;i<MODS;i++) Bike[i]) ;
    printf("The projected revenue is %.0f\n",Z);

    detailed_output();        // Left to reader
}
```

The program produces a plan that projects revenue of 141014 based on the manufacture of 1217 motorcycles. Note that these numbers are not obtained by rounding off the fractional value of each wp(Bike[i]) but rather by rounding down a sum of terms. This is a situation where the scale of operations is such that the fractional solution represents a fitting approximation to the process.

End of Model 3.3

We also make use of the basic technique we have set up for handling sparse data in the next model. The model is an example of a classic application known as the *set covering problem*.

Model 3.4 The Committee

The board of a large school district is organizing a committee of citizens to prepare a complete report on all aspects of education, its quality, its costs, its administration, and so on. Many different brands of expertise will have to be represented on this committee. A long list has been prepared of people who are willing to serve together with their areas of expertise. However, it is important to keep the size of the committee down.

In the Table 3.7, the rows represent the different kinds of expertise required. The second column lists the citizens with that expertise, and

3.3 Sparse data

TABLE 3.7 Committee Data

Expertise	Citizens with Expertise	Size of Set	Expertise	Citizens with Expertise	Size of Set
0	2,4,7,18,22	5	21	12,14,17,8,2	5
1	8,0,6,19,32	5	22	28,11,6,9,2	5
2	7,3,6,13,25	5	23	27,23,26,3,5	5
3	4,5,14,29	4	24	14,15,4,9	4
4	10,18,27	3	25	0,8,17	3
5	11,12,20,23	4	26	1,19,10,13	4
6	12,14,7,22	4	27	2,24,7,12	4
7	18,1,16,28	4	28	8,0,16,18	4
8	17,13,16,28	4	29	7,13,6,28	4
9	12,4,17,18,22	5	30	2,24,17,8,12	5
10	18,14,6,19,32	5	31	28,31,26,9,32	5
11	17,3,26,13,25	5	32	7,23,6,31,15	5
12	4,5,14,29	4	33	24,5,14,19	4
13	16,18,27	3	34	21,28,7	3
14	11,31,20,23	4	35	11,31,30,3	4
15	12,14,7,22	4	36	18,4,27,32	4
16	18,32,16,28	4	37	8,32,17,25	4
17	17,13,32	3	38	7,23,32	3
18	19,30,26,28	4	39	29,30,3,28	4
19	15,32,16,18	4	40	5,30,13,18	4
20	17,22,26,28	4	41	7,32,26,31	4

the third column lists the size of that set. The number of types of expertise is 42, the number of citizens is 33, and the number of entries in the second column is 174. The task at hand is to determine a committee and to keep its size under control.

Let us make the definitions

```
#define NUM 33
#define EXPS 42
#define TOTNZ 174
```

With each expertise is associated the set of citizens having that expertise. Let us declare an array `people[TOTNZ]` to hold the elements of these sets and an array `sz_of_set[EXPS]` to list the size of each set. Both of these arrays are of type `int`. The data can be entered by a call to a procedure such as

```
data()
{
    people = {
            2, 4, 7, 18, 22,     // Citizens with expertise
                 ...
            7,32,26,31
    };
    sz_of_set = { 5,..., 4 };  // Size of set
}
```

The unknown here is whether or not an individual citizen is to be on the committee. Let us introduce a continuous variable Citizen[i] for each citizen to represent membership on the committee. The idea will be for this variable to be fixed at 1.0 if the citizen is to serve and at 0.0 otherwise. The declaration of the array of citizens and a variable for the objective function will be

```
continuous Citizen[NUM], Z;
```

Now to express that a sufficient number of committee members have a given type of expertise j, we can require that the sum of the Citizen[i], where i is in the set of those with this type of expertise, must be greater than or equal to 1. With these constraints all the variables Citizen[i] have coefficient equal to 1. So to make use of the generic routine sparsgteq, the >= analog of sparslteq, we introduce local array rhs and cffs of type double and set all their entries to 1.0. This can be done in a setup routine as follows:

```
setup()
{
double cffs[TOTNZ], rhs[EXPS];

    and(int j=0;j<NUM;j++)
        Citizen[j] <= 1.0;              // Upper bounds

    and(int i=0;i<TOTNZ;i++)            // All 1's
        cffs[i] = 1.0;

    and(int i=0;i<EXPS;i++)             // All 1's
        rhs[i] = 1.0;

    // Generic >= routine
    sparsgteq(Citizen,sz_of_set,people,cffs,rhs,EXPS);

    // Objective function
    Z == sigma(int k=0;k<NUM;k++) Citizen[k];
}
```

3.3 Sparse data

Though not critical, it is a good idea to bound the variable `Citizen[i]` by `1.0`, since this is better programming practice. It also emphasizes that a person can only appear once on the committee; if the model were changed so that several members of the committee were to have some given expertise, these bounding constraints would then be necessary.

To determine an acceptable committee, our code must decide who exactly is to serve. For this a solution to the constraints that have been set up must be found where each citizen is either on the committee or not. In Model 2.3, Call 911, we used progressive roundoff in a similar situation. This technique can also be used here, since fixing `Citizen[i]` at the ceiling of its witness point value will not violate the constraints: in this model all the constraints have the form

```
sigma(int i=0;i<=N;i++) a[i]*X[i] >= b ;
```

with all coefficients nonnegative. Constraints of this type are demand constraints and are not violated when a solution is rounded upwards. Using the progressive roundoff method to find a solution leads to the code

```
21p_main()
{
    data();
    setup();

    and(int i=0;i<NUM;i++) {    // Progressive roundoff loop
        min: Z;                 // Optimize the linear relaxation
        Citizen[i] == ceil(Citizen[i]);   // Round up
    }
    output();                   // Left to reader
}
```

This program yields a good solution to the school board's problem, keeping the size of the committee down to 10.

The term "set covering" derives from the fact that this application can be viewed as constructing a collection of sets whose union covers a given set. Each citizen can be thought of as a set of expertises, and the committee can be thought of as a collection of these sets whose union includes all the expertises. So a citizen is a set and an expertise is an element. Therefore the continuous variable `Citizen[j]` of the model represents the set of expertises of the jth citizen. The left-hand side of constraint i represents the ith expertise, and the constraint itself expresses the fact that the expertise has to be represented on the committee. So the variables represent sets, and the left-hand sides of the constraints represent elements. In the progressive roundoff strategy, we proceed through the list of sets and decide whether or not to place a set in the cover. There is a dual way of looking at this situation: One can cycle through the elements to ensure that some set goes into the cover that contains that element. This alternative way of building a solution is taken up in Section 12.3.

This model is similar to Model 2.3, Call 911. However, the progressive roundoff heuristic works much more effectively in Model 2.3. The reason is that the scheduling in Model 2.3 has a special cyclic structure that makes the problem close to one for which the heuristic is exact. The problem in the current model is very general and does not have this kind of special structure.

The progressive roundoff heuristic used for this model leads to a valid solution because of the special form of the constraints. For example, if we add the requirement that the size of the committee be limited to at most a certain number of members, then things change dramatically. For one thing, there is no *a priori* guarantee that a solution does exist. This form of the model is known as the *minimum cover problem*. Methods for addressing this version of the problem will be developed in later chapters, starting in Chapter 8. A point to retain, however, is that techniques for solving the simpler version of this and other problems can serve as heuristics for solving the more complex versions.

<u>End of Model 3.4</u>

Data sets for the set covering problem can be obtained on the internet from the OR-Library via anonymous `ftp` to `mscmga.ms.ic.ac.uk` or at the web page `http://mscmga.ms.ic.ac.uk/`.

Exercises

3.3.1. Write a greedy discrete algorithm for the set covering problem. Try it on the data of Model 3.4.

3.3.2. Suppose that the number of citizens needed in Model 3.4 with certain skills is greater than 1. Will progressive roundoff still be guaranteed to find a solution if there is one?

3.3.3. Obtain data sets for the set covering problem from the internet; test your skill.

3.3.4. A company has 10 resources and it uses them to make 12 products. Table 3.8 lists the amount of each resource consumed in the production of a unit of each product and the revenue brought in by the sale of a unit of each product. Table 3.9 gives the amount available of each resource in hundreds of units. Write a linear program to find a production strategy which optimizes revenue.

3.4 Stochastic programming

TABLE 3.8 Resource Requirements

Product Number	Number of Resources Required	Resources Required	Quantity of Resource Required by Product	Revenue per Unit of Product
0	3	0,2,4	22,23,12	40
1	4	0,1,4,5	30,14,44,9	80
2	4	1,2,6,11	21,24,11,22	73
3	3	3,4,5	33,10,21	80
4	4	4,6,7,8	11,14,16,32	40
5	3	6,7,8	42,52,12	88
6	5	3,5,6,7,8	12,26,9,8,5	90
7	5	3,4,9,10,11	5,29,9,7,25	110
8	6	1,4,6,7,8,9	4,30,6,4,7,9	220
9	4	3,4,10,11	24,14,19,31	145

TABLE 3.9 Resource Data in Hundreds

Resource	0	1	2	3	4	5	6	7	8	9	10	11
Available	14	25	36	22	40	38	20	11	22	45	26	55

3.4 Stochastic programming

In this section we develop a method for dealing with uncertainty in linear programming models. The method will be based on some basic ideas from probability theory.

In Model 3.2, British Cooking, prices for oils to be purchased in the following months were treated as known in advance because futures contracts could be obtained to guard against price swings. So the program could be written with a fixed view of what the future had in store. A given view like this of future events is called a *scenario*. In a multiperiod situation, the decision-making process splits naturally into two parts. There are the decisions that are made at the outset and those that are to be made later. The decisions to be made at the outset are called *stage one decisions*. In this section we develop another approach to multiperiod planning that integrates different future scenarios into the determination of the course of action to take at the first stage. The idea of the method is to formulate each of the scenarios independently and then to merge the scenarios by having them share the first-stage decisions. Each of the future scenarios must be assigned a probability and these must sum to 1.0. In the merged model the objective function can

TABLE 3.10 Interest Rate Scenarios

	Now	In 3 Months	In 6 Months	In 9 Months
Scenario 1: Down				
3 Month CDs	1.02	1.0175	1.01625	1.015
6 Month CDs	1.0425	1.040	1.3875	NA
12 Month CDs	1.09	NA	NA	NA
Scenario 2: Steady				
3 Month CDs	1.02	1.02	1.02	1.02
6 Month CDs	1.0425	1.0425	1.0425	NA
12 Month CDs	1.09	NA	NA	NA
Scenario 3: Up				
3 Month CDs	1.02	1.0225	1.02375	1.03
6 Month CDs	1.0425	1.045	1.04625	NA
12 Month CDs	1.09	NA	NA	NA

then be set to weigh the effect of the future scenarios according to their likelihoods.

We will illustrate these ideas by going back to Model 1.2. There, the treasurer had to insure that investment of the library fund in CDs generated a certain cash flow every 3 months and achieved a certain end of year result. The assumptions were that interest rates would not change in the months ahead; however, these assumptions have been challenged.

Model 3.5 The Library Fund Reprised

After meeting with the committee, the treasurer consulted the library's accountant who did not think it reasonable to suppose that interest rates would stay more or less fixed. In fact, the accountant sketched three reasonable scenarios: interest rates could drift up, they could stay the same, or they could drift down. Rates of total return on the different CDs for these scenarios are listed in Table 3.10. In the accountant's considered opinion, these scenarios have the probabilities .25, .35, and .40, respectively.

The treasurer's task in planning the investment of the fund is now more complex. The challenge is to formulate an initial investment strategy that will leave the treasurer in a good position to adapt to changing interest rates at later breakpoints. Here as in Model 1.2, the key things to consider are the breakpoints that occur at 3-month intervals, the CDs that can be purchased at each breakpoint, and the CDs purchased earlier that can yield cash flow at a given breakpoint. Cash flow at a breakpoint is the amount of liquid capital that is available and the cash

3.4 Stochastic programming

flow from the CDs must meet the liquidity requirement of $75,000 set by the library committee. So let us first set up machinery to handle a given interest rate scenario, effectively rewriting Model 1.2 in a more structured way.

We begin by defining some relevant symbolic constants

```
#define SC 3        // Number of interest rate scenarios
#define BKS 4       // Number of breakpoints at which
                    // investments can be made
#define NUM 3       // Number of different kinds of CDs
#define NA 0.0      // Not applicable
```

Since we are dealing with 3-, 6- and 12-month CDs, we can record those that mature at a breakpoint and those that can be purchased at a breakpoint in an arrays mtrs of dimension NUM by 1+BKS and an array prchsbl of dimension NUM by BKS as follows:

```
mtrs = {     // CD i matures at breakpoint j if entry is 1
    0, 1, 1, 1, 1,
    0, 0, 1, 1, 1,
    0, 0, 0, 0, 1
};

prchsbl = {          // CD i can be purchased at
    1, 1, 1, 1,      // breakpoint j if entry is 1
    1, 1, 1, 0,
    1, 0, 0, 0
};
```

Each CD has a time to maturity that can be stored in an array ttm[NUM] of type int:

```
ttm = {3/3,6/3,12/3}; // Time to maturity in breakpoints
```

The entries $1, 2, 4$ reflect the facts that 3-month CDs mature in one breakpoint, 6-month CDs in two breakpoints, and 12-month CDs in four breakpoints.

Next we can access the interest rates of three scenarios as the slices of a master three-dimensional array r[SC][NUM][BKS] that will be initialized as

```
r = {
    // Total rate of return in scenario k of
    // CD i at breakpoint j
    1.02,    1.01785, 1.01625, 1.015,
    1.0425,  1.040,   1.03875, NA,      // Scenario 0
    1.09,    NA,      NA,      NA,
```

3 Structured Linear Programming

```
        1.02,   1.02,   1.02,    1.02,
        1.0425, 1.0425, 1.0425,  NA,       // Scenario 1
        1.09,   NA,     NA,      NA,

        1.02,   1.0225, 1.02375, 1.03,
        1.0425, 1.045,  1.04625, NA,       // Scenario 2
        1.09,   NA,     NA,      NA
    };
```

We can record the probabilities associated with each scenario in a global array `prob` initialized as

```
    prob = { .25, .35, .40 }; // Probabilities of scenarios
```

Two final data items are the initial value of the fund and the minimum amount of cash that is to be on hand at each breakpoint; in the notation of Model 1.2, these are `infd` and `lq`, and they are initialized to `100000` and `75000`, respectively. All of these initializations can be grouped in a routine `data`.

The unknowns are the amounts to be invested in each type of CD at each breakpoint and the cash flow that the investment stream generates at each breakpoint. These are the unknowns for each scenario. Let us make the declarations

```
continuous Cd[SC][NUM][BKS],CashFlow[SC][BKS+1];
```

For each scenario s, the initial cash flow is equal to `infd`, and since everything is to be in cash at the end, the value of the fund at the end of the year will be the final cash flow `CashFlow[s][BKS]`. For each scenario the current and future total returns on the CDs are represented by the two-dimensional slice `Cd[s][NUM][BKS]`, which is an array of dimension NUM by BKS. The analysis of a scenario can be encapsulated in a procedure call

```
scenario(continuous CashFlow[],Cd[][BKS],double r[][BKS])
{
    CashFlow[0] == infd;          // Initial fund

    and(int j=0;j<BKS;j++)        // Cash going out
        CashFlow[j] ==
        sigma(int i=0;i<NUM;i++) prchsbl[i][j]*Cd[i][j];

    and(int j=1;j<=BKS;j++)       // Cash coming in
        CashFlow[j] ==
            sigma(int i=0;i<NUM;i++)
                mtrs[i][j]*r[i][j-ttm[i]]*Cd[i][j-ttm[i]];
```

3.4 Stochastic programming

```
    and(int j=0;j<=BKS;j++)    // Liquidity constraints
        CashFlow[j] >= lq;

    and(int i=0;i<NUM;i++)     // Good programming practice
        and(int j=0;j<BKS;j++)
            if prchsbl[i][j] == 0;
            then Cd[i][j] == 0.0;

    max: CashFlow[BKS];        // Optimize worth

    printf("The fund will be worth %.2f\n",
                  CashFlow[BKS]);
}
```

Note that in this call, the scalars `lq`, `infd`, and the arrays `mtrs` and `prchsbl` are all global, while the continuous variables and the array of rates of total return are passed to this routine as parameters. Naturally all of these data structures could be passed as parameters, but for what we do next, it will be important that the array of rates and the continuous variables be parameters.

The three models can be run side by side by means of the code

```
    and(int s=0;s<SC;s++) {    // For each scenario
        printf("In scenario %d\n",s);
        scenario(CashFlow[s],Cd[s],r[s]);    // Run it
    }
```

However, we want to formulate the scenarios together and merge them into one super model.

We have before us three scenarios for future interest rate movement UP, STEADY, and DOWN. The decisions that are made at stage one will commit investment capital and limit the latitude in decision making at subsequent stages. The three models give three very different investment plans. Therefore what is wanted now is a stage one decision that somehow takes into account the likelihood of UP, STEADY and DOWN and that will tend to put the investor in a good position at the next breakpoint. To achieve this, we link the three scenarios together in a *stochastic model* by running the three scenarios while having them share the stage one continuous variables; it is these variables that determine the investments to be made at the start, at breakpoint zero. We make the objective function equal to the weighted sum of the objective functions of the three scenarios, where the weights are given by the probabilities. This has the effect of making the stage one decisions simultaneously sensitive to the three scenarios. To achieve the sharing of variables, for each CD i, we alias the variables `Cd[s][i][0]` for all

94   3 Structured Linear Programming

the scenarios s. This is done by means of equality constraints between pairs of continuous variables:

```
alias_first_stage_variables(continuous Cd[][NUM][BKS])
{
    and(int s=1;s<SC;s++)
        and(int i=0;i<NUM;i++)
            Cd[s][i][0] == Cd[s-1][i][0];// Share stage one
}
```

The new objective then is

```
max: sigma(int s=0;s<SC;s++) prob[s]*CashFlow[s][BKS];
```

This objective function measures the *expected value* of the fund. Because the first stages have been merged, this is very different from taking the weighted average of the results of the three scenarios independently.

Adding output routines and putting things together, we have

```
2lp_main()
{
continuous Cd[SC][NUM][BKS],CashFlow[SC][BKS+1];
string duration[NUM];            // For I/O

    data();
    duration = // Duration of CDs
        {"3 month","6 month","12 month"};

    and(int s=0;s<SC;s++)  // Create each scenario
        scenario(CashFlow[s],Cd[s],r[s]);

    alias_first_stage_variables(Cd);   // Merge stage ones

    max: sigma(int s=0;s<SC;s++) prob[s]*CashFlow[s][BKS];

    output_stage1(Cd[0],duration);

    and(int s=0;s<SC;s++)
        later_stages(s,CashFlow[s],Cd[s],duration);

}

output_stage1(continuous Cd[][BKS],string dur[])
{
    printf("\nAt stage one the plan is:\n\n");
    and(int i=0;i<NUM;i++)
```

3.4 Stochastic programming

```
            printf("Invest %.2f in a %s CD now\n",
                                            Cd[i][0],dur[i]);
}

later_stages(int s,continuous Flo[],Cd[][BKS],string dur[])
{
    printf("\nThe plan in scenario %d is then:\n\n",s);
    and(int j=1;j<BKS;j++)
        and(int i=0;i<NUM;i++) {
            printf("Invest %.2f ",Cd[i][j]);
            printf("in a %s CD in %d months\n",
                                            dur[i],3*j);
        }
    printf("The fund will be worth %.2f\n",Flo[BKS]);
}
```

Analyzing the output for the different scenarios, both when run independently and when merged into the stochastic model, can prove instructive. So far we have not taken into account the committee's requirement that the fund be worth at least $108250 at year's end. In fact, the output shows that this can simply not be achieved if interest rates go down for in this case the best possible result would be $107,621. Moreover, in this case, this best possible result is achievable only if the treasurer follows the optimal plan for this scenario from the start. However, if interest rates do go down, then the best possible result after the stage one plan recommended by the stochastic program would be $107,589. As a hedge against so-called *downside risk*, one could try to guarantee a result as good as $107,621 in the DOWN situation by adding the constraint `CashFlow[0][BKS] >= 107621` to the merged stochastic model. Rerunning the stochastic model with this additional constraint leads to a different stage one investment policy that offers greater protection in the case interest rates go down while not changing the final value of the fund in the STEADY case and only slightly reducing the best possible final value of the library fund in the UP case.

End of Model 3.5

In the above model we had a small number of scenarios, and each scenario had a small number of breakpoints. In complex models there can easily be a very large number of scenarios, each being a significant linear program in its own right.

Two important early papers on stochastic linear programming are Dantzig (1955) and Beale (1955). Two recent texts to consult are Hall and Wallace (1994) and Censor and Zenios (1996). The field is a growing one.

TABLE 3.11 New Interest Rate Scenario

	Now	In 3 Months	In 6 Months	In 9 Months
Scenario 4: Wayup				
3 Month CDs	1.02	1.0275	1.02825	1.033
6 Month CDs	1.0425	1.0450	1.0475	NA
12 Month CDs	1.09	NA	NA	NA

Exercises

3.4.1. To Model 3.5, the accountant has added a fourth scenario WAYUP which has interest rates given in Table 3.11. Suppose the probabilities of the four scenarios are 18%, 30%, 33% and 19%. Redo the model with these data. Then add the same hedge against downside risk as in Model 3.5.

3.4.2. Here is a one stage stochastic program. Consider the following change to Model 3.3, Motorcycles Inc. Suppose that the prices that the different models will fetch are not firm as is assumed there. Instead, one must take into account the fact that these prices will be a function of weather conditions. In dry weather motorcycles sell well; in normal weather they bring in the prices given in Table 3.6; in wet weather they sell poorly. The dry and wet weather prices are given in Table 3.12. The almanac states that the coming driving season will be dry with a probability of 53%, normal with a probability of 20%, and wet with a probability of 27%. The same upper limits on production apply as in Model 3.3. Plan a production strategy which maximizes the expected income.

TABLE 3.12 Dry and Wet Weather Cycle Prices

Cycle	0	1	2	3	4	5	6	7	8	9	10	11
Dry	70	120	80	170	80	140	190	180	150	260	160	170
Wet	45	85	50	130	45	114	150	130	118	184	100	125

4

Conjunction and Implication

In mathematical logic the four basic propositional connectives are conjunction, disjunction, implication, and negation. The passage from logic to computer programming alters the roles and interpretations of these logical connectives. A program has aspects such as time, sequence and state that are absent from mathematics. The presence of continuous variables and constraints affords an opportunity to generalize the semantics of the logical connectives in the programming language context in order to provide more powerful problem-solving tools.

So an important theme in this and succeeding chapters is how to extend the programming semantics of the logical connectives to encompass constraints. Another theme is the relation between the programming use of these connectives and their meaning in logic and mathematics.

4.1 Conjunction

In symbolic logic, conjunction is expressed by means of the connective $\wedge$. Thus to assert that p and q are both true, one writes $p \wedge q$. The individual elements of the conjunction, in this case p and q, are called *conjuncts*. The fundamental properties of conjunction are commutativity

$$p \wedge q \text{ is equivalent to } q \wedge p$$

and idempotence

$$p \wedge p \text{ is equivalent to } p$$

With the shift from mathematics to programming, procedural elements such as state and change of state are introduced, and the meaning of conjunction changes.

As we have seen, in 2LP conjunction is expressed either by simple juxtaposition

```
p(); q();
```

or by an and loop

```
and(int i=0;i<N;i++)
    p(i);
```

In 2LP as in other programming languages, the commutativity of conjunction cannot be assumed. For example, if a is currently set to 0, the two code segments

```
a == 1; a = 1;
```

and

```
a = 1; a == 1;
```

yield very different results: The first fails, while the second succeeds. However, if the statements are constraints, then they do commute in that the conjunction will succeed or fail independently of the order in which the constraints are called. Moreover, if they succeed, the feasible region is also the same. This extends to and loops of constraints. Thus, for example, the two statements

```
and(int i=0;i<N;i++)
    X[i] == i;
```

and

```
and(int i=N-1;i>=0;i--)
    X[i] == i;
```

will either both succeed or both fail. If they succeed, the two loops will lead to the same feasible region. However, the intermediate feasible regions generated during the loop will not usually be the same. Let us look now at an example where this consideration is important. The model is based on "what-if" analysis, a sampling technique that tests the effect that changes in data will have on the recommended solution.

Typically activities that bring a high level of reward carry a higher rate of risk than do less ambitious activities. Maximizing reward and minimizing risk are opposing objectives. A decision will usually be based on a consideration of the trade-off between the two. In this model we

4.1 Conjunction

consider an example of this situation where we can draw a graph to visualize this trade-off. To simplify things, the notation we use in this model is based on that of Model 2.4, The Linear Knapsack.

Model 4.1 Risk vs Reward

Fashion accessories are to be shipped from Miami to Montréal in a truck that has capacity M. There are N kinds of accessories in the shipment and the order calls for at least lo[i] of item i and at most hi[i]. Item i takes up w[i] space in the truck and brings in an expected profit of p[i]. Letting the continuous variable X[i] represent the quantity of item i in the shipment, we have the single resource constraint

```
sigma(int i=0;i<N;i++) w[i]*X[i] <= M;
```

So if all items were guaranteed to be purchased, we could then determine the most profitable shipping strategy by maximizing the objective function

```
sigma(int i=0;i<N;i++) p[i]*X[i];
```

as we did in Model 2.4.

However, consumer taste in fashion accessories as in many things is driven by fads. So with each item i is associated a fickle factor ff[i], which is a number between 100 and 1000 that measures the likelihood that this item will *not* sell in Montréal this time of year. In fact, the items that are more profitable are the ones that typically have a higher fickle factor ff[i]. So shipments with a higher hoped-for profit also have a greater likelihood of not being profitable at all.

The average risk that the company associates with a shipment is the ratio of

```
sigma(int i=0;i<N;i++) ff[i]*X[i];
```

to

```
sigma(int i=0;i<N;i++) X[i];
```

Therefore the risk of a shipment can be bounded by a given level, r, by means of the additional constraint

```
sigma(int i=0;i<N;i++) ff[i]*X[i]
    <=  r*sigma(int i=0;i<N;i++) X[i];
```

Optimizing the objective function with this constraint enforced yields a solution that is the best possible when the risk is restricted this way. To encapsulate this process, we introduce a procedure `evaluate_risk` as follows:

```
evaluate_risk(double ff[],w[],p[],r, continuous X[])
{
    sigma(int i=0;i<N;i++) ff[i]*X[i]
        <= r*sigma(int i=0;i<N;i++) X[i];

    max: sigma(int i=0;i<N;i++) p[i]*X[i];
}
```

To record the evaluation, we can use a routine such as

```
record_evaluation(double p[],r, continuous X[])
{
    printf("At risk %.0f the expected profit is %.2f\n",
           r,sigma(int i=0;i<N;i++) p[i]*X[i]);
}
```

The fashion company would like to get an idea of the trade-off between risk and reward for this shipment. The goal is to determine a situation they are comfortable with. To do this, we can make `r` a loop variable that goes from high to low risk, and we can compute the best possible reward when the risk is limited to each value of `r` in turn. The loop will go from a high fickle factor to a very low one. Let us set the symbolic constants HIRISK to 800 and LORISK to 200, and decrement the bound on the average fickle factor by 50 each time through the loop. So the loop will be

```
    and(int r=HIRISK;r>=LORISK;r=r-DCRMNT){
        evaluate_risk(ff,w,p,r,X);
        record_evaluation(p,r,X);
    }
```

where DCRMNT is a symbolic constant for the amount `r` is to be decremented each time through the loop, namely 50.

Incorporating the data and a routine to set up the linear knapsack problem, we have the code

```
#define M 800
#define N 12
#define HIRISK 800
#define LORISK 200
#define DCRMNT 50
```

4.1 Conjunction

```
21p_main()
{
double p[N],w[N],lo[N],hi[N],ff[N];
continuous X[N];

    data(p,w,lo,hi,ff);
    knapsack(lo,hi,w,p,X);

    and(int r=HIRISK;r>=LORISK;r=r-DCRMNT){
        evaluate_risk(ff,w,p,r,X);
        record_evaluation(p,r,X);
    }
}

data(double p[],w[],lo[],hi[],ff[])
{
    p  = {  28.8, 24.7, 19.7, 14.75, 10.7, 9.0,
            8.05, 7.0, 6.0, 5.0, 5.0, 4.0 };
    lo = {  0, 0, 10, 12, 15, 22, 50, 10, 8, 22, 5, 27 };
    w  = { .5, .4, .8, .7, .4, .5, .6, .8, .4, .3, .6, 1 };
    hi = {  290, 200, 200, 200, 240, 250,
            100, 230, 215, 300, 300, 280};
    ff = {  1000, 950, 950, 900, 400, 400, 200,
            190, 120, 110, 100, 100 };
}

knapsack(double lo[],hi[],w[],p[], continuous X[])
{
    and(int i=0;i<N;i++) {
        lo[i] <= X[i];
        X[i]  <= hi[i];
    }
    sigma(int i=0;i<N;i++) w[i]*X[i] <= M;
}
```

The output is

```
With risk 800 the expected profit is 25335.10
With risk 750 the expected profit is 25335.10
With risk 700 the expected profit is 25308.41
With risk 650 the expected profit is 25232.45
With risk 600 the expected profit is 24964.64
With risk 550 the expected profit is 24653.48
With risk 500 the expected profit is 23990.83
With risk 450 the expected profit is 22858.52
With risk 400 the expected profit is 20821.71
```

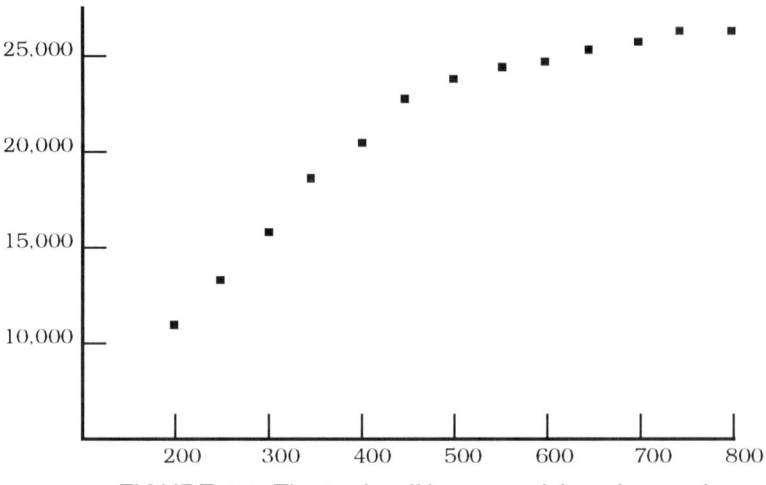

FIGURE 4.1 The trade-off between risk and reward.

With risk 350 the expected profit is 18147.40
With risk 300 the expected profit is 15576.06
With risk 250 the expected profit is 13098.41
With risk 200 the expected profit is 10651.46

These numbers lead to the graph in Figure 4.1 that relates risk and hoped-for profit. We see that assuming a risk between 500 and 600 nearly optimizes the expected profit.

In this model the loop header

```
and(int r=HIRISK;r>=LORISK;r=r-DCRMNT)
```

cannot be replaced by

```
and(int r=LORISK;r<=HIRISK;r=r+DCRMNT)
```

since this would have the effect of restricting the average fickle factor to LORISK at the very outset, and the later constraints bounding the risk would be redundant and have no effect on the feasible region. In other words, a sequence of constraints can only make the feasible region successively smaller, and a region can never be increased in size by further constraints. Thus the order in which the and loop generates the constraints is important.

The fickle factors in this model are drawn from some *ad hoc* measure of consumer caprice. In the literature such measures are called *uncertainty factors* (or dually, *certainty factors*); another term used is *confidence factor*.

4.2 Implication

In this model, if the constraints at some risk level r should prove infeasible, the program will simply halt and print a message to that effect. In the sections to follow, we will treat methods for detecting infeasibility and continuing the program.

End of Model 4.1

Languages such as C, C++, and LISP will only evaluate as many conjuncts as are required to determine that one of them fails; once a conjunct fails, evaluation ceases, and the conjunction as a whole fails. This strategy is known as *lazy* evaluation. It is also used in 2LP. Other languages such as Pascal evaluate all the conjuncts whether or not one or more of them fail; this is called *strict* evaluation.

As for the idempotence of conjunction, it is straightforward to come up with examples that show how it is violated by a mix of tests and assignment statements. When only constraints are involved, a form of idempotence remains because repeating a constraint does not change the feasible region.

Exercises

4.1.1. Give an example involving tests and assignment statements to show that conjunction is not necessarily idempotent.

4.1.2. To achieve market share in a product a sufficient number of units must be produced. Producing more units of a particular product in a product mix problem will have an effect on the objective function. Track the trade-off between market share and revenue in Models 3.1 and Model 3.3 by successively increasing the lower bond on a given product.

4.2 Implication

In mathematical logic, implication is expressed by means of the connective $\to$. Thus, to assert that p implies q, one writes $p \to q$. The proposition p is called the *antecedent* of the implication and q is called the *consequent*. We also read $p \to q$ as "if p then q."

In logic, the interpretation of $p \to q$ is that either p is false or both p and q are true. In classical programming languages, the interpretation of "if-then" does not maintain the link between the antecedent and the consequent. For example, suppose that a is currently set at 1, then the following C code segment cannot assign 0 to a and keep the antecedent true:

```
if (a == 1) a = 0;      /* C code */
```

In working with constraints, 2LP supports a version of the mathematical semantics of implication. Thus

```
if X <= 6;
then Y >= 5;
```

succeeds if either X <= 6 fails or both X <= 6 and Y >= 5 succeed together. If both X <= 6 and Y >= 5 succeed, the feasible region is changed by the addition of these constraints. If X <= 6 fails, the program continues without changing the feasible region. On the other hand, this if/then statement will itself fail if X <= 6 succeeds but then Y >= 5 fails. Another way of putting this is: Either the first constraint fails, or they succeed together; this is an analog of "either p is false or both p and q are true."

Formally, the syntax of the implication statement in 2LP is

```
if <statement>
then <statement>
```

Reiterating, the semantics are as follows: If the antecedent statement succeeds, the consequent statement is called; if the consequent fails, the if/then statement fails, otherwise it succeeds. If the antecedent statement fails, the if/then statement succeeds without calling the consequent statement.

In programming, one typically wants to provide an alternative course of action in the case the antecedent of an if/then construct fails. In this case we use the if/then/else statement whose syntax is

```
if <statement>
then <statement>
else <statement>
```

In this construct, if the antecedent fails, the statement following else is called. By way of example, consider the statement

```
if X >= Y;
then Z <= 100;
else Z >= 150;
```

If the constraint X >= Y is consistent with the current feasible region, then the conditional reduces to the conjunction X >= Y; Z <= 100. If the constraint X >= Y is not consistent with the current polyhedron, Z <= 100 is not invoked, but the constraint Z >= 150 is tried; in this case, if it is not feasible, the entire if/then/else statement fails.

The if/then and if/then/else statements are often called *conditional statements*. When constraints are not involved, the interpretation of implication reduces to the standard programming

4.2 Implication

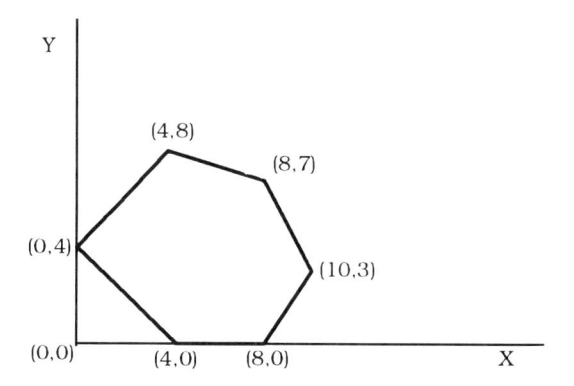

FIGURE 4.2 The original constrained region.

semantics for if/then/else. Thus, for example, if a and b are of type int or double, the code segment

```
if a > b;
then printf("Good morning world");
else printf("Good evening world");
```

will print "Good morning world" only if the test a > b succeeds; otherwise, it prints "Good evening world."

Let us illustrate the use of constraints in a conditional by stepping through a geometric example. In the code below the first five constraints are taken from Figure 1.2. They yield the region shown again in Figure 4.2. Now, in the code, we have added an if/then/else statement that as a result changes the polygon to that of Figure 4.3

```
21p_main()
{
continuous X,Y;

    -X + Y <= 4;
    X + 4*Y <= 36;
    2*X + Y <= 23;
    3*X - 2*Y <= 24;
    X + Y >= 4;

    // At this point the polyhedron is that of Figure 4.2

    if Y <= X;
    then Y >= 3.0;
    else Y <= 2.0;
}
```

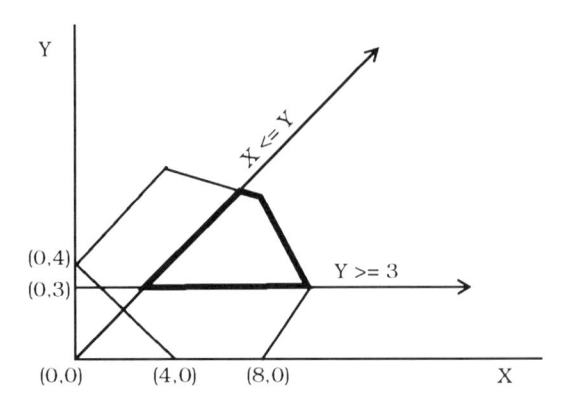

FIGURE 4.3 The constrained region further reduced by the addition of a conditional statement.

Since the constraint `X[1] <= X[0]` in the conditional succeeds, the polygon defined by the constraints is changed accordingly. Then the constraint `X[1] >= 3.0` is added to the polyhedral set, resulting in the graph in Figure 4.3.

As we have seen, statements can be grouped in a block by enclosing them within a matching pair of braces. This is especially useful if several constraints are to be generated in some part of an `if/then/else` statement. For example, if we change the `if/then/else` in the above code to

```
if Y <= X;
then {
     Y >= 3.0;
     X >= 8.0;
}
else {
     Y <= 2.0;
     X >= Y - 0.5;
}
```

we obtain the region drawn in Figure 4.4.

A block may be used in the `if` part of an `if/then/else` statement as well as in the `then` or `else` part. When a block is used in the `if` part, the entire block must succeed in order for the `then` branch to be executed. If any of the constraints in the `if` statement block are infeasible, then the entire block fails and only the `else` branch is executed. As an example, the conditional statement

4.2 Implication

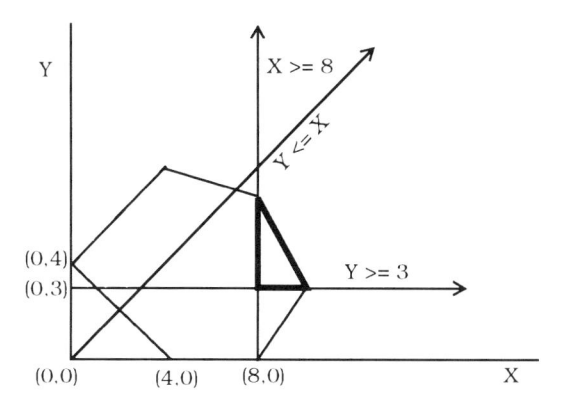

FIGURE 4.4 The constrained region generated by a different conditional

```
if {
    X <= Y;
    X == Y + 1.0;
}
then Y >= 3.0;
else Y <= 2.0;
```

will generate only the constraint `Y <= 2.0` of the `else` branch, since the constraints of the statement block necessarily fail, although the first one `X <= Y` by itself succeeds.

In the next model we illustrate the use of the `if/then/else` construct in a program to determine how quickly a mortgage can be paid off if the monthly payment is increased. For this we revisit Model 2.1, The Fixed Rate Mortgage, and consider a question that the new homeowners are now asking themselves.

Model 4.2 The Fixed-Rate Mortgage Reviewed

Through some fortunate developments just after closing on the house, the new homeowners now can afford to pay up to $1200 a month on the mortgage. So they want to know how long it will take to pay it off if they increase their monthly payment accordingly.

To compute this, we will first constrain the monthly payment to be less than or equal to the maximum the home-buyers can afford

```
Pymt <= PAYMAX;
```

where `PAYMAX` is a symbolic constant for `1200.0`. What we want is to run the loop of equations of Model 2.1,

```
    and(int k=0;k<MONTHS;k++)
        Bal[k+1] == (1 + RATE/12)*Bal[k] - Pymt;
```

stopping at the first k where `Bal[k] == 0` is possible. When this value of k is reached, we want to print the result and exit the loop. For this we will use the `break` statement, which exits the `and` loop with success.

Adding a symbolic constant for the amount of the mortgage, as well as one for the maximal payment that the homeowners can make,

```
#define MORTGAGE 100000.0       // Initial balance
#define PAYMAX 1200.0           // Maximum monthly payment
#define MONTHS 180              // Number of months
#define RATE .09                // 9% annual rate
```

we can redo the main routine to read as follows:

```
2lp_main()
{
continuous Pymt,Bal[MONTHS+1];

    Bal[0] == MORTGAGE;
    Pymt <= PAYMAX;
    and(int k=0;k<MONTHS;k++)
        if Bal[k] == 0.0;   // Mortgage can be paid off ?
        then {
            printf("Make %d payments of $%.2f\n",k,Pymt);
            break;  // Then exit loop
        }           // Else make another payment
        else Bal[k+1] == (1 + RATE/12)*Bal[k] - Pymt;
}
```

The model will print

Make 132 payments of $1196.08

Note that we print the results before we exit from the `and` loop. If we had placed the `printf` statement after the `and` loop, we would not have been able to use the loop control variable k in the print statement.

End of Model 4.2

Since procedures and compound statements either succeed or fail, they too can be used in the `if` part of a conditional statement as well as in the `then` or `else` part. Thus the `if/then/else` statement can be used to deal more smoothly with situations where a program can reach an infeasible region. For example, in Model 4.1 nothing guarantees that the constraint

4.2 Implication

```
sigma(int i=0;i<N;i++) ff[i]*X[i]
    <= r*sigma(int i=0;i<N;i++) X[i];
```

is consistent for all values of the loop variable r. So to detect this and continue the code without having the program as a whole fail, we can rewrite the loop to read

```
and(int r=HIRISK;r>=LORISK;r=r-DCRMNT)
    if evaluate_risk(ff,w,p,r,X);
    then record_evaluation(p,r,X);
    else {
        printf("Risk %d is not feasible\n",r);
        break;   // Exit the loop and continue program
    }
```

This kind of precaution gives the programmer more control over the run of the program.

As noted in Section 2.3, it is very simple to write code for the linear knapsack problem without using continuous variables, making it a discrete application. One reason that knapsack models are interesting is that they are a point where continuous and discrete methods coincide. We will make use of this fact later in Chapter 7 and Chapter 11. Here we present code for this application that does not use continuous variables, but which does illustrate use of the if/then/else construct. We will also break the code out into procedures.

The continuous variables X[N] become discrete:

```
double x[N];
```

The main procedure has a straightforward mission.

```
2lp_main()
{

    data();              // As before
    initialize();        // Updated
    // Test for feasibility
    sigma(int k=0;k<N;k++) w[k]*x[k] <= M;
    greedy_strategy();
    output();            // Left to reader
}
```

The requirement that at least lo[i] of item i be placed in the knapsack becomes an initialization rather than a lower bound on a continuous variable:

```
initialize()
{
    and(int i=0;i<N;i++)
        x[i] = lo[i];
}
```

The greedy strategy is to use as much as possible of each item in turn:

```
greedy_strategy()
{
    and(int i=0;i<N;i++)
        if w[i]*(hi[i] - lo[i]) <=
            M - sigma(int k=0;k<N;k++) w[k]*x[k];
        then x[i] = hi[i];       // Use all available x[i]
        else x[i] = x[i] +       // Use as much as needed
            (M - sigma(int k=0;k<N;k++)w[k]*x[k])/w[i];
}
```

Having done the discrete version of the maximization form of the linear knapsack problem, let us do the minimization version also. Here we are given an amount M and the total weight of the articles placed in the knapsack must be at least M. Again we are given upper and lower bounds on the amounts of each article to be used. This time we think of the objective function coefficients as costs rather than profits, and the goal is to reach total weight M at the least cost. We consider the items in nondecreasing order of cost to weight, placing the item that has the lowest ratio c[i]/w[i] first. This puts the less costly items ahead of the more costly items. Interestingly the greedy strategy for the minimization problem is exactly the same as for the maximization problem, since in both cases the aim is to fill the knapsack exactly to level M. So except for the reversal of data and the switch from <= to >= in the test for feasibility, the code is identical to that for the maximization problem:

```
2lp_main()
{
    reverse_data();      // Previous data entered backward
    initialize();        // Same as for maximization
    // Test for feasibility
    sigma(int k=0;k<N;k++) w[k]*hi[i] >= M;
    greedy_strategy();   // Same as for maximization
    output();
}
```

For completeness, let us list the data entry code:

4.2 Implication

```
reverse_data()
{
    c  = {  5.0, 4.75, 12.5,  9.0, 7.0, 10.75, 12.25, 9.5 };
    w  = {  9.5, 6.25, 13.75, 9.0, 6.5, 6.0,   7.0,   5.0 };
    lo = {  1.2, 0.4,  1.4,   2.1, 0.0, 0.2,   0.1,   0.2 };
    hi = {  4.2, 2.4,  3.4,   5.1, 0.2, 1.2,   0.4,   1.0 };
}
```

Testing and exception handling lie at the heart of programming. By way of example, in Section 3.3, we discussed a way of setting up constraints whose coefficients were sparse. With each index i, we supposed that there was associated a constraint. However, it might well happen that there is no constraint that is to be associated with a given index, i_0, and this can be reflected by the fact that nnz[i_0] is equal to 0. In this case we will have end equal to begin in the loop that sets up the constraints and the sigma operator will return 0. So it is a good precaution to skip what reduces to the unwanted test 0 <= rhs[i_0] by first testing whether the constraint is nonempty; this can be done by asking if begin < end before adding the constraint. Inserting this test yields the revised code:

```
sparslteq(continuous X[],int nnz[],indices[],
                    double cffs[],rhs[],int m)
{
int begin,end;

    end = 0;
    and(int i=0;i<m;i++) {
        begin = end;
        end = end + nnz[i];
        if begin < end;      // Check constraint not vacuous
        then
        sigma(int j=begin;j<end;j++) cffs[j]*X[indices[j]]
        <= rhs[i];
    }
}
```

This kind of elementary precaution together with good error messages and error-handling routines are basic ingredients of programming.

Exercises

4.2.1. A group of friends wins the lottery. They will receive $1M now and $1M per year for the next 20 years. The lottery people tell them they have won $21M. However, one can purchase annuities from an insurance company that pay 9.2% annual interest for 20 years; this money is

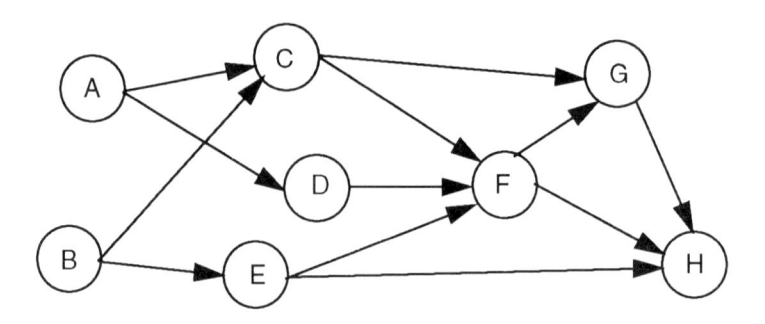

FIGURE 4.5 A Network.

paid out in equal annual payments at the end of each year, the same as a reverse mortgage. How much has the group really won? That is, what is the present value of the income stream they have won?

4.2.2. Prove that the greedy algorithm yields the optimal solution to the minimization form of the linear knapsack problem.

4.3 Network models

Networks are the source of an important and extensive class of linear programming applications. Networks are naturally represented pictorially; as an example, in Figure 4.5 we have a diagram where the circles denote nodes of the network and the arrows denote the arcs of the network. A typical application represented by a network such as this one is shipping a quantity of material across a transportation system; material arrives at a node and then is divided up to flow across the arcs that lead to succeeding nodes. The quantity of material that moves along an arc is called its *flow*.

In a typical application one would want to find a minimal cost way to route material from the start nodes A and B to the terminal node H. For example, let us suppose that 100 packages are to be shipped from node A and 150 from node B across the network of Figure 4.5. In Table 4.1 we have both the unit cost for moving flow across each arc and the capacity each arc can carry. The goal is to have the packages reach the terminal node H at minimal cost. This is an example of a *transshipment problem*. Putting aside for now the concern that packages are to survive the journey intact, we can formulate this as a linear program.

First we have symbolic constants for the arcs

```
#define ARCS 12
#define AC 0
#define AD 1
```

4.3 Network models

TABLE 4.1 Network Costs and Capacities

Arc	Unit Cost	Capacity	Arc	Unit Cost	Capacity
AC	2	125	DF	5	75
AD	4	100	EF	2	100
BC	3	150	EH	1	120
BE	5	125	FG	4	65
CG	2	85	FH	3	90
CF	2	95	GH	3	100

```
#define BC 2
  ...
#define FH 10
#define GH 11
```

For each arc we introduce a continuous variable to represent the amount of flow across the arc:

```
continuous F[ARCS];
```

The data are easily stored in integer arrays:

```
int cost[ARCS],capacity[ARCS];
```

And the data from Table 4.1 can be recorded therein:

```
data()
{
    cost = { 2,4,...,3 };
    capacity = { 125, 100, ... ,100 };
}
```

With each node in the network we associate a constraint to the effect that the flow out of the node is equal to the flow into the node. Let us envelop this information in a procedure:

```
node_constraints()
{
    F[AC] + F[AD] == 100;       // Flow from supply at A
    F[BC] + F[BE] == 150;       // Flow from supply at B
    -F[AC] - F[BC] + F[CF] + F[CG] == 0;  // Flow through C
    -F[AD] + F[DF] == 0;                   // Flow through D
    -F[BE] + F[EF] + F[EH] == 0;           // Flow through E
    -F[CF] - F[DF] - F[EF] + F[FG] + F[FH] == 0;// through F
```

```
    -F[CG] - F[FG] + F[GH] == 0;         // Flow through G
    -F[EH] - F[FH] - F[GH] == -250;      // Flow into H
}
```

The following procedure handles the capacity constraints:

```
capacity_constraints()
{
    and(int a=0;a<ARCS;a++)
        F[a] <= capacity[a];
}
```

Finally, the objective function is to minimize the cost of the operation. So we can summarize with a main procedure such as

```
21p_main()
{
    data();
    node_constraints();
    capacity_constraints();
    min: sigma(int a=0;a<ARCS;a++) cost[a]*F[a];
    output();
}

output()
{
string arcs[ARCS];

    arcs = { "AC", ... ,"GH" };

    printf("The optimal linear solution is %f\n",
                sigma(int a=0;a<ARCS;a++) cost[a]*F[a]);
    printf("The plan is to ship\n");
    and(int a=0;a<ARCS;a++)
        if wp(F[a]) > 0.0; then
        printf("\t %.2f packages on %s\n",F[a],arcs[a]);
}
```

If we run this linear program, we find that the solution is

```
The optimal linear solution is 1660.000000
The plan is to ship
        100.00  packages on AC
         30.00  packages on BC
        120.00  packages on BE
         40.00  packages on CG
```

4.3 Network models

    90.00 packages on CF
  120.00 packages on EH
    90.00 packages on FH
    40.00 packages on GH

This solution has a remarkable property: All flows are integral, and the packages arrive whole without having to be reassembled at the terminus.

To show where the integrality of the solution is coming from, we need to look at the structure of the constraints in some detail. Let us recall some basic notions and some mathematical terminology from Section 1.4. A constraint has the form

<affine expression> <closed relational operator> <affine expression>

For mathematical purposes it is convenient to represent constraints in the special form where the expression on the left is a linear combination of continuous variables and the one on the right is a constant term. Thus the usual notation for a $\leq$ constraint is

$$a_1 x_1 + \ldots + a_n x_n \leq b$$

The left-hand side is also conveniently denoted as the scalar product of a row vector of coefficients $\mathbf{a} = [a_1,\ldots,a_n]$ and a column vector of continuous variables, $\mathbf{x} = [x_1,\ldots,x_n]^T$. Thus the above constraint can be written simply as

$$\mathbf{a}\mathbf{x} \leq b$$

With this notation, an objective function

$$c_1 x_1 + \ldots + c_n x_n$$

can be written concisely as $\mathbf{cx}$. Naturally the mathematical notation does not need to distinguish between dense and sparse data and represents everything in dense form.

In our formulation of the network program, we wrote the constraints in equality form. A set of m equality constraints on n variables can be given as

$$a_{11} x_1 + \ldots + a_{1n} x_n = b_1$$
$$\vdots$$
$$a_{m1} x_1 + \ldots + a_{mn} x_n = b_m$$

We write this in matrix form as **Ax** = **b**. More explicitly

$$\mathbf{Ax} = \begin{bmatrix} a_{11} & \cdots & a_{1n} \\ \vdots & & \vdots \\ a_{m1} & \cdots & a_{mn} \end{bmatrix} \begin{bmatrix} x_1 \\ \vdots \\ x_n \end{bmatrix} = \begin{bmatrix} b_1 \\ \vdots \\ b_m \end{bmatrix} = \mathbf{b}$$

In the example above, the network constraints are defined in the procedure `node_constraints`. Filling in the nonzeros only from this example, the matrix **A** is

$$\begin{bmatrix} 1 & 1 & & & & & & & \\ & & 1 & 1 & & & & & \\ -1 & & -1 & & 1 & 1 & & & \\ & -1 & & & & & 1 & & \\ & & & -1 & & & & 1 & 1 \\ & & & & -1 & & -1 & -1 & & 1 & 1 \\ & & & & & -1 & & & & -1 & & 1 \\ & & & & & & & & & -1 & & -1 & -1 \end{bmatrix}$$

The critical properties of this matrix are that (1) every nonzero entry is either +1 or -1, (2) no column has more than two nonzero entries, and (3) no two nonzero entries in a column have the same sign. Matrices with this form are called *network matrices*. The +1 coefficients represent *supply*, and the -1 coefficients represent *demand*. Therefore each arc can occur at most once as a consumer of supply and at most once as a source of supply.

Assuming that the right-hand side only has integer entries and likewise for all bounding constraints, the fundamental mathematical property assured by network matrices is that all vertices of the feasible region are integer-valued. This is shown in Chapter 15. For this to hold, the relational operators in the constraints can be any of =, ≤, or ≥ . Whence our confidence that the solution to the problem above would not require that packages be broken up along the way. Linear programming applications with this kind of structure are called *network problems*.

Let us make one cautionary remark. The special network structure is lost if additional constraints are added to a network problem. For example, if we add the requirement that a certain proportion of the packages shipped pass through nodes AC, CF, and EF, say,

```
11*Flow[AC]  +  15*Flow[CF]  +  5*Flow[EF]  == 2100;
```

4.3 Network models

then the output would read

The optimal linear solution is 1660.000000
The plan is to ship
      100.00   packages on AC
       30.00   packages on BC
      120.00   packages on BE
       63.33   packages on CG
       66.67   packages on CF
         ...
       63.33   packages on GH

The wonderful connection between network matrices and integer-valued solutions also underlies the many special algorithms for network problems; these algorithms replace the simplex method in these situations and can be more efficient by orders of magnitude. For more on this topic, see Bertsekas (1994). This means that very effective discrete engines can be built for a wide range of important applications that have network structure in all or in part.

Schematically the network linear programming problem can be formulated as follows. There are n nodes labeled $0,\ldots,n-1$; an arc is an ordered pair (i,j) with $i < j$, and the sets of arcs is Arc. With each arc is associated a cost or value c_{ij}. For each arc (i,j) we have a continuous variable x_{ij}.

For each node $i = 0,\ldots,n-1$, the flow constraint is

$$\sum_{(j,i) \in Arc} x_{ji} - \sum_{(i,j) \in Arc} x_{ij} = b_i$$

When the constraint for a node is an equality constraint, the right-hand side tells whether the node is a supply node or a demand node: A supply node i adds a positive amount b_i to the flow, while a demand node adds a negative amount b_i. A similar interpretation applies to the other two closed relational operators.

There will be bounding constraints on the flows:

$$x_{ij} \leq f_{ij}, \ (i,j) \in Arc$$

When the goal is to minimize the cost of the flow, we have the objective function:

$$\min \sum_{(i,j) \in Arc} c_{ij} x_{ij}$$

When the c_{ij} represent value rather than cost, we have a maximization problem:

$$\max \sum_{(i,j) \in Arc} c_{ij} x_{ij}$$

Network problems appear in many situations. In fact, the reason that the progressive roundoff heuristic of Model 2.3, Call 911, is so effective is that the related problem where there are no lunch breaks can be shown to be equivalent to a network problem, Bartholdi, Orlin, and Ratliff (1980). Moreover, some special cases of the network linear programming problem come up again and again. Above we saw a transshipment problem. A transshipment problem with no intermediate nodes is called a *transportation problem*. In this problem there are m supply nodes $i = 0,...,m-1$ with supply a_i and n demand nodes $j = 0,...,n-1$ with demand b_j. The arcs are the shipping routes from each supply point to each demand point. The unit cost for supplying j from i is c_{ij}. We can set x_{ij} to be the flow from supply point i to demand point j. The situation is as follows:

For each $i = 0,...,n-1$, the supply constraint is

$$\sum_{j=0}^{n-1} x_{ij} = a_i, \quad i = 0, 1, ..., m-1$$

For each demand node the consumer constraint is

$$\sum_{i=0}^{m-1} -x_{ij} = -b_j, \quad j = 0, 1, ..., n-1$$

Finally, the objective function is

$$\min \sum_{i=0}^{m-1} \sum_{j=0}^{n-1} c_{ij} x_{ij}$$

Examples of this structure range from warehouse scheduling to computing capital equity requirements on options positions, Rudd and Schroeder (1982). In the next model we have an example where a single machine scheduling application turns out to be a transportation problem.

Model 4.3 Batch Scheduling

A server on a computer network must be scheduled to run eight large jobs. Each job will take a certain processing time or duration to

4.3 Network models

TABLE 4.2 Times for Preemptive Single Machine Scheduling

job number	0	1	2	3	4	5	6	7
ready time	0	1	1	11	11	5	11	7
due time	5	5	8	26	15	10	24	22
duration	2	3	3	1	4	2	2	5

complete. Also each job cannot begin before a certain *ready* time and must be finished by a *due* time. All of these times are given in Table 4.2.

The server runs the jobs in such a way that a job can be stopped, put on a queue, and then put back on without any increase in processing time. In other words, the jobs can be *preempted*. This can occur at the discrete times 0,1,2,3,... . The time to preempt a job and put another in its place is negligible. What is needed is a schedule that will process the jobs and finish the last job at the earliest possible time.

To model this situation, we can think in terms of a network. From each unit interval of time, we have potential flow to each of the corresponding jobs that can be running at that time. Each time unit can provide one unit of run time; each job requires its processing time from the sources.

Since the dimensions of this problem are small, we need not concern ourselves with sparse data considerations. Instead, we can define symbolic constants for the number of time slices and the number of jobs

```
#define T 26
#define J 8
```

Then, thinking in terms of network flow, we introduce an array of continuous variables for the flow from time slice i to job j:

```
continuous X[T][J];
```

The idea is that X[i][j] will be 1 if job j is running in the time interval [i,i+1). In the network model this will mean that there is a unit of flow from source i to consumer j. The flow from each source will be bounded by 1, which will ensure that no two jobs can be running at the same time.

The data can be stored in integer arrays

```
int ready[J], due[J], dur[J];
```

and can be easily entered in a data routine. To ensure that each job j is run for dur[j] time units, we stipulate that the total flow from time i to job j with i in the interval [ready[j],due[j]-1] must be equal to dur[j]:

```
    sigma(int k=ready[j];k<due[j];k++) X[k][j] == dur[j];
```

The object is to minimize the finishing time of the last job to be run. The following objective function will try to do this and a little more. Minimizing the objective function comes down to minimizing the last index i so that some `X[i][j]` is not 0:

```
min: sigma(int i=0;i<T;i++)      // Charge i for active
         sigma(int j=0;j<J;j++) i*X[i][j]; // unit i
```

The "little more" referred to is that this objective function will not only try to minimize the final finishing time but will also strive to finish the earlier jobs as soon as possible.

The constraints can be assembled in a routine to set up the network structure:

```
network()
{
    and(int i=0;i<T;i++)    // At most one job per time unit
        sigma(int j=0;j<J;j++) X[i][j] <= 1;
    and(int j=0;j<J;j++)    // Each job must be completed
        dur[j] ==
        sigma(int k=ready[j];k<due[j];k++) X[k][j];

    min: sigma(int i=0;i<T;i++)
            sigma(int j=0;j<J;j++) i*X[i][j];
}
```

The main routine needs only to call `network` and to see to the output.

```
21p_main()
{
    data();
    network();
    output();
}

output()
{
    ...            // To list sequence of scheduled jobs
}
```

The output is interesting.

At time 0 run job 0
At time 1,2,3 run job 1
At time 4 run job 0
At time 5,6,7 run job 2

4.3 Network models

At time 8,9 run job 5

At time 10 run job 7

At time 11,12,13,14 run job 4

At time 15,16,17,18 run job 7

At time 19,20 run job 6

At time 21 run job 3

The solution found has a pattern. At each point in time, an available job is running that has the earliest finishing time. This strategy for preemptive single machine scheduling is known as *Jackson's preemptive schedule* (Jackson 1955). The linear programming formulation of the model has discovered, so to speak, this strategy. It should be said, however, that the linear programming solution can well find an alternative optimal solution that does not follow the Jackson strategy. In the exercises, the reader is asked to prove directly that the Jackson strategy is optimal.

The network structure in this application would not be there if preemption were not allowed. That is, single machine scheduling without preemption is a problem cut from an entirely different cloth. It is an example of a *disjunctive programming problem*, a topic discussed in Chapter 9. Let it be said, however, that for finding good solutions to these more difficult problems, Jackson's preemptive schedule is an important heuristic (see Carlier and Pinson 1990).

End of Model 4.3

The special case of the transportation problem where the number of supply nodes is equal to the number of consumer nodes and where the right-hand side terms are all 1 is called the *assignment problem*. Here the idea is to match n sources with n consumers so as to optimize an objective function. This problem is usually represented schematically in the following way:

For every i there is a matching j:

$$\sum_{j=0}^{n-1} x_{ij} = 1, \quad i = 0, 1, \ldots, n-1$$

For every j there is a matching i:

$$\sum_{i=0}^{m-1} x_{ij} = 1, \quad j = 0, 1, \ldots, n-1$$

Typically the objective is to minimize the cost of the match

$$\min \sum_{i=0}^{m-1} \sum_{j=0}^{n-1} c_{ij} x_{ij}$$

Strictly speaking, this does not yield a network matrix because all coefficients are +1. However, the model can be put in network form by multiplying the consumer constraints through by -1:

$$\sum_{i=0}^{m-1} -x_{ij} = -1, \quad j = 0, 1, ..., n-1$$

This was a rather trivial operation. However, detecting matrices that have the properties of network matrices is a rich area; it includes mathematical gems such as the Decomposition Theorem of Seymour (1980). For the assignment problem, the network structure guarantees that the optimal linear programming solution will make exactly one x_{ij} have value 1 for each i, ensuring the desired match. Naturally, in writing a network model one doesn't have to wrench the constraints into a special form, since the geometric property of having only integer-valued vertices is not lost so long as the constraint matrix is equivalent to a network matrix.

In assignment problems the fact that the solution will be integer-valued is critical, for often one is matching people with jobs, airplanes with passenger gates, and the like. When the matching is subject to additional constraints, then the problem is no longer a network problem, and other methods have to be used to obtain integer-valued solutions. Methods for this are discussed from Chapter 8 on.

In our formulation of these network problems, we have maintained a duality between arcs and nodes, representing arcs as variables and nodes as left-hand sides of constraints. However, one can also explicitly include continuous variables for nodes in a network problem, and this will preserve the network structure of the constraint matrix. This can be convenient in formulating the problem. For example, if costs are associated with each node rather than with each arc, it is convenient to have names for the nodes for the objective function. We will see this in the next model.

So far we have assumed that flow across an arc is conservative and that the flow exiting the arc is equal to the flow entering the arc. When flow can be dampened, the matrix will no longer have 0, +1, −1 entries, although each column will still have at most two nonzero entries and no two such of the same sign. In this case we have a *generalized network* or a *network with gains*. In fact, Model 1.2, The Library Fund, can be seen to have this kind of structure. With generalized networks, the wonderful property of integer-valued solutions is lost; however, special purpose

4.3 Network models

algorithms can still exploit the matrix structure and outperform general purpose linear programming engines on these applications. The next model is an example of a generalized network. In the last model we sidestepped the issue of sparse data because of the simplicity of the problem. In network applications of any size this issue cannot be avoided, and much of the problem formulation becomes an exercise in manipulating sparse data.

Model 4.4 The Grid

A utility provides power for the people in the valley by producing electricity at a central power plant and distributing it through a network of high-tension lines. The network is diagramed in Figure 4.6. The high-tension lines form the arcs of the network that fork or amalgamate at various nodes. The lines link the plant to substations that change voltage and deliver electricity to the public. Each arc along the network has a capacity as does each substation. Substations disperse the electricity they receive to consumers, and they are the terminal nodes on the network. The other nodes amalgamate and/or split the high-tension lines that come into them. Each of the substations has a certain demand for power that is determined by its client base; the intermediate nodes have a threshold requirement of 110 megawatts, which is considered the minimum traffic that a node of this type should support for maintenance reasons. Each arc has a certain conservation coefficient, which accounts for the power loss to be expected by transmitting electricity across this line. Thus a conservation factor of .96 means that 4% of the electricity that enters that arc is lost in transmission.

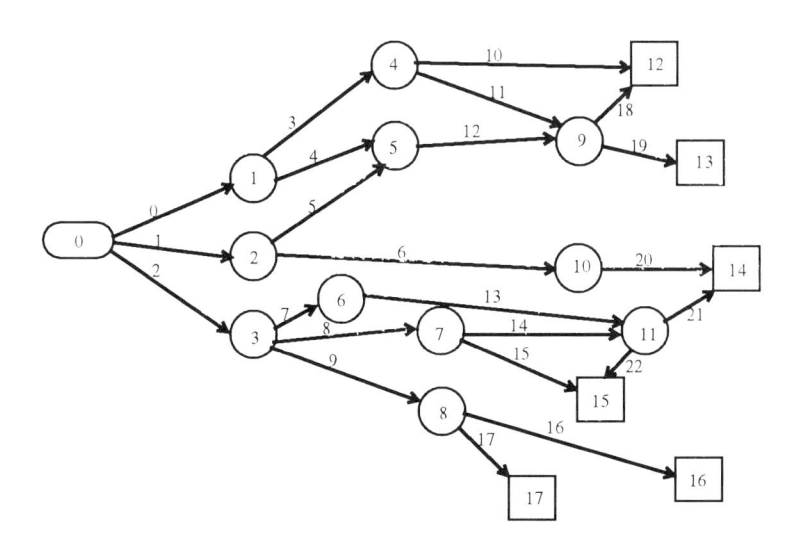

FIGURE 4.6 The network of substations and high-tension lines.

TABLE 4.3 High-Tension Line Data

Arc	From/To	Capacity	Conservation Coefficient	Arc	From/To	Capacity	Conservation Coefficient
0	0,1	400	.88	12	5,9	250	.98
1	0,2	400	.9	13	6,11	300	.99
2	0,3	550	.9	14	7,11	300	.96
3	1,4	330	.92	15	7,15	250	.98
4	1,5	230	.88	16	8,16	500	.96
5	2,5	340	.89	17	8,17	220	.92
6	2,10	610	.96	18	9,12	311	.88
7	3,6	330	.99	19	9,13	233	.89
8	3,7	250	.96	20	10,14	190	.97
9	3,8	320	.92	21	11,14	278	.99
10	4,12	140	.88	22	11,15	175	.92
11	4,9	350	.89				

The task is to route the electricity through the network so as to meet demand while minimizing the amount of electricity that has to be produced at the central plant.

In Table 4.3 the rows represent the high-tension lines between the stations. In Table 4.4 we have the capacity and the demands of the nodes in megawatts. Nonterminal nodes all have the same threshold demand of 110 megawatts in this scheme of things.

To be able to enter the data, let us declare appropriate symbolic constants and arrays:

```
#define ARCS 23
#define NODES 18

int from_to[ARCS][2];
double arc_cap[ARCS];
double cons[ARCS];

double node_cap[NODES];
double node_demand[NODES];
```

The unknowns are the amount of power that will enter and then flow through each arc and the amount of power that will arrive at each node. So let us introduce continuous variables to represent these quantities:

```
continuous Flow[ARCS], Power[NODES];
```

4.3 Network models

TABLE 4.4 Substation Data

Node	Capacity	Demand	Node	Capacity	Demand
0	1200	0	9	621	110
1	700	110	10	632	110
2	700	110	11	798	110
3	650	110	12	223	222
4	600	110	13	310	100
5	400	110	14	421	120
6	520	110	15	300	120
7	625	110	16	254	111
8	775	110	17	285	107

We need to have a list of the arcs that come into each node and a list of the arcs that leave each node. For both the incoming and outgoing arcs, this information is implicit in the data in the From/To column of Table 4.3. In tabular form, for the arcs entering nodes we want the information set forth in Table 4.5. This information has the typical structure of sparse data. Following the notation used in the previous sparse models, we can store this information in arrays

```
int in[ARCS],nnz_in[NODES];
```

Similarly, we can set up corresponding arrays to keep track of the arcs leaving a node:

```
int out[ARCS],nnz_out[NODES];
```

TABLE 4.5 Entering Arcs

Node	Entering Arcs	Size of Set	Node	Entering Arcs	Size of Set
0	None	0	9	11,12	2
1	0	1	10	6	1
2	1	1	11	13,14	2
3	2	1	12	10,18	2
4	3	1	13	19	1
5	4,5	2	14	20,21	2
6	7	1	15	15,22	2
7	8	1	16	16	1
8	9	1	17	17	1

The information from Table 4.3 will be stored in the array from_to. From this, we can build the arrays in, nnz_in, out, and nnz_out. The following code segment will do the job:

```
tables(int dir,indices[],nnz[])    // dir will be 1 for in
{                                  // and 0 for out
int end;
    end = 0;
    and(int n=0;n<NODES;n++)
        and(int arc=0;arc<ARCS;arc++)
            if from_to[arc][dir] == n;
            then {
                indices[end] = arc;
                nnz[n] = nnz[n]+1;
                end = end + 1;
            }
}
```

Now the model basically reduces to replicating Figure 4.6. To start, we must state that the amount of electricity that reaches a node is equal to the sum of the flow along the arcs that enter the node. Since Power[n] represents the flow that reaches node n, the power that reaches the node will be given by the sum of the flow that comes in on the arcs reaching the node. For each arc, this will be the power that entered the arc multiplied by the conservation factor. These equations will be generated by the following routine:

```
flows_in()
{
int begin,end;

    end = 0;
    and(int n=0;n<NODES;n++) {
        begin = end;
        end = end + nnz_in[n];
        if begin < end;
        then     // Flow in is sum of flow along incoming arcs
            Power[n] ==
            sigma(int k=begin;k<end;k++)
                        cons[in[k]]*Flow[in[k]];
    }
}
```

A similar routine flows_out can be written for the flows out of the nodes, but this time we will not have to multiply by the conservation factor, since that has already been accounted for. The structure of the program will be to enter the data, build the tables, set up the represen-

4.3 Network models

tation of the grid, and then to minimize the power produced at the central plant. Therefore the main routine will be

```
2lp_main()
{
    data();                    // From Tables 4.3 and 4.4
    tables(1,in,nnz_in);       // Build auxiliary tables
    tables(0,out,nnz_out);
    setup();                   // Node constraints
    min: Power[0];             // Minimize power produced at plant
    output();                  // Left to reader
}
```

For completeness, let us write the `setup` procedure in detail:

```
setup()
{
    and(int n=0;n<NODES;n++) {
        Power[n] <= node_cap[n];
        Power[n] >= node_demand[n];
    }

    and(int arc=0;arc<ARCS;arc++)
        Flow[arc] <= arc_cap[arc];

    flows_in();
    flows_out();
}
```

The program determines that the plant should produce 1117.75 megawatts of power; at the substation end, this translates into 861.26 megawatts to meet a demand of 780 megawatts.

Note that we introduced explicit variables `Power[n]` for the intermediate nodes of the network in this model; the flows in and out of the nodes could have been linked directly as the two sides of an equality constraint. However, introducing the variables for the nodes does help to formulate the objective function and other aspects of the model; also it does not change the generalized network structure of the constraint matrix.

End of Model 4.4

Exercises

4.3.1. Prove that Jackson's preemptive schedule is optimal. Hint: Think greedy.

TABLE 4.6 Time in Minutes for Scouts to Do Tasks

	Task 0	Task 1	Task 2	Task 3	Task 4	Task 5	Task 6	Task 7	Task 8	Task 9
Scout 0	22	33	44	42	35	67	81	23	44	52
Scout 1	32	43	45	32	25	47	61	43	48	22
Scout 2	25	47	55	62	33	26	42	36	19	33
Scout 3	42	23	35	37	29	27	48	29	41	44
Scout 4	52	23	33	66	15	67	39	23	38	62
Scout 5	22	33	44	42	35	67	81	23	44	52
Scout 6	12	37	66	21	59	41	41	43	29	17
Scout 7	35	37	65	52	43	16	2	36	19	33
Scout 8	47	26	30	17	20	77	38	38	24	47
Scout 9	72	63	43	50	17	47	19	63	58	11

4.3.2. Reorganize Model 1.2, The Library Fund, so that its constraint matrix is that of a network with gains. Rewrite Model 3.5, The Library Fund Reprised, by having only one set of stage one variables and eliminating the need to alias three sets of variables. Comment on the resulting generalized network structure.

4.3.3. Write the routine `flows_out` that handles the flow out of the nodes in Model 4.4.

4.3.4. A troop of 10 scouts has 10 tasks to perform. In Table 4.6, the amount of time it takes each scout to perform each of the tasks is given. Assign each scout a task to do individually so as to minimize the average time spent working by the scouts.

4.4 Smoothing out extreme solutions

As we know, another name for vertex is extreme point and linear programming solvers maintain the witness point as an extreme point of the polyhedron defined by the constraints. On some occasions, to refine a solution to a constraint problem, it is desirable to balance the values to give a satisfactory solution.

By way of example, consider the situation where 36 tons of gold bullion are to be stored in four locations, three of which are capable of holding 12 tons and one of which is capable of holding 6 tons. A vertex solution to the constraints can yield 12,12,12, and 0 as the allocations, and the vertex found depends on the details of the implementation of the linear programming solver that is used.

4.4 Smoothing out extreme solutions

One approach to balancing the values in a solution is first to *make the largest value as small as possible*. This can be done by setting a variable Z to be a simultaneous upper bound on all the variables in question and to minimize Z. Let's see what this smoothing technique does in the example of the gold bullion. The code would be

```
#define N 4

2lp_main()
{
    continuous X[N];
    continuous Z;

    X[3] <= 6;
    and(int i=0;i<3;i++)
        X[i] <= 12;

    sigma(int i=0;i<N;i++) X[i] == 36;

    and(int i=0;i<N;i++)
        X[i] <= Z;
    min: Z;

    printf("Z is %f\n",Z);
    and(int i=0;i<N;i++)
        printf("X[%d] is now %f\n",i,X[i]);
}
```

The output is

Z is 10.000000
X(0) is now 10.000000
X(1) is now 10.000000
X(2) is now 10.000000
X(3) is now 6.000000

So the value of Z at the witness point is 10.0, and in effect we have added the constraints X[i] <= 10 which force the witness point to be one that balances the values of the X[i].

What is needed, in general, is a way of iterating this step. That is, once the largest value is made as small as possible, how do we make the second largest value as small possible, and so on. The following theorem provides the basis for implementing such a strategy:

Theorem (Flat Covering Theorem) *Let P be a polyhedral set, and let $H_0,...,H_{n-1}$ be hyperplanes such that*

$$P \subseteq \bigcup_{i<n} H_i$$

Then for some $j < n$, we have $P \subseteq H_j$.

The proof is given in Chapter 15. The geometric intuition behind the proof is that a full dimensional feasible region cannot be covered in a nontrivial way by flat hyperplanes. Thus in two dimensions a square cannot be covered by a finite set of straight lines; on the other hand, a line segment can be covered by a finite set of straight lines, but then it must be completely included in one of them.

To see how to employ the Flat Covering Theorem in a program, suppose that we have continuous variables `X[0]`, ... , `X[N-1]` and possibly other continuous variables as well, all constrained to form a feasible region. Let `U` be a new continuous variable, and set `X[i] <= U` for all `i`. Now minimize `U` and fix `U` at its minimum value; so in 2LP we would call

```
min: U;
b = wp(U);
U == b;
```

recording the witness point value of `U` in the identifier `b` of type `double`. So at this point we have `X[i] <= U` for all `i` and `U == b` in force. We also have `X[i] <= b`, since `b` is equal to `wp(U)`. Denote the current feasible region by F, and denote the hyperplane `X[i] == b` by H_i. Since `b` is the minimum value of `U` and `U` is the least upper bound on the `X[i]`, every point in F has at least one coordinate `X[i]` with witness point value `b`. Thus we have $F \subset \cup H_i$. Then, by the Flat Covering Theorem, for at least one j, $F \subset H_j$: for this `X[j]` the witness point value `wp(X[j])` is henceforth fixed at `b`. Having fixed at least one of the `X[i]`, the strategy can be iterated and applied again to the variables that are not yet fixed. Since at least one new variable is fixed each time, the process will terminate by fixing all the variables `X[i]`.

In employing this strategy, after minimizing `U`, we would check if the variable `X[i]` is now at its minimum by calling

```
min: X[i];
if fabs(X[i] - U) < EPSILON;
then X[i] == wp(X[i]);
```

where `EPSILON` is a symbolic constant for a value such as 1e-6 (in decimal notation .000001). We use `EPSILON` in the test to see if `wp(X[i])` is the same as `wp(U)` because these floating-point numbers are equal for all practical purposes if they are within `EPSILON` of one another.

4.4 Smoothing out extreme solutions

In the discussion thus far, we have been applying this compression process from above in the sense that we have been compressing the upper bounds on continuous variables. In the situation where the continuous variables X[i] are all bounded from above, this process can be applied from below to smooth out the solution by pushing variables up to maximum values rather than down to minimum ones.

In the following example, we apply this compression process from above to smooth the solution to a set of linear constraints. The model is based on work of Sankaran (1989); see also Shrage (1991, chapter 17).

Model 4.5 Fairer Co-op Tax

A agricultural cooperative consisting of six individual family farms is forced to levy a tax of $7.5 thousand dollars on its members for barn roof repairs. The tax on the different individual family farms in the cooperative is based on a formula devised by an accountant back in the days when the cooperative was established. Written into the bylaws are linear constraints that relate the tax levies of the different participating family farms. These constraints take into account such critical factors as proximity to the road, acreage, and size of livestock facilities. Given the amount required for the repairs, what is needed is a levy on each family farm so that these taxes sum to the needed amount and meet the constraints of the bylaws.

However, among the many different possible assessments of tax liabilities that meet these requirements, some will appear less fair than others. In fact, if linear programming is used to compute the levies, an extreme point will be found that might make some people extremely unhappy. So the cooperative wants to use a smoothing strategy to equalize the tax assessments as much as possible. In terms of the strategy derived from the Flat Covering Theorem, this means the levies should be determined so that the highest tax paid among all the family farms is the smallest highest tax possible and, given that, that the second highest tax is the smallest possible, and so on.

To program this strategy, we introduce continuous variables Tax[N] to represent the tax in thousands of dollars to be paid by the family farms that make up the cooperative, where N is a symbolic constant for the number of family farms, in this case 6. The program will first set up the constraints of the bylaws and a constraint to raise the required amount of tax money:

```
sigma(int i=0;i<N;i++) Tax[i] == TXBILL;
```

where TXBILL is a symbolic constant for 7.5.

To carry out the smoothing strategy, we will introduce additional continuous variables Z[j] which will be used to serve as uniform bounds on sets of the Tax[i]. To keep track of the variables Tax[i] that are already determined to be at their final value, we can use some standard programming techniques. We introduce an array `flag[N]` of type `int`. The elements of this array will initially be 0. When the final value of the variable `Tax[i]` is determined, we will set `flag[i]` to 1. To count the number of `Tax[i]` that have been finalized this way, we use an identifier `count` of type `int`, which is also initially 0, and which will be incremented by 1 each time some `flag[i]` is set to 1. The code will complete when `count` is equal to N and all `flag[i]` are equal to 1. So we will need the following definitions and declarations:

```
#define N 6              // Number of individual family farms
#define EPSILON .000001
#define TXBILL 7.5       // In thousands of dollars

continuous Tax[N];
continuous Z[N];    // Z[j] is the bound for the jth appli-
                    // cation of the Flat Covering Theorem
int count;
int flag[N];
```

The work of smoothing out the `Tax[i]` can be done by means of an and loop that will terminate when `count` is equal to N. Each time through this loop, additional variables `Tax[i]` will be finalized. In doing this, the body of the loop will take the next available variable `Z[j]` and use it to smooth out the solution following the strategy above. Let us encapsulate this task in a procedure:

```
#define OFF 0
#define ON 1
body_of_the_loop(continuous Z)
{
    and(int i=0;i<N;i++)
        if flag[i] == OFF;
        then Z >= Tax[i];

    min: Z;
    Z == wp(Z);
    and(int i=0;i<N;i++)
        if flag[i] == OFF;
        then {
            min: Tax[i];
            if fabs(Tax[i]-Z) < EPSILON;
            then {
                    flag[i] = ON;
                    count = count + 1;
            }
        }
}
```

4.4 Smoothing out extreme solutions

This loop will be invoked from the main procedure.

```
21p_main()
{
    by_laws();

    // Funds required for repairs
    sigma(int i=0;i<N;i++) Tax[i] == TXBILL;

    printf("With the naive solution we have\n");
    and(int i=0;i<N;i++)
        printf("Tax on family farm %d is $%.2f\n",
                                        i,1000*Tax[i]);

    printf("\n\n");

    // Even out tax levies
    and(int j=0;count!=N;j++)
        body_of_the_loop(Z[j]);

    printf("After smoothing we have \n");
    and(int i=0;i<N;i++)
        printf("Tax on family farm %d is $%.2f\n",
                                        i,1000*Tax[i]);
}
```

Finally, let us code the bylaws:

```
by_laws()
{
    // Relations among tax levies dictated by bylaws

    // Proximity to road
        Tax[0] + 2*Tax[1] + 4*Tax[2] + 7*Tax[4] >= 16;

    // Proximity to silos
        2.5*Tax[0] + 3.5*Tax[1] + 5.2*Tax[4] >= 17.5;

    // Dwelling size
        0.4*Tax[1] + 1.3*Tax[3] + 7.2*Tax[5] >= 12;
    // Size of acreage
        2.5*Tax[1] + 3.5*Tax[2] + 5.2*Tax[4] >= 13.1;

    // Size of livestock facilities
        3.5*Tax[0] + 3.5*Tax[3] + 5.2*Tax[5] >= 18.2;
}
```

The output shows that the smoothing strategy has a significant effect on the taxes levied on the individual family farms:

With the naive solution we have
Tax on family farm 0 is $0.00
Tax on family farm 1 is $0.00
Tax on family farm 2 is $634.62
Tax on family farm 3 is $0.00
Tax on family farm 4 is $3365.38
Tax on family farm 5 is $3500.00

After smoothing we have
Tax on family farm 0 is $1794.64
Tax on family farm 1 is $1051.79
Tax on family farm 2 is $325.26
Tax on family farm 3 is $739.03
Tax on family farm 4 is $1794.64
Tax on family farm 5 is $1794.64

As one can see, the process of making the maximum tax levy as small as possible and then proceeding iteratively does tend to equalize the amount each group has to pay.

End of Model 4.5

Another way of looking at the method employed in the last model is as a progressive way of minimizing an error term. Think of it this way. Nobody wants to pay taxes. So in any solution to the constraints of the bylaws and the need to raise $7500, the vector of values

$$(\text{wp}(\text{Tax}[0]) - 0.0, \ldots, \text{wp}(\text{Tax}[N-1]) - 0, 0)$$

is a measure of how far the solution is from the ideal solution $(0.0, \ldots, 0.0)$. The vector $(\text{wp}(\text{Tax}[0]), \ldots, \text{wp}(\text{Tax}[N-1])$ is the error term associated with the solution. An error term like this is a vector and not just a single number. The size of a single number is measured by its absolute value; there are many ways of measuring the size of an error term given as a vector.

The measure of the size of a vector is called a *norm*. For vectors of real numbers $\mathbf{x} = (x_1, \ldots, x_n)$, the standard measures are the L_p norms, which are given by

$$\|x\|_p = \left(\sum_i |x_i|^p\right)^{1/p}, \quad 1 \le p < \infty$$

4.4 Smoothing out extreme solutions

For $p=2$ we find the usual Euclidean norm. For $p=1$, the norm reduces to the sum of the absolute values of the components of the vector. Since

$$\lim_{p \to \infty} \|x\|_p = \max\{|x_i|\}$$

the L_∞ norm of **x** is defined by the equation

$$\|x\|_\infty = \max\{|x_i|\}$$

The L_1, L_2, and L_∞ norms are used extensively. The measure to use in a given situation depends very much on the nature of the phenomenon being measured and on the algorithmic tools available. The L_1 and L_∞ norms have robust mathematical properties and can be easily computed. In fact, in the model just considered, with the first pass through the loop, we minimized the maximal entry in a vector of nonnegative entries and restricted the error term accordingly. In other words, with the first call to the routine `body_of_the_loop`, we restricted ourselves to solutions to the constraints, which are minimal in the sense of the L_∞ norm. By the Flat Covering Theorem, we know that certain components of the error term cannot be reduced any further in size. Successive passes through the loop then minimized the L_∞ norm of the subvector of remaining components. Thus what we have done in Model 4.5 is to carry out a progressive minimization of the L_∞ measure of the error term.

It is interesting to note that the first use of the simplex method by Fourier was to minimize the L_∞ norm of an error term. For discussion and textual fragments, see Chvátal (1983) and Schrijver (1986). Minimizing this norm is called *Chebychev approximation*.

The L_1 norm is not very useful as a measure of error in the situation of Model 4.5 because the constraint

```
sigma(int i=0;i<N;i++) Tax[i] == TXBILL;
```

fixes the L_1 norm of the vector `(wp(Tax[0]),...,wp(Tax[N-1]))` once and for all. Although the L_1 norm is not of use here, there are many situations where it is the measure of choice.

The following model will illustrate the use of the L_1 and L_2 norms. First we will need to recall some elements of linear algebra. A vector **a** is called a *unit vector* if its L_2 norm is equal to 1. If $\mathbf{a} = (a_1,...,a_n)$ is a unit vector, then the scalar product $\mathbf{av} = a_1 v_1 + ... + a_n v_n$ of **a** with **v** measures the projection of **v** onto the axis defined by **a**. If $\mathbf{ax} \leq b$ is a constraint, the vector **a** is perpendicular to the hyperplane $\mathbf{ax} = b$ and points away from the half space defined by $\mathbf{ax} \leq b$. Therefore, if **a** is a unit vector and if $\mathbf{av} \leq b$ holds, then $b - \mathbf{av}$ measures the Euclidean or L_2 distance from **v** to the hyperplane $\mathbf{ax} = b$; see Figure 4.7.

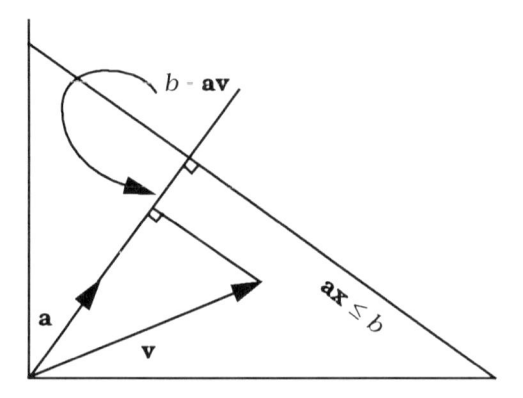

FIGURE 4.7 A half space defined by a unit vector.

Model 4.6 Extreme Vertices

Suppose that we have a feasible region such as that of Figure 1.5 (b). Consider the constraint X + Y + Z <= 1. If a point in the region lies on the boundary defined by this constraint, then it satisfies its equality form X + Y + Z == 1. So for a point in the region with coordinates X,Y,Z, the difference 1 - (X + Y + Z) is measure of the discrepancy between the boundary defined by the constraint and the point. The significance of this discrepancy will depend on the application being modeled, and this measure can be applied to anything from chickens to CDs. If it is the geometry of the region that we are interested in, then the natural thing to measure is the Euclidean distance of a point in the region to the hyperplane defined by the bounding constraint. To have the discrepancy measure this, we must normalize the left-hand side of the constraint. In the case at hand, it means computing the L_2 norm of the vector (1,1,1) and then dividing the coefficients and constant term by this quantity. Since the norm of this vector is equal to the square root of 3, or 1.733, this process yields a new version of the constraint X + Y + Z <= 1, namely

```
1/1.733*(X + Y + Z)  <= 1/1.733 ;
```

The Euclidean distance between a point in the feasible region and the hyperplane defined by the equality form of this constraint is

```
1/1.733 - 1/1.733*(X + Y + Z)
```

or, more simply,

```
1/1.733*( 1 -   (X + Y + Z) )
```

4.4 Smoothing out extreme solutions

For the constraint X >= 0, which is also a boundary constraint of the region, the L_2 norm of the vector (1,0,0) is already 1. In the first quadrant the difference X - 0 measures the Euclidean distance from a point to the YZ plane. Similar remarks apply to the boundary constraints Y >= 0 and Z >= 0.

The average distance from a point to the boundary constraints is a measure of how far the point is from lying on all the boundaries of the region. A point in the region that minimizes the average distance is in some sense a typical point of the feasible region. Minimizing the average distance is, of course, the same as minimizing the sum of the distances; this is attained at a point which minimizes the L_1 norm of the vector of distances from the boundary constraints. In other words, this point will minimize the L_1 norm of the 4 dimensional vector of L_2 distances to the boundary. Such a point can be found by making the sum of the distances the objective function and minimizing it. In the current situation the 2LP code is

```
2lp_main()
{
continuous X,Y,Z;

    X + Y + Z <= 1;

    min: 1/1.733*(1 - (X + Y + Z)) + X + Y + Z ;

    printf("X is %.1f\n",X);
    printf("Y is %.1f\n",Y);
    printf("Z is %.1f\n",Z);
}
```

The point found is (0,0,0) which is a vertex of the region that yields an objective function value of 1/1.733. This minimizing vertex is "least extreme" in that it comes closest in the sense of the L_1 norm to satisfying the equality form of all the constraints that define the region. If we change min: to max: we find the "most extreme" vertex, which is at (1,0,0). This is not the only solution to the maximization problem, since the vertices (0,1,0) and (0,0,1) clearly yield the same value for the objective function. In fact every point on the face of the solid defined by the equality constraint X + Y + Z == 1 is an optimal solution. The geometric explanation of this is that the level surfaces

```
    c == 1/1.733*(1 - (X + Y + Z)) + X + Y + Z;
```

of the objective function are all planes parallel to the plane 1 == X + Y + Z.

End of Model 4.6

In the next chapter, we will see more examples where the L_1 norm is used as the measure of the size of the error term.

Exercises

4.4.1. Determine "least extreme" and "most extreme" vertices for the the feasible region of Figure 1.2.

4.4.2. In Model 4.6, try to determine a point that will minimize the L_2 norm of the 4 dimensional vector of L_2 distances to the boundary.

4.4.3. Prove that

$$\lim_{p \to \infty} \|x\|_p = max\{|x_i|\}$$

5

Conditional Disjunction

In this chapter we begin a treatment the propositional connective disjunction. In the first section we study the form of disjunction most commonly used in programming and extend it to programming with constraints. Next we deal with the issue of competing objectives in linear programming problems, and go on to apply linear programming to pattern recognition.

5.1 A programming interpretation of disjunction

In symbolic logic, disjunction is denoted by the binary connective $\vee$. The notation $p \vee q$ expresses that at least one of p or q is true; the terms of a disjunction are called *disjuncts*.

In mathematics, a disjunction can be known to be true without determining the truth or falsity of its disjuncts. Thus for example if x and y vary over the nonnegative real numbers and we have shown that $xy \geq 100$, the statement $(x \geq 10) \vee (y \geq 10)$ can be assumed to be true without first proving x to be ≥ 10 or else first proving y to be ≥ 10. The mathematical problem solver can employ the strategy of considering the case $x \geq 10$ and then coming back later to consider the case $y \geq 10$. This interpretation of disjunction is known as *classical disjunction*; it is, in fact, a very powerful logical principle. We will return to this point in Section 9.1. Disjunction is called *intuitionistic* when one of the disjuncts must be determined to be true in order for the disjunction to be evaluated as true. This is the starting point of an alternative mathematical logic championed by Brouwer and the intuitionistic school; see Troelstra (1977).

In traditional programming languages, the analog of classical disjunction is not available; in order for the code to proceed through a disjunction, at least one of the disjuncts must first be determined to hold, as in intuitionistic logic. Let us illustrate this point using the C/C++ notation for disjunction. If a,b are of type `double` or `int`, to evaluate the disjunction

```
            (a >= 10 || b >= 10)
```

the interpreter first will perform the test `a >= 10`: If this succeeds, the second test is skipped and the program proceeds. If the first test fails, the second test will be performed. The system must determine explicitly that one of the disjuncts succeeds. If the first test succeeds, the second test is not executed. This interpretation of disjunction is called the *conditional or* or *c-or*; see Gries (1981). Languages such as C, C++ and LISP employ this strategy and only evaluate as many disjuncts as are required to determine that one of them succeeds. Once a disjunct succeeds, the program proceeds without leaving the option of coming back and considering an alternative disjunct. This is an example of a lazy evaluation strategy. Other languages such as Pascal follow a strict evaluation strategy and evaluate all the disjuncts whether or not one of them succeeds. But, in all of these languages, if the disjunction succeeds or fails, the program proceeds without leaving open the possibility of coming back and considering an alternative branch of the disjunction.

In Chapters 8 and 9 a programming approximation to classical disjunction will be studied, but first let us consider the extension of conditional disjunction to programming with constraints. For conditional disjunction the 2LP syntax is

```
c_either <statement>
or <statement>
   . . .
or <statement>
```

Conditional disjunction is close to `if/then/else` in spirit and in semantics. In fact,

```
c_either <statement1>
or <statement2>
```

is equivalent to

```
if <statement1>
then ;              // Null statement
else <statement2>
```

For example, as noted above in Section 4.2, nothing guarantees that the call `evaluate_risk(ff,w,p,r,X)` will succeed for all values of the loop control variable `r` in Model 4.1, Risk vs Reward. In Section 4.2 we saw how to deal with this using the `if/then/else` construct in the loop. Alternatively, we can rewrite the loop to read

5.1 A programming interpretation of disjunction

```
and(int r=HIRISK;r>=LORISK;r=r-DCRMNT)
    c_either evaluate_risk(ff,w,p,r,X);
    or {
        printf("Risk %d is not feasible\n",r);
        break;  // Exit the loop, continue program
    }
```

As with the coding in terms of if/then/else in Chapter 4, this code allows the program to detect that new constraints can no longer be added consistently and to go on to the next phase of execution without abruptly ending.

As is the case for conjunction, two basic properties of mathematical disjunction are commutativity and idempotence. However, it is quite easy to construct situations where conditional disjunction fails to satisfy commutativity or idempotence. For the former, we have the following simple example:

```
a = 0;
c_either {
    a = 1;
    a == 0;
}
or a == 0;
```

With similar ingenuity one can construct a statement to illustrate that idempotence can also fail for conditional disjunction. However, when the disjuncts are constraints or conjunctions of constraints, something positive can be said. Trivially a conditional disjunction of the same conjunction of constraints is idempotent, for if the first disjunct fails, the feasible region is restored and so the same constraints will fail again. As for commutativity, the compound statement

```
c_either <conjunction_of_constraints1>
or <conjunction_of_constraints2>
```

will succeed or fail the same as the compound statement

```
c_either <conjunction_of_constraints2>
or <conjunction_of_constraints1>
```

although the feasible regions are not necessarily the same. Thus, for example, assuming there are no constraints yet on the continuous variable X, the code segments

```
X >= .5;
c_either { X >= .75; X <= 1.0; }
or X == 0;
```

and

```
X >= .5;
c_either X == 0;
or { X >= .75; X <= 1.0; }
```

both succeed, although through different paths. In this case the two forms of the disjunction lead to the same feasible region .75 <= X <= 1.0. In cases where the feasible regions are different, further developments in the program might be affected. By way of example, the code segment

```
c_either X == 0;
or X >= 1;
X == 1;
```

behaves quite differently from

```
c_either X >= 1;
or X == 0;
X == 1;
```

In fact, the first segment fails while the second succeeds.

As an illustration of the usefulness of the conditional disjunctive construct, we can employ it to try a different approach to Model 3.4, The Committee. Let us apply a simple local search algorithm, known as *Drop/Add*. The idea is to start with some solution and successively to improve it by dropping and adding elements, but always maintaining a solution. For the model at hand, we start by adding everyone to the committee and then loop through the list of citizens dropping from the committee anyone not needed. So in this case the Add portion gives us an initial solution; then we use the Drop portion of the technique. Implementing this heuristic certainly doesn't require continuous variables. However, since we have already have the machinery in place, let us use the setup of Model 3.4 and apply conditional disjunction to encode this new heuristic. In the place of the loop to carry out the progressive roundoff, we substitute

```
and(int i=0;i<NUM;i++)
    c_either Citizen[i] == 0.0;  // Try to drop citizen i
    or Citizen[i] == 1.0;        // Citizen i needed
```

For this code we are using the fact that if Citizen[i] == 0.0 is consistent with the linear relaxation, then there is a solution with this citizen off the committee. With the data of Model 3.4, this strategy finds a committee of 11 members, not quite as good a result as the one obtained

5.1 A programming interpretation of disjunction

with progressive roundoff. In Section 6.3, we discuss a way of improving the performance of the Drop/Add heuristic.

There is a loop construct for conditional disjunction, the c_or loop, that is analogous to the and loop. The syntax of the c_or loop is

    c_or <loop header> <statement>

Again, the control variable may not be assigned in the body of the loop or accessed after the loop is exited. Also, if the control variable is passed as a parameter to a procedure in the body of the loop, this must be a pass by value.

As an illustration of the use of the c_or loop construct, let us look at a situation encountered in work on compilers for parallel computation. It can be important to know if complex expressions for indices in nested loops can become equal as in Bixby, Kennedy and Kremer (1994). This leads to the task of determining whether a set of linear equations has an integer solution. This can be a daunting problem. As a very simple example, the following code checks to see if a pair of linear equations has integer-valued solutions in the interval [1,100]:

```
c_or(int i=1;i<=100;i++)
    c_or(int j=1;j<=100;j++) {
        2*i + 7 == 3*j + 1;
        3*i + 5 == 4*j + 1;
        printf("A solution is (%d,%d)\n",i,j);
    }
```

Let us step through this code segment. Although the first test succeeds with i=3,j=4 and again with i=6,j=6, the second test fails in both cases and the search continues until the mutual solution i=12,j=10 is found. Thus for each i=1,...,11, the inner loop is entered 100 times only to fail each time.

Let us see what happens when continuous variables are used. Assuming that X and Y are otherwise unconstrained, the following fragment finds the same solution at i=12,j=10:

```
2*X + 7 == 3*Y + 1;
3*X + 5 == 4*Y + 1;
c_or(int i=1;i<=100;i++) {
    X == i;
    c_or(int j=1;j<=100;j++){
        Y == j;
        printf("A solution is (%d,%d)\n",i,j);
    }
}
```

With the continuous variables, the constraint X == i fails for i<12, and so the inner loop is not even entered until i reaches 12. Then the constraint Y == j fails for all values of j<10 and finally succeeds at j=10. This remarkable ability of continuous variables and constraints to foresee inconsistency is the basis of the important problem solving paradigm known as *constrain-and-generate*; these matters are taken up in Section 8.3.

In the next model both the c_either/or construct and its loop variant c_or will be used. The model deals with *graph coloring*, a mathematical problem that can come up in many situations. A graph is given by a set *V* and a set *E* of 2 element subsets of *V*; the elements of *V* are called *vertices* and the elements of *E* are called *edges*. If {u,v} is an edge, the vertices *u* and *v* are said to be *connected*. A *coloring* of the graph is a mapping on *V* that assigns distinct values to vertices that are connected. This will be a discrete model in which continuous variables are not used; we program it in 2LP to avoid changing programming languages and to illustrate the syntax for conditional disjunction in 2LP.

Model 5.1 Parallel Sessions

An important international medical conference will be attended by high-ranking officials from public health ministries. The conference will last one day and be held at a facility with a large number of meeting rooms. During the day there will be workshops on public health issues led by leading specialists. All together 19 sessions are planned. In order to keep the conference shorter and to allow for more time for each workshop, there will be parallel sessions. There is no real limit on the number of sessions that can be held in any one time slot because of the size of the facility. However, the organizers do not want to schedule sessions at the same time if someone would like to attend both of them. So the attendees have been polled and asked to indicate which sessions they definitely plan to attend. Their responses have been tabulated and a list of all conflicting pairs of sessions in given Table 5.1; for the table

TABLE 5.1 Data for Parallel Sessions

0, 8	0, 3	0, 5	0, 4	0, 9
0, 12	0, 13	1, 10	1, 7	1, 11
1, 6	1, 4	1, 12	1, 5	2, 6
2, 10	2, 13	2, 4	2, 9	2, 7
2, 12	3, 14	3, 8	3, 11	3, 16
3, 12	4, 15	4, 8	4, 17	5, 10
5, 6	5, 15	5, 16	6, 7	6, 15
7, 9	8, 11	9, 18	10, 18	11, 14
13, 17	14, 17	16, 18		

5.1 A programming interpretation of disjunction 145

the sessions have been numbered from 0 through 18. The organizers' goal is to schedule the workshops in a small number of time slots.

To model this, let each session be a vertex in a graph, and place an edge between two vertices if there is at least one official who intends to attend both sessions. In fact, Table 5.1 is simply a list of the edges of this graph. If we think of a time slot as a color, our task is to color the graph with a small number of colors. This way sessions of the same color do not conflict and can be scheduled in parallel. (The graph itself is not easy to draw because it is not planar; if it were, we could invoke the famous theorem that 4 colors suffice to have an upper bound on the number of colors needed.)

Let us consider how to find a quick-and-dirty solution to this graph-coloring problem. The palette of available colors consists of the integers from 0 to V-1, where V is the number of vertices in the graph. In the worst case, if all vertices are connected to one another, then all V colors will be needed. Since the vertices are listed in numerical order, a simple approach is go down the list and color each vertex with the first available color; that is, assign to vertex v the first color that has not yet been assigned to a vertex vv < v that shares an edge with v. A natural way to enhance this strategy is to consider the more difficult vertices first. For this the vertices must be listed in the order determined by the number of neighbors each vertex has in the graph, starting from the highest number and going down to the lowest. In fact, the vertices are given so that the most popular vertices appear first in numerical order. Thus 0 is the vertex with the greatest number of neighbors, and so on. If the data are entered in some other order, subroutines can be written to rearrange the data.

Let us develop the code to carry out this strategy for coloring the graph. For the data, we have the declarations:

```
#define V 19    // Number of vertices or sessions
#define E 43    // Number of edges or session conflicts

int edges[E][2];
```

This is another situation where the programmer has to set up the data structures needed. The vertices are listed as 0,...,V-1. For each vertex v, what we need is a list of the vertices u < v that are connected to v. For each edge in the graph connecting u with v, either u < v or v < u. Therefore, the size of the array needed to enumerate the lists of earlier vertices connected to each vertex is E itself. Hence, we create an array of size E to hold the lists we want and an array of size V to index into this set of lists

```
int earlier[E];    // Array of lists of earlier neighbors
int nghbrs[V];     // Indexes into the array earlier
```

```
                    earlier neighbors of vertex v
earlier
array      ┌──┬──┬──┬──┬──┬──┬──┬──┐
           │  │  │  │  │  │  │  │  │
           └──┴──┴──┴──┴──┴──┴──┴──┘
                    ▲              ▲
                     ╲              ╲
                      ╲              ╲
nghbrs            ┌──┬──┬──┬──┬──┐
array             │0 │  │  │  │  │
                  └──┴──┴──┴──┴──┘
                          v-1  v
```

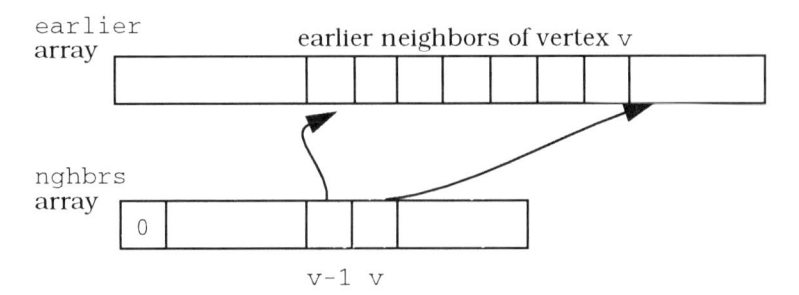

FIGURE 5.1 A List of lists

We will fill these arrays so that the earlier neighbors of vertex v are all listed from `earlier[nghbrs[v-1]]` through `earlier[nghbrs[v]]-1`. This is illustrated in Figure 5.1. Note that vertex 0 has no earlier neighbors, so `earlier[0]` and `nghbrs[0]` are 0, which is the default initialization of a global variable of type `int`. For the rest of the vertices, we cycle through and place in the array `earlier` all connected smaller vertices. During this pass we can count the number of earlier neighbors of each vertex in the array `ngbhrs`. Finally, we incrementally move everything up in the `nghbrs` array so that the interval in the array `earlier` where the neighbors of vertex v are found is `[nghbrs[v-1],nghbrs[v])`. The code for building these arrays follows:

```
build_data_structures()
{
int cnt,ncnt;

    cnt = 0;

    and(int v=1;v<V;v++)
        and(int k=0;k<v;k++)
            if connected(k,v);
            then {         // k is an earlier neighbor of v
                earlier[cnt] = k;    // Enter neighbor
                nghbrs[v] = nghbrs[v]+1; // Count neighbors
                cnt = cnt + 1;       // Increment placeholder
            }

    and(int v=1;v<V;v++)   // Readjust indexing scheme
        nghbrs[v] = nghbrs[v] + nghbrs[v-1];
}
```

Checking whether vertex v and vertex w are connected by an edge in the graph means looping through the edges until a connecting edge is found:

5.1 A programming interpretation of disjunction

```
connected(int v,w)
{
    c_or(int e=0;e<E;e++)
        c_either {
            edges[e][0] == v;
            edges[e][1] == w;
        }
        or  {
            edges[e][0] == w;
            edges[e][1] == v;
        }
}
```

For the coloring itself, we will need an array to hold the color assigned to each vertex and a counter to keep track of the number of distinct colors being used in the coloring process:

```
int color[V];      // To enter coloring
int nc;            // To count number of colors used
```

This means that the first vertex will be assigned color 0 and at this point nc, the number of colors used, will be 1:

```
color_first_vertex()
{
    color[0] = 0;   // Actually the default
    nc = 1;         // One color has been used
}
```

As for the other vertices, testing that color c is available for vertex v is straightforward, since it reduces to checking that no earlier neighbor of this vertex has the color c:

```
conflict_free(int v,c)
{
int begin,end;

    begin = nghbrs[v-1];               // v >= 1
    end = nghbrs[v];

    and(int k=begin;k<end;k++)
        color[earlier[k]] != c;
}
```

When coloring a vertex, we can also keep track of the number of colors used thus far:

```
color_the_vertex(int v,c)
{
    conflict_free(v,c);// Check color c available

    color[v] = c;         // Color vertex v with color c
    if nc == c;           // If color c is new
    then nc = c + 1;      // then increment number of colors
}
```

To color the entire graph, we color all the vertices:

```
color_the_graph()
{
    color_first_vertex();
    and(int v=1;v<V;v++)
        c_or(int c=0;c<=nc;c++)
            color_the_vertex(v,c);
}
```

Putting things together, we have a main procedure:

```
2lp_main()
{
    data();            // Entries from Table 5.1
    build_data_structures();
    color_the_graph();
    output();
}
```

The `output` routine is left to the reader to formulate. This code does find that the meeting can be run with four parallel sessions, which will make the organizers happy.

With this strategy for coloring the graph, we have arranged the vertices in a meaningful order; however, we try to assign colors to a vertex v in the fixed order 0,...,nc. The search could perhaps be made more intelligent by sorting the colors and trying to assign one to this vertex according to a criterion such as the number of earlier vertices already assigned this color.

An alternative is to organize the program from the dual point of view and to assign vertices to colors rather than colors to vertices. To do this we need only replace the call to `color_the_graph` with a call to the following routine:

```
#define UNCOLORED -1
dual_color_the_graph()
{
int vcnt;        // Number of vertices colored
```

5.1 A programming interpretation of disjunction

```
    vcnt = 0;    // None colored yet
    and(int v=0;v<V;v++)    // Initialize vertices
        color[v] = UNCOLORED;
    color[0] = 0;    // Assign first color to first vertex
    and(int c=0;c<V;c++) {    // Cycle through colors
        and(int v=1;v<V;v++)    // Apply color where possible
            if  {
                color[v] == UNCOLORED;
                conflict_free(v,c);
            }
            then {
                color[v] = c;
                vcnt = vcnt+1;
            }
        if vcnt == V; // All vertices colored ?
        then {
            printf("It took %d colors\n",c+1);
            break;
        }
    }
}
#undef UNCOLORED
```

This strategy for coloring the graph also finds a way to do it with only four colors. Clearly heuristics could be developed to work with this strategy as well.

We will have several occasions to revisit this model, in particular, in Section 10.2. The question of finding an optimal coloring that employs a minimal number of different colors is the subject of a significant literature; for further reading, one can start with Berge (1991).

<u>End of Model 5.1</u>

In this last model, we followed the strategy of coloring the more difficult vertices first. This approach is characteristic of a large family of heuristics. We will see other examples, especially in Section 10.2 and Section 12.1.

Exercises

5.1.1. Apply the Drop/Add heuristic to Model 2.3, Call 911.

5.1.2. Write routines to rearrange the data for Model 5.1 in the case the data are not ordered according to the "popularity" of the vertices. Experiment with different data sets to check the importance of the heuristic of working through the vertices in this order. Consider other heuristics for solving the problem both by looping through vertices and by looping through colors.

5.1.3. Code Model 5.1 in a traditional programming language.

5.2 Goal programming

In the models considered so far, constraints have been "hard" in the sense that they are enforced as written. The term *goal constraint* is used for a condition that is desirable but is allowed to be violated to some extent. When an application involves goal constraints, developing a program for it is called *goal programming*. For example, suppose that we are interested in the value of a continuous variable Quantity and would like to make this value reach or surpass some target goal, if possible; failing that, we would like to come as close to the target as possible. The following routine can be used to carry this out:

```
try_to_stay_above_goal(continuous Quantity, double goal)
{
    c_either Quantity >= goal;    // Goal is met
    or {              // Do the best you can given that
                      // goal can't be reached
        max: Quantity;
        Quantity == wp(Quantity);
    }
}
```

A similar routine can be used to try to keep a continuous variable below some target level. Trying to make a continuous variable be as close as possible to a goal is also easily encapsulated:

```
try_to_stay_at_goal(continuous Quantity, double goal)
{
    if wp(Quantity) <= goal;    // Currently beneath goal
    then {
        Quantity <= goal;       // Stay beneath goal
        max: Quantity;          // Get as close as possible
        Quantity == wp(Quantity);  // Fix variable
    }
    else {  // Currently above goal
        Quantity >= goal;       // Stay above goal
        min: Quantity;          // Get as close as possible
        Quantity == wp(Quantity);  // Fix variable
    }
}
```

5.2 Goal programming

An application will often have many goal constraints. Some of these goals may be of equal importance, others may be more or less important. Arranging goals in order of importance is called *prioritizing*. As an illustration of this, let us suppose that in Model 3.2, British Cooking, optimization of the objective function Revenue is too simple a formulation and the following goals are to be satisfied if possible:

1. Achieve a 6-month revenue of at least $90,000.
2. Aim to use about 30 tons of sesame oil in the blend every month.
3. Keep the 6-month total usage of Corn oil under 60,000 tons.
4. Given all this, still maximize revenue.

Goal 2 is really six goals, one for each month. Let us prioritize these subgoals simply by placing the goals in order from the first month to the sixth month. We change the main procedure to

```
continuous Plan[MONTHS][MODES][KINDS];
continuous Revenue,TotalCorn;
2lp_main()
{
    data(price,hardness);      // Unchanged from Model 3.2
        ...
    and(int i=1;i<MONTHS;i++)            // Ditto
        link(Plan[i-1][STOR],Plan[i][PURC],
            Plan[i][PROC],Plan[i][STOR]);

    TotalCorn
    == sigma(int i=0;i<MONTHS;i++) Plan[i][USED][CORN];
    try_to_stay_above_goal(Revenue,90000);    // Goal 1
    and(int i=0;i<M;i++)         // Goals 2
        try_to_stay_at_goal(Plan[i][USED][SESAME],30000);
    try_to_stay_below_goal(TotalCorn,60000);  // Goal 3
    try_to_stay_above_goal(Revenue,1e6);      // Goal 4
}
```

Note that we have formulated the fourth goal of maximizing revenue as the goal of making revenue as close to a million as possible, illustrating the fact that optimization itself is a form of trying to achieve a goal.

In this last example, we were able to prioritize the goals in a strict order. When two or more goals have equal priority, things are more complicated. Let us look at ways of handling this.

Suppose that we have the twin goals of making X close to xgoal and Y close to ygoal at the same time. Mathematically speaking, we want a solution to our constraints that minimizes the size of the error vector (fabs(X-xgoal),fabs(Y-ygoal)). Let us consider ways of minimizing the size of this vector according to the L_∞ and L_1 norms.

For the former, we can introduce a new continuous variable, New, and add the constraints

```
New >= X - xgoal; New >= xgoal - X;
New >= Y - ygoal; New >= ygoal - Y;
```

These constraints imply the relations

```
wp(New) >= fabs(X - xgoal);
wp(New) >= fabs(Y - ygoal);
```

that are not in and of themselves linear constraints. Then, if we call

```
min: New;
```

the minimization of New forces one of the following two equalities to hold:

```
wp(New) == fabs(X - xgoal);
```
or
```
wp(New) == fabs(Y - ygoal);
```

At this point, if we fix the variable New

```
New == wp(New);
```

the error vector is constrained to have minimal L_∞ size. We also know from the Flat Covering Theorem of Section 4.4 that at least one of the goals cannot be approached any closer; it might still be possible to get closer to the other goal.

Now let us turn to the L_1 norm. A constraint of the form

```
sigma(int i=0;i<N;i++) a[i]*X[i] <= b;
```

can be turned into an equality constraint by the addition of a new continuous variable, called the *slack variable*, which will measure the gap from the sum of the a[i]*X[i] to b:

```
sigma(int i=0;i<N;i++) a[i]*X[i] + Slack == b;
```

Similarly a constraint of the form

```
sigma(int i=0;i<N;i++) a[i]*X[i] >= b;
```

can be turned into an equality constraint by the addition of a new continuous variable, called the *surplus variable*, which will measure the gap from the sum of the a[i]*X[i] down to b:

5.2 Goal programming

```
sigma(int i=0;i<N;i++) a[i]*X[i] - Surplus == b;
```

These variables are useful measures of how far the witness point is from satisfying the equality form of a constraint. In fact, in Model 4.6, Extreme Vertices, we used the slack as a measure of the discrepancy between a point in the feasible region and the boundary constraint for ≤ constraints, and we used the surplus for ≥ constraints.

Now let's consider the case where the witness point can lie on either side of a constraint. For example, suppose once more that we are interested in the continuous variable X and would like to measure the discrepancy between this variable and the target xgoal. A way to express the relation between the continuous variable and the target is by using both a slack and a surplus variable

```
X + UnderX - OverX == xgoal;
```

To achieve the effect of the `try_to_stay_above_goal` routine above, we would call

```
min: UnderX;
UnderX == wp(UnderX);
```

or, alternatively,

```
try_to_stay_below_goal(UnderX,0);
```

When slack and surplus variables are used this way in goal programming, they are called *deviation variables* because they measure the deviation of the variable from the target. Deviation variables can also be used in a program to help keep the solution close to the target. Thus, for example, when a goal is expressed with deviation variables, the way to force the witness point to come as close as possible to the target is to minimize the sum of the deviations. With a new variable, say New, the code would be

```
X + UnderX - OverX == xgoal;
New == UnderX + OverX;
try_to_stay_below_goal(New,0);
```

or, alternatively,

```
min: UnderX + OverX;
UnderX + OverX == wp(UnderX + OverX);
```

Let us see what we are measuring here. When the call to

```
min: UnderX + OverX;
```

is made, at least one of these two deviation variables must be driven to 0.0, because their sum is minimized. Therefore, after the call to minimize the sum, the sum of the deviation variables is measuring the absolute value of the difference between the value of X and the goal; this value is fabs(xgoal - X); in other words, after minimization we have the relation

```
wp(UnderX + OverX)  ==  fabs(xgoal - X);
```

Now, if we have a second continuous variable Y and target goal ygoal, we can introduce more deviation variables and write

```
Y + UnderY - OverY  ==  ygoal;
```

If the two equality goals for X and Y have equal priority, a way of treating them as a unit is to minimize the sum of their deviations

```
min: UnderX + OverX + UnderY + OverY;
```

Again, after minimization of the sum of the deviation variables, at least one variable in each pair will be driven to 0.0, and so we are minimizing the sum of the absolute values of the differences between X and Y and their goals xgoal and ygoal. In other words, we are minimizing the L_1 norm of the error term. That done, we could then simply freeze the values of X and Y:

```
X == wp(X);
Y == wp(Y);
```

Alternatively, we could freeze the expression for the L_1 norm of the error term, UnderX + OverX + UnderY + OverY, and then carry out a progressive minimization of the L_∞ size of the error vector. For yet another alternative, we could first minimize the L_∞ size of the error vector and then minimize its L_1 size. Then a progressive minimization of the L_∞ norm could be carried out. The interesting thing is that in interleaving these strategies, the minimization of the L_1 size of the error vector can only be done once and cannot be iterated the way minimization of the L_∞ norm can. In a similar way, deviation variables and the L_∞ and L_1 norms can be used if the equally important aims are to have X reach or surpass xgoal and Y stay below or at ygoal, and so on.

Deviation variables can also be employed when priorities are given to the goals; this is accomplished by assigning larger coefficients in the objective function to the variables that measure the deviation from the more important goals. By making the coefficients much greater for the more important goals, the effect of strictly prioritized goals can be achieved. However, if this is done too naively, numerical problems can arise which impede the work of a linear programming solver. On the

5.3 Pattern recognition

other hand, by modulating the coefficients, a continuum of effects can be achieved.

Exercises

5.2.1. A small plant produces 3 products, known by the code names Alpha, Beta, and Gamma. Each week, the operation is capable of production at the rate of 10 units of product Alpha per hour, 15 units of Beta or 12 units of Gamma. 250 hours of labor time have been allotted and any overtime will cost $15 per hour. The union contract limits overtime to 100 hours. The company cannot sell more than 3000 units of all products combined in a single month. The products bring in revenue of $20, $28, and $40 each for Alpha, Beta, and Gamma, respectively. After production each unit of product must be painted and decorated. This work is done by an outside contractor; the painting and decoration time required is 20 minutes for each unit of Alpha, 24 minutes for each unit of Beta, and 30 minutes for each unit of Gamma. The agreement with the contractor provides for up to 1400 hours of work but things run more smoothly if this is kept down to around 1200 hours. The plant's priorities, in order of importance, are (1) to generate monthly revenue of at least $90,000, (2) to keep overtime for the production crew as close to 0 as possible, (3) to limit the total amount of painting and decorating time to 1200 hours, and (4) to produce as close to 200 units of Beta as possible. Formulate a plan to meet these goals in a reasonable way.

5.2.2. Suppose the goals in Exercise 5.2.1 are not prioritized, but are all of equal importance. Formulate a plan for this situation.

5.3 Pattern recognition

The term *pattern* is used to denote a collection of values given to features of an object. Patterns that are similar constitute a class. A *pattern recognition* algorithm is a method for determining which class a given pattern belongs to. To classify an object, measures are taken of features of the object such as shape or color. These measures constitute a pattern, and the object can be assigned a class based on this. For our purposes, we can consider a pattern as a point in n-dimensional space. In Figure 5.2 (a) we have the situation where a pattern is represented as a point in two-dimensional space; the patterns are divided into two classes designated as xs and os.

To fix ideas, in this section we will only consider the case where the patterns are divided into two classes. A *discriminating function* is a map that is strictly positive on one class of patterns and strictly negative on the other. In Figure 5.2 (b) the two classes are separated by a hyperplane. This means that they are distinguished by a linear function $ax +$

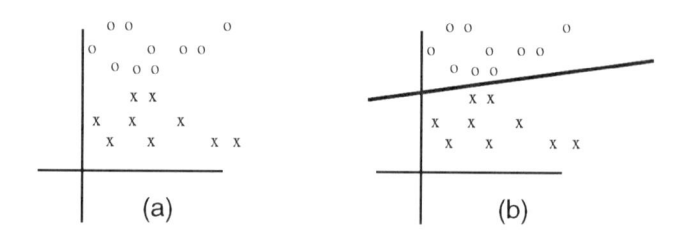

FIGURE 5.2 (a) Two classes of patterns; (b) the patterns separated by the level curve of a linear function.

$by + c$; the hyperplane is the set of zeroes of the discriminating function. A linear discriminating function is an example of a *perceptron*, a concept introduced in the pioneering work on machine learning of Rosenblatt (1962); see Minsky and Papert (1988).

In Figure 5.3 (a), we have the Exclusive-Or relation which is represented by two xs, the points (0, 0) and (1, 1), and two os, the points (1, 0) and (0,1). This is the classic example of two classes of patterns that cannot be separated by a hyperplane.

The sets of Figure 5.3 (a) can, however, be discriminated by a quadratic function. Linear programming can be used to look for both linear and quadratic discriminators because polynomial functions are linear in their coefficients. By this we mean that if we reverse the role of the coefficients and the variables, a quadratic expression such as c*X*Y becomes x*y*C and if it is the coefficient that is the unknown, linear programming can be used to compute it. This is another example of looking at a situation from a dual point of view.

Let us look at the case of the Exclusive-Or more closely. We seek a quadratic function $q(x,y) = ax^2 + bxy + cy^2 + dx + ey + f$ such that $q(0,0) \geq \varepsilon$, $q(1,1) \geq \varepsilon$ and $q(0,1) \leq -\varepsilon$, $q(1,0) \leq -\varepsilon$. The values of x, y that interest us are given. What we have to determine are the coefficients a, b, c, d, e, f. Reversing roles and looking at the dual situation, we consider the function $x^2 a + xy\, b + y^2 c + x\, d + y\, e + f$, making the terms a,b,c,d,e,f the variables. Since there is no reason for these terms to be nonnegative, we need to simulate free continuous variables to determine them. For that we can use the device of Model 2.5, The Paris Bourse, where we split a free variable into a pair of nonnegative continuous variables. The program can be written as follows:

```
continuous Ap,Am,Bp,Bm,Cp,Cm,Dp,Dm,Ep,Em,Fp,Fm;
#define EPSILON 1e-2

2lp_main()
{
    0*(Ap-Am) +0*(Bp-Bm) +0*(Cp-Cm) +0*(Dp-Dm) +0*(Ep-Em)
        + Fp-Fm >= EPSILON;
```

5.3 Pattern recognition

```
1*(Ap-Am)+1*(Bp-Bm)+1*(Cp-Cm)+1*(Dp-Dm)+1*(Ep-Em)
   + Fp-Fm >= EPSILON;
0*(Ap-Am)+0*(Bp-Bm)+1*(Cp-Cm)+0*(Dp-Dm)+1*(Ep-Em)
   + Fp-Fm <= -EPSILON;
1*(Ap-Am)+0*(Bp-Bm)+0*(Cp-Cm)+1*(Dp-Dm)+0*(Ep-Em)
   + Fp-Fm <= -EPSILON;

printf("a is %f\n",Ap-Am);
printf("b is %f\n",Bp-Bm);
printf("c is %f\n",Cp-Cm);
printf("d is %f\n",Dp-Dm);
printf("e is %f\n",Ep-Em);
printf("f is %f\n",Fp-Fm);
}
```

Printing out the witness point yields the solution $a = -.02$, $b = .04$, $c = -.02$, $d = 0.0$, $e = 0.0$, $f = .01$. So we have

$$q(x,y) = -.02\, x^2 + .04\, xy + -.02\, y^2 + .01$$

To represent the way this function separates the two classes geometrically, we draw the level curve $q(x,y) = 0$. Using the quadratic formula to solve this equation, we find that $y = x \pm \sqrt{2}/2$. As sketched in Figure 5.3 (b) the level curve of the discriminating function separates the two classes by means of a degenerate conic section consisting of two straight lines. So we have found the next best thing to a separating hyperplane, namely a separating pair of hyperplanes.

The method is perfectly general and will find a discriminating function if one exists. In fact, this extends the Perceptron Learning Theorem of Rosenblatt (1962) which deals with the case of a linear discriminating function. In this example, because of the simplicity of the situation, we could be sure that the constraints to determine the coefficients of the quadratic discriminator would prove feasible. When things are more

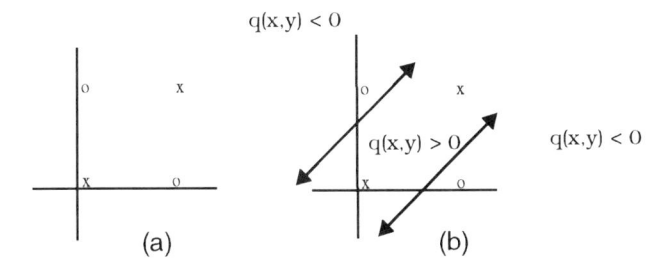

FIGURE 5.3 (a) The exclusive or relation; (b) the level curve of a discriminating function for the exclusive or.

complex, the constraints have to be treated as goals. The next model illustrates this.

Model 5.2 Cell Discrimination

Lab scientists have developed a scheme for classifying cells into two categories called benign and malignant. At the start of the classification process each cell is given nine nonnegative numbers that measure different features. Assigning these numbers to a cell from a sample is quite easy and, in fact, can be done automatically during a scan of the area under examination. There is certainly a relationship between the nine numbers that evaluate the features of a cell and determine whether it is benign or malignant. In fact, by looking at the data in Tables 5.2 and 5.3, one can see that malignant cells tend to have higher-valued entries. The question arises whether there is a simple mathematical expression, say a polynomial, that defines a function that will discriminate between these benign and malignant cells with a high degree of accuracy, a positive value being returned for benign cells and a negative value for malignant cells.

The data from Tables 5.2 and 5.3 consist of 50 items for each case and there are nine fields for each data item. Each field contains the evaluation of one of the nine features. The evaluation is an integer in the interval [0, 9]. The data can be entered in two-dimensional arrays:

```
#define N 9          // Number of data fields
#define SZ 50        // Size of training sets
int be[SZ][N], ma[SZ][N];
```

Each benign data item can be accessed as a vector be[n], and each malignant one by ma[n].

Let us first set up the machinery for finding a quadratic function to discriminate between the two classes of cells. For this we need coefficients for the quadratic and linear terms and a constant term. We will try to have the function return a positive value in the benign case and a negative value in the other case. Since the coefficients we are seeking may well take on negative values, we again need to simulate free variables by splitting a free variable into a pair of nonnegative continuous variables. For the constant term this means a pair of continuous variables Cp,Cm; for the linear terms, we need 9 such pairs Lp[i],Lm[i] for 0<=i<N, and for the quadratic terms, 81 pairs of continuous variables Qp[i][j],Qm[i][j] for 0 <= i,j < N.

For a vector be[n] with entries from a benign cell, we want

```
    sigma(int i=0;i<N;i++)      // Sum of quadratic terms
        sigma(int j=0;j<N;j++)
            be[n][i]*be[n][j]*(Qp[i][j]-Qm[i][j]))
  + sigma(int k=0;k<N;k++)      // Sum of linear terms
        be[n][k]*(Lp[k]-Lm[k])
```

5.3 Pattern recognition

TABLE 5.2 Malignant Data Set

7,9,9,7,6,9,8,6,7	4,2,2,2,1,2,3,3,0	6,3,5,3,5,0,3,2,0
9,6,6,5,3,9,3,0,1	9,4,4,2,5,6,6,9,0	7,3,4,0,1,0,6,2,0
9,6,6,2,7,4,6,3,2	9,9,9,7,5,0,7,8,0	1,4,2,2,5,6,6,4,0
9,3,2,0,2,2,5,4,1	4,5,4,5,9,0,2,0,0	9,9,9,3,7,0,7,9,0
6,7,6,1,3,7,2,7,1	8,4,7,0,1,2,1,0,4	9,2,5,1,2,4,3,9,1
4,4,4,7,9,7,6,2,6	9,5,5,2,3,4,2,5,0	7,9,9,0,2,5,2,8,0
4,1,2,0,5,9,4,0,0	8,4,4,1,1,1,4,0,0	8,9,9,0,9,7,2,2,0
5,2,3,0,4,1,2,8,0	4,2,3,0,7,9,3,8,0	7,2,7,2,3,8,7,8,7
8,3,4,9,5,9,3,7,0	9,5,3,0,2,3,2,1,2	4,9,5,0,9,3,3,9,9
2,2,5,3,4,7,3,3,0	8,5,8,1,9,5,1,8,9	6,4,5,9,4,9,6,8,3
1,2,3,3,1,4,1,4,0	7,1,2,0,5,2,6,0,0	6,2,3,3,2,2,2,1,6
9,9,9,7,1,9,3,0,0	5,4,3,3,2,8,6,7,2	7,5,3,2,4,8,2,0,0
9,9,9,2,9,7,7,0,0	3,4,4,9,3,9,6,4,7	4,2,4,0,7,9,4,2,0
4,3,5,6,8,6,7,9,0	7,2,4,3,4,9,0,5,1	4,9,7,9,7,9,2,5,2
2,3,4,1,5,7,3,0,0	7,7,6,3,9,9,6,7,6	9,9,7,5,3,4,7,9,0
4,4,4,5,2,9,2,0,0	9,6,6,3,4,9,4,6,1	5,9,9,9,7,9,9,9,6
4,9,9,2,7,0,4,9,2	7,6,5,3,3,9,4,0,0	

```
+   (Cp-Cm)              // Constant term
>=  EPSILON ;            // Small positive quantity
```

And similarly for the malignant cells but with `<= -EPSILON` instead of `>= EPSILON`.

One can observe that the quadratic terms can be assembled as monomials in products `be[n][i]*be[n][j]` with `i <= j`. Therefore, we can fix at `0.0` all the unneeded variables `Qp[i][j]` and `Qm[i][j]` with indices such that `j < i`. This remark nearly halves the number of continuous variables needed to represent a quadratic discriminating function. This can be done in the procedure that sets up the constraints of the model.

Since we cannot be sure that a discriminator of the type sought exists, we will seek a function that discriminates well overall but that might err in one direction or the other on occasion. An item which is incorrectly classified by a labeling scheme is called an *outlier*. Detecting outliers that resist an attempt at classification is important, for these cells might well point to interesting clues. Even if there is no perfect discriminating function of the type sought, we want to flag those data items that cause the breakdown in the classification.

One thing we can do is to loop through the data items and try add the above constraint for `be[n]` to ensure that this item is classified correctly; if this is not feasible, no constraint would be added and the item becomes an outlier. However, this approach will favor the earlier entries

TABLE 5.3 Benign Data Set

4,0,0,0,1,0,2,0,0	4,3,3,4,6,8,2,1,0	5,7,7,0,2,3,2,6,0
3,0,0,2,1,0,2,0,0	1,0,1,0,1,0,2,0,0	1,0,0,0,1,0,0,0,4
0,0,0,0,0,0,2,0,0	1,0,0,0,1,0,1,0,0	3,0,0,0,1,0,1,0,0
3,0,0,0,1,0,2,0,0	2,0,0,0,1,0,1,0,0	0,0,0,0,1,0,2,0,0
4,0,0,0,1,0,1,0,0	1,0,0,0,1,0,1,0,0	2,0,0,0,0,0,1,0,0
1,0,0,0,1,0,2,0,0	2,0,1,0,1,0,1,0,0	1,0,0,0,1,0,1,0,0
5,5,5,8,5,0,6,7,0	0,0,0,0,1,0,1,0,1	3,0,0,2,1,0,2,0,0
0,0,0,0,1,1,1,0,0	3,0,0,0,1,0,2,0,0	0,0,0,0,1,0,2,1,0
0,2,2,1,1,0,6,1,0	0,0,1,0,1,1,3,1,0	4,2,0,1,1,0,1,0,0
2,0,0,0,1,2,2,0,0	1,1,1,0,0,0,6,0,0	3,0,0,1,1,0,1,0,0
2,0,0,0,1,1,6,0,0	3,0,0,0,1,0,2,0,0	0,0,0,0,1,0,2,0,0
2,0,0,1,1,0,0,0,0	0,0,0,0,1,0,1,0,0	1,0,0,0,1,0,2,0,0
1,0,0,1,1,0,0,0,0	4,0,0,0,1,0,2,0,0	0,0,0,0,1,0,1,2,0
0,2,0,1,1,1,4,2,1	0,0,0,0,1,4,0,0,0	7,2,2,0,1,1,2,1,0
2,1,0,0,1,1,2,0,0	0,0,1,1,1,0,2,0,0	0,0,0,0,1,0,1,0,0
2,0,0,0,1,0,2,0,0	4,0,2,0,1,0,1,0,0	2,0,0,0,2,0,1,0,0
2,0,0,0,1,0,0,0,0	2,0,0,0,1,0,0,0,0	

in the data stream. Since the data items are given in no particular order, we want a strategy where all the items are treated equally and those that are exceptional relative to all the rest are denoted outliers. Another reason for emphasizing the global structure of the data is that the discriminating function will subsequently be used on many new data items. For this reason the data we employ to construct the discriminating function are called the *training set*. A way to find a solution in this situation is to make the constraints for all of the data items goal constraints of equal priority and to use deviation variables to measure the discrepancy in satisfying a goal constraint exactly.

Let us introduce a continuous variable Under[n] to measure the gap for a benign cell in the training set. That is, if the discriminating function assigns the benign cell be[n] a value below EPSILON, this variable will represent the error term. For this we change the last term in the constraint to

```
>=  EPSILON - Under[n];    // Deviation
```

A similar thing can be done for the malignant case with a variable Over[n], where Over[n] measures the excess over -EPSILON. All this can be encapsulated in a procedure to which we pass a flag value of +1 in the benign case and -1 in the malignant case:

5.3 Pattern recognition

```
approx(double c[],continuous Dev,int sign)
{
    sign*( // sign is +1 in benign case, -1 otherwise
        sigma(int i=0;i<N;i++)
            sigma(int j=i;j<N;j++)
                c[i]*c[j]*(Qp[i][j]-Qm[i][j])
        + sigma(int k=0;k<N;k++) c[k]*(Lp[k]-Lm[k])
        + (Cp-Cm)
    )
    >=   EPSILON - Dev;
}
```

The parameter Dev will be Over, and the vector c will be ma[n] in the malignant case; they will be Under and be[n] in the other case.

This setup routine takes the form:

```
#define BENIGN 1
#define MALIGNANT -1

setup()
{
    and(int i=0;i<N;i++)    // Zero out unneeded variables
        and(int j=0;j<i;j++) {
            Qp[i][j] == 0;
            Qm[i][j] == 0;
        }

    and(int n=0;n<SZ;n++) {  // SZ is training set size
        approx(be[n],Under[n],BENIGN);
        approx(ma[n],Over[n],MALIGNANT);
    }
}
```

Since for each benign data item we have a continuous variable Under[n], and for each malignant data item a variable Over[n], the vector of these variables represents the error term in the discriminating function. The aim is to minimize the L_1 norm of this error term. In other words, we seek to minimize the sum of the deviations:

```
min: sigma(int n=0;i<SZ;n++) (Under[n] + Over[n]);
```

Since it is interesting to know if there is a linear discriminating function, let us first try to find one. Moreover, if there is no linear discriminator, it will still be of interest to flag the data items that are outliers for this attempt at classification. Since the linear case is an instance of the quadratic one obtained by setting all the Qp[i][j] and Qm[i][j] to 0.0, the machinery we have set up thus far can be used.

The main procedure will of course enter the data and set up the constraints for finding the quadratic discriminating function. Then the attempt can be made to find a linear discriminating function. If that fails, a quadratic alternative can be tried. Putting things together, the declarations and main procedure can take the form

```
#define EPSILON 1e-3

double be[SZ][N];
double ma[SZ][N];

continuous Qp[N][N],Qm[N][N];
continuous Lp[N],Lm[N],Cp,Cm;
continuous Under[SZ],Over[SZ];

2lp_main()
{
    data();
    setup();

    c_either linear();      // Try for a linear discriminator
    or quadratic();         // Else do the best you can
}
```

It remains to write the routines for the linear and the quadratic cases. We can distinguish between them for reporting purposes by means of symbolic constants.

```
#define LINEAR 0        // For i/o
#define QUADRATIC 1     // For i/o
```

For the linear case, we will zero out the quadratic terms and minimize the L_1 norm of the deviation

```
linear()
{
int cnt;    // Storeback variable
    and(int i=0;i<N;i++)    // Zero out quadratic terms
        and(int j=0;j<N;j++) {
            Qp[i][j] == 0.0;
            Qm[i][j] == 0.0;
        }
    min:  sigma(int i=0;i<SZ;i++) (Under[i] + Over[i]);
    cnt = 0;                // Initialize count of outliers
    report(cnt,LINEAR);     // List outliers if any, etc
    cnt == 0;   // Does linear discriminator exist?
}
```

5.3 Pattern recognition

For the quadratic case, we need only minimize the objective function and issue a report:

```
quadratic()
{
int cnt;      // Storeback variable
    min:    sigma(int i=0;i<SZ;i++) (Under[i] + Over[i]);
    cnt = 0;                  // Initialize count of outliers
    report(cnt,QUADRATIC); // List outliers if any, etc
}
```

The role of the reporting routine will be to flag outliers, to record the coefficients determined, and to return the number of outliers. To record this last piece of information, the reporting routine will be passed the storeback variable cnt by reference:

```
report(int & cnt, int case)
{
    if case == LINEAR;
    then printf("Linear report:\n");
    else printf("Quadratic report:\n");
    and(int i=0;i<SZ;i++) {
        if wp(Under[i]) > EPSILON;
        then {
            printf("benign %d is an outlier\n",i);
            cnt = cnt + 1;
        }
        if wp(Over[i]) > EPSILON;
        then {
            printf("malignant %d is an outlier\n",i);
            cnt = cnt + 1;
        }
    }
    printf("In all %d outliers found\n",cnt);
}
```

On the given training set, the linear discriminating function flagged six outliers. With the quadratic discriminating function there were no outliers.

One more point: Some programmers may find it unaesthetic to declare 72 useless variables $Qp[i][j]$ and $Qm[i][j]$ with $j < i < N$. A two-dimensional array that is not rectangular is called a *ragged array*, and that is the kind of data structure needed here. Nonstandard data structures such as these have to be designed manually by the programmer; this approach is left as an exercise.

<u>*End of Model 5.2*</u>

The above model is based in large part on Ferris and Mangasarian (1995). In Mangasarian (1995) linear programming is used to build sophisticated discriminating functions for this kind of appplication known as *neural networks* or *neural nets*.

For other work on neural nets and linear programming, see Roy and Mukhopadhyay (1991) and Ignizio and Cavalier (1994). Another view of neural net construction uses ideas from nonlinear optimization; the principal technique in this approach is called *back-propagation*, (Rumelhardt and McClelland 1987). Neural net pattern recognition is called a *connectionist* method. Connectionist approaches to machine learning fell out of favor in the 1970s, but the discovery of back-propagation methods relaunched interest in this approach to problem solving. Neural networks are an example of how computational power can add a fundamentally new method for solving problems to one's arsenal in addition to extending the reach of existing methods.

Data and information on this kind of problem can be found on the internet. Let us cite two web addresses:

```
http://www.ics.uci.edu/AI/ML/MLDBRepository.html
http://www.cs.wisc.edu/~shavlik/uwml.html
```

Exercises

5.3.1. Replace the arrays Qp[N][N] and Qm[N][N] by one-dimensional arrays of length N(N+1)/2 to represent ragged arrays. Rewrite the model using this representation of the ragged arrays.

5.3.2. Download data sets from the internet sites above. Build and test discriminating functions.

6

Negation

In this chapter we treat the propositional connective negation. We introduce negation-as-failure, a method for dealing with negation in computational logic pioneered in Prolog. This construct is then used in some important local search algorithms.

6.1 Negation-as-failure

In classical logic the fundamental property of negation is involution

$$\sim \sim p \text{ is equivalent to } p$$

However, this property of negation has stirred controversy among mathematicians. For example, it is not valid in intuitionistic logic and other branches of constructive mathematics (see Fitting 1983). In traditional programming languages, negation simply inverts the truth value returned and therefore the involutive property of negation is maintained. Something else can happen to the numerical values returned; for example, in C and C++, the expression !!7 evaluates to 1.

For working with constraints, the semantics of negation are based on *negation-as-failure*, a concept used in database work and in logic programming. This is a generalization of the semantics of negation of traditional programming languages. Formally, negation in 2LP has the syntax

```
not <statement>
```

and the semantics of not <statement> are that it succeeds if <statement> fails and conversely, that it fails if <statement> succeeds. Operationally, not <statement> is equivalent to

```
if <statement>
then 0 == 1;   // Always fails
```

In other words, when the code not <statement> is reached, <statement> itself is executed, and then its truth value is reversed and the feasible region is restored to its previous state. However, variables of type int, double, and string keep the values assigned to them by the execution of <statement>.

Since negation does not change the feasible region, it can be used to test the consistency of sets of constraints and of procedures. For example, the code segment

```
if not X <= Y;
then printf("The constraint would fail\n");
else printf("The constraint would succeed\n");
```

will determine if the constraint X <= Y could be successfully added to feasible region and the code segment does this without changing the current feasible region. This is quite different from the code

```
if X <= Y;
then printf("The constraint has succeeded\n");
else printf("The constraint has failed\n");
```

which changes the feasible region by adding the constraint X <= Y if it is possible to do so.

For a procedure p() the code

```
if not p();
then printf("A call to p() can be successful\n");
else printf("A call to p() cannot be successful\n");
```

determines if p() can be called successfully; again, this is done without changing the current feasible region. In the following model we use this functionality to run the same routines on different sets of data.

Model 6.1 Double Data

A professor is preparing a linear programming assignment for her class and is planning to use two data sets that she compiled when giving this course last semester. She will give out the two data sets but wants to be sure that one of them leads to an infeasible problem. The assignment is straightforward, two maximization problems involving 8 constraints on 7 continuous variables. The data are dense and are given in Tables 6.1 and 6.2. The objective function is given in the row labeled -1; all the constraints are given in ≤ form. The professor wants to run the problem with both data sets, determine which set is inconsistent, and note the value of the objective function in the consistent case.

We can run the linear program with the different data sets on the same continuous variables X[i] by wrapping the setup constraints and

6.1 Negation-as-failure

TABLE 6.1 First Data Set

Rows	Columns							
-1	3	55	22	18	7	9	6	
0	15	6	-18	-3	22	11	-6	<=	43
1	-33	-4	6	7	-8	9	0	<=	32
2	5	10	8	-11	-65	0	21	<=	-28
3	5	66	-88	33	-21	-56	0	<=	41
4	19	-32	-36	531	-46	99	16	<=	-77
5	38	-55	0	5	0	17	0	<=	0
6	10	37	92	-84	0	59	3	<=	2
7	22	0	-5	12	41	0	56	<=	71

TABLE 6.2 Second Data Set

Rows	Columns							
-1	2	34	12	28	11	13	16	
0	5	-6	8	-3	22	11	-6	<=	43
1	7	-4	6	7	-8	9	0	<=	-32
2	5	0	8	-11	-65	44	21	<=	28
3	4	66	-88	33	-21	-56	73	<=	41
4	9	-32	136	531	-46	99	16	<=	-77
5	38	-55	0	5	0	17	0	<=	0
6	0	37	92	-84	0	59	3	<=	2
7	0	0	0	12	41	0	56	<=	71

the optimization in the if/not construct. Since this construct will undo the constraints and the optimization, we will print out the value of the objective function when it is determined. The professor can check her data with the following code:

```
#define M 8        // Number of constraints
#define N 7        // Number of variables
#define K 2        // Number of data sets
double a[M][N],rhs[M],c[N];
continuous X[N],Z;    // Z is the objective function
2lp_main()
{
    and(int k=0;k<K;k++)
        if not version(k);
        then printf("Version %d is inconsistent\n",k);
}
```

```
version(int k)
{
    data(k);                          // New data set
    and(int i=0;i<M;i++)              // New constraints
        sigma(int j=0;j<N;j++) a[i][j]*X[j] <= rhs[i];
    printf("Version %d is consistent\n",k);
    Z == sigma(int j=0;j<N;j++) c[j]*X[j];// New objective
    max: Z;                           // Optimize
    printf("With maximum value %f\n",Z);
}

data(int k)
{
    if k == 0;
    then {
        a = { 15, ..., 56 };          // From Table 6.1
        rhs = { 43, ..., 71 } ;
        c = { 3, ..., 6 };
        return;        // Built-in command
    }

    if k == 1;
    then {
        a = { 5, ..., 56 };           // From Table 6.2
        rhs = { 43, ..., 71 };
        c = { 3, ..., 6 };
        return;        // Built-in command
    }
}
```

The command return used in the data routine simply successfully exits the procedure in which it is called. Running the program reveals that the first data set is consistent, while the second indeed leads to infeasible constraints.

End of Model 6.1

To reiterate, a key fact about negation-as-failure is that the feasible region is never changed by a call to a negated statement. This functionality will prove very useful. For instance, it will be used presently to develop heuristic search strategies of artificial intelligence where future possibilities must be enacted, evaluated, and then undone.

Another example of the utility of negation-as-failure is that it can be used to do "what-if" analysis for goal programming. One can simply place if/not constructs in front of the code for differing lists of goals and priorities and record the results. Negation-as-failure will also prove

6.1 Negation-as-failure

useful in Chapter 7 for parametric analysis of linear programming models.

When constraints are involved, the semantics of the double negation

> not not <statement>

are different from those for the statement itself. Firstly, a double negation is a negation and so will not affect the feasible region. Now, the double negation will succeed if and only if the negation not <statement> fails; that is, the double negation succeeds if and only if <statement> succeeds, but the double negation leaves the feasible region unchanged. So from the point of view of the constraints and the feasible region, the double negation tests the consistency of the statement without committing to it. Naturally, if the <statement> fails, then the double negation also fails.

By way of example, in Model 3.5, The Library Fund Reprised we ran the same routine scenario on different sets of continuous variables. The declarations of the continuous variables in 2lp_main were

```
continuous Cd[SC][INS][BKS],CashFlow[SC][BKS+1];
```

where SC was a symbolic constant for the number of scenarios. For each scenario a complete set of continuous variables is declared. If we only want to run the three separate scenarios without merging them into the stochastic model, we can reuse the same continuous variables by calling the double negation not not scenario rather than scenario:

```
2lp_main()
{
continuous Cd[INS][BKS],CashFlow[BKS+1];   // Only one set

    data();
    and(int s=0;s<SC;s++) {
        printf("In scenario %d\n",s);
        not not scenario(CashFlow,Cd,r[s]);
    }
}
```

If <statement> does not involve constraints, then the negation not <statement> and the double negation not not <statement> have their usual programming language meaning. Thus

> not not {a <= 5;b == 3;c = 0;}

is equivalent to

> {a <= 5;b == 3;c = 0;}

and

```
not {a <= 5;b == 3;c = 0;}
```

is equivalent to

```
{c_either a > 5; or b !=3; or not c = 0;}
```

where, of course, the identifiers a, b, and c are of type int or double. Note that

```
not c = 0;
```

will fail, since an assignment statement always succeeds. Let us also note that in C and C++ the expression

```
!(c = 0)
```

evaluates to 1, since assignment is interpreted as returning the value assigned. But such subtleties aside, the semantics of negation-as-failure coincide with the usual programming language semantics of negation when constraints are not in the picture.

Exercises

6.1.1. Use negation-as-failure to run the loop in Model 4.1, Risk vs Reward, in the direction that increases the loop variable r.

6.2 Evaluation functions and lookaheads

The term *heuristic* is used in different ways in the literature. On the one hand, the term denotes a search strategy that yields an acceptable solution to a difficult application, though not necessarily the best possible solution. On the other hand, a heuristic is a guide used to determine which action to take next. Typically, when used as a guide, the heuristic assigns a measure to each possible next action; the function used to compute this measure is called an *evaluation function*. When an application is modeled so as to have a linear relaxation, then the optimum of the linear relaxation provides a natural evaluation function.

Since negation does not change the feasible region, it can be used to peer into the future. For looking ahead, the term *ply* is used to denote a step forward, and an *n-ply lookahead* is one that peers n layers into the future. We next consider some techniques for lookaheads of this

6.2 Evaluation functions and lookaheads

kind. We limit ourselves to one-ply situations to keep the discussion simple.

By way of illustration, suppose that future1() and future2() are routines that develop alternative possibilities in a model. Suppose that Z is an objective function and that we are seeking a solution that makes Z large. Then at any point in the search, the maximum value of Z over the current linear relaxation provides us with a natural evaluation function. So confronted with two possible choices for the future, the following code enables us to look into each of these possible worlds and to compare them:

```
look_into_the_future(double & cache1,& cache2)
{
    if not {
        future1();          // Go into the first future
        max: Z;             // Optimize
        cache1 = wp(Z);     // Record evaluation
    }
    then cache1 = -INFINITY; //Executed if future1() fails

    if not {
        future2();          // Go into second future
        max: Z;             // Optimize
        cache2 = wp(Z);     // Record evaluation
    }
    then cache2 = -INFINITY; //Executed if future2() fails
}
```

When this code returns, an evaluation of the two futures will be contained in cache1 and cache2, and the current feasible region will not have changed. Now one of the futures can be selected based on these evaluations. Note that we are passing the variables cache1 and cache2 by reference and using them to communicate the result of the computations made in the routine look_into_the_future. Variables used this way are called *storeback variables* or *storeback parameters*.

To describe how to develop heuristic search strategies that look one level into the future, let us fix ideas and consider the set covering problem in the form of Model 3.4.

Model 6.2 The Committee Realigned

People find that the committee of Model 3.4 still has too many members for efficient operation. The Drop/Add heuristic of Section 5.1 did not lead to an improvement over the results of the progressive roundoff method. Let us return to the progressive roundoff technique and see if we can improve it. For the progressive roundoff, the main routine, with some comments added, is

```
21p_main()
{
    data();
    setup();
    and(int i=0;i<NUM;i++) {   // Loop over NUM citizens
        min: Z;
        Citizen[i] == ceil(Citizen[i]);   // Will succeed
    }
    output();
}
```

In the central loop, in the case where the witness point value of Citizen[i] is a fraction strictly between 0 and 1, this value is rounded up, and the continuous variable is fixed at 1. Thus, if the linear optimum places a fractional part of the citizen on the committee, this is upgraded to full membership. But in this fractional case we could probably improve things if we could look one-ply ahead into the future and compare the effects of both putting the citizen on the committee and of not putting the citizen on the committee individually before deciding which alternative to choose. We can do this by using negation-as-failure. We will look ahead at each of the alternative nodes Citizen[i] == 1 and Citizen[i] == 0. For each alternative we will cache the minimum value of the objective function and return to where we were without changing the feasible region. We will then choose the node with the better evaluation.

To evaluate the effect of placing Citizen[i] on or off the committee, we can use an auxiliary procedure.

```
evaluate(int on_or_off,continuous Citoyen,double & value)
{
    Citoyen == on_or_off;
    min: Z;
    value = wp(Z);
}
```

Then to evaluate the effect of placing this citizen on the committee we can make the call

```
    not not evaluate(1,Citizen[i],min_when_on);
```

This code assigns the minimum value of the objective function when Citizen[i] is placed on the committee to the storeback parameter min_when_on of type double. The not not construct removes the constraint Citizen[i] == 1 from the feasible region but does not change the value assigned to min_when_on in the execution of the call to evaluate(1,Citizen[i],min_when_on). Here we are using the fact that we know that it is always consistent to put someone on the committee

6.2 Evaluation functions and lookaheads

because there is no constraint limiting the number of committee members. Thus the `not not` must succeed.

The situation is quite different for evaluating the effect of keeping someone off the committee because the constraint `Citizen[i] == 0` could fail. For this reason we cannot duplicate the above code. However, we can use implication and negation together and test for the success or failure of this constraint. The following code segment will do this:

```
if not evaluate(0,Citizen[i],min_when_off);
then min_when_off = NUM;   // Total number of citizens
```

Now if the constraint `Citizen[i] == 0` is inconsistent, the `if` part of the conditional will succeed without changing the feasible region; the identifier `min_when_off` will then be assigned the value `NUM`, which is the worst possible evaluation. If the constraint is consistent, the effect will be the same as with `not not`; the minimum value of the objective function will be cached in the storeback variable `min_when_off`, the feasible region will not be changed, and the second part of the conditional will not be executed.

To test if the witness point has `Citizen[i]` at an integer value, we can run a test such as `wp(Citizen[i]) == nint(Citizen[i])`. Alternatively, we can call `integral(Citizen[i])`, using the built-in procedure `integral`. This procedure call is equivalent to the test `fabs(wp(Citizen[i]) - nint(Citizen[i])) < EPSILON`, using 1e-6 as EPSILON. So to make use of the one-ply lookahead in the fractional case, we can change the central loop in of the main routine to read:

```
lookahead_code()
{
    and(int i=0;i<NUM;i++) {   // Revised loop
        min: Z;
        if integral(Citizen[i]);
        then Citizen[i] == wp(Citizen[i]);
        else one_ply(i);
    }
}
```

where the `one_ply` routine is

```
one_ply(int i)
{
double min_when_on,min_when_off;   // Storeback variables

    not not evaluate(1,Citizen[i],min_when_on);

    if not evaluate(0,Citizen[i],min_when_off);
    then min_when_off = NUM; // Total number of citizens
```

```
    bold_decision(i,min_when_off,min_when_on);
}
```

A search strategy that uses an evaluation function to select one alternative while discarding the rest is called a *hill-climbing* strategy. In this example, we can use the evaluation function to decide which alternative to pursue:

```
bold_decision(int i, double off,on)    // Hill_climbing
{
    if off < on;
    then Citizen[i] == 0;
    else Citizen[i] == 1;
}
```

The main routine is simply

```
2lp_main()
{
    data();
    setup();
    lookahead_code();
    output();
}
```

This version of the code is more effective than the version of Model 3.4; a solution is found with only 8 people on the committee. By considering the optimal solution to the linear relaxation, one can see that a solution with 8 people is the best possible.

<u>End of Model 6.2</u>

We will have recourse to the lookahead method several times in order to obtain a quick-and-dirty solution to get a good start in an optimization problem. The hill-climbing strategy itself is a fundamental technique, which we will have reason to use many times.

Exercises

6.2.1. Extend the lookahead in Model 6.2 to a 2-ply lookahead.

6.3 Randomization and local search

The study of random phenomena in science goes back to Aristotle. In *The Physics*, Aristotle analyzes causality in terms of four basic forms: material, efficient, formal, and final. This is followed by a commentary on the role of randomness in causality. The modern mathematical analysis of randomness was begun by Fermat and Pascal. Today randomness is a fundamental principle in subjects ranging from thermodynamics to evolutionary biology, (e.g. see Dennett 1995). The use of randomness as a tool in mathematical algorithms themselves, however, is a product of the computer age.

Intuitively a long sequence of numbers is *random* if there is no *a priori* way to generate the sequence by means of a compact program or algorithm. Computer-generated sequences of their very nature cannot be random. However, for applications we can make do with "sufficiently random" sequences, which are known as *pseudorandom sequences*. These sequences are generated by clever numerical algorithms, and their development is an important research field.

Typically a pseudorandom number generator starts with a seed provided by the programmer; it outputs a sequence of real numbers in the interval (0, 1). If a different seed is given, a different sequence is produced. For development and debugging, one will use the same seed again and again; for production runs, one will use varying seeds. The pseudorandom number generator of 2LP is a function with signature

```
double random( )
```

The seed is provided by a call to the procedure `seed` which has the signature

```
seed( int )
```

The parameter passed to this procedure serves as the seed for the pseudorandom sequence. The 2LP system uses the generator of Percus and Kalos (1989) for its built-in function. This is a linear congruential pseudorandom number generator.

The simple use we will make of the pseudorandom number generator is to scramble a sequence of integers 0,...,n-1. For that, we will employ a classic shuffling algorithm; see Knuth (1964). An integer array and a integer are passed to a subroutine, as follows:

```
find_next_random_sequence(int q[], int n)
{
int temp, rand;

    and(int i=0;i<n;i++)
        q[i] = i;
```

```
    and(int i=0;i<n;i++) {
        rand = random()*(n-i);  // random integer
                                // between 0 and n-i-1
        temp = q[i+rand];
        q[i+rand] = q[i];
        q[i] = temp;
    }
}
```

In Models 3.4 and 6.2 we used progressive roundoff strategies to find a solution to a set covering problem. In both cases we looped through the continuous variables in the given order $0,...,N-1$. This order can be replaced by any permutation $i_0,...,i_{N-1}$. In general, by applying a heuristic to a considerable number of different sequences, we can hope to obtain a richer family of solutions, including solutions with better objective function values and solutions with other desirable features.

As an example, let us scramble the order in which the decisions are made in the Drop/Add approach to the problem of Model 3.4, The Committee, and Model 6.2, The Committee Realigned. The thing to remember is to use negation-as-failure to be sure that the feasible region is restored after each randomized run of the central loop of the program. In fact, in order to allow the heuristic to fail, we will wrap the runs in an `if/then` construct with the null statement as the `then` clause; this combination itself can never fail and is coded as part of the main procedure below.

```
#define ITERATIONS 30            // Number of randomized runs

2lp_main()
{
double rand, best;
int permu[NUM];      // NUM is number of citizens

    data();
    setup();

    seed(59038);     // Seed random number generator
    best = 1e6;      // Initialize to very high value

    and(int k=0;k<ITERATIONS;k++) {
        find_next_random_sequence(permu,NUM);
        if not {     // Undo variable fixings when done
            Z <= best-1;   // Force the issue
            and(int i=0;i<NUM;i++)      // Heuristic
                c_either Citizen[permu[i]] == 0.0;// Drop
                or Citizen[permu[i]] == 1.0;      // Add
            if wp(Z) < best;
```

6.3 Randomization and local search

```
                then best = wp(Z); // Does not get undone
        }
        then ;          // Do nothing
    }
    printf("The best solution found was %.0f\n",best);
}
```

This time the heuristic finds the optimal solution, as did the one-ply lookahead of Model 6.3.

Note that we have pushed things along in this code: The constraint `Z <= best-1` challenges the code to find better and better solutions. The initial value of `best` is so large that the heuristic will necessarily find a first solution, since, as decisions are made, the linear relaxation will stay feasible. However, as realistic values of `best` are obtained, they are small enough that the heuristic might well reach an impasse and not be able to continue. On the other hand, the presence of the constraint `Z <= best-1` might perturb things in such a way as to improve the performance of the heuristic. The effect can be studied by removing the constraint, or by making it more aggressive, say, `Z <= best-2`.

Randomization can be employed to bias a decision that is based on information returned by an evaluation function. Such methods are called *probabilistic*. Here's a very simple example. In the code for the one-ply lookahead for the set covering problem of Model 6.2, the two future possibilities are evaluated and the one with the better evaluation is chosen all the time. We can make this code probabilistic by preferring the future with the better evaluation with a probability of 75% rather than 100%. For that we replace the routine `bold_decision` with the following one:

```
probabilistic_bold_decision(int i, double off,on)
{
double factor;
    factor = .75
    if {
        off < on;
        random() > factor;
    }
    then Citizen[i] == 0;
    else Citizen[i] == 1;
}
```

This code can be run once, or run many times, and the order in which the citizens are treated can be permuted randomly as well. Yet another idea is to run the code many times, changing the value of the parameter `factor`.

Now let us apply randomization to a classic scheduling problem, known as the *job shop problem*. The data are from Fischer and Thompson (1963). The model doesn't require continuous variables in an essential way, but we will make use of them to simplify the programming.

Model 6.3 The Body Shop

An automobile body shop has to perform 6 jobs using 6 machines. Each job is composed of 6 tasks, one for each machine. For each job the tasks must be done in a certain order. Also, when a task is being run, it cannot be preempted and taken off the machine it is on. In Table 6.3 the order in which these tasks must be performed is given as well as the time each task takes and the machine on which it must be run. The goal is to schedule the machines and the tasks so as to finish processing all the jobs in a timely manner. The time it takes to finish all the jobs is called the *makespan*. Ideally one would find a schedule that provided the shortest makespan. This is an issue we address in later chapters; here we will seek a quick-and-dirty solution.

There is a natural evaluation function given by the relaxation obtained by supposing that there are 6 copies of each machine for the yet unscheduled tasks. This evaluation function is weak because it completely ignores the fact that no two tasks can be running on a given machine at the same time. Still, using this evaluation function, we can employ a hill-climbing strategy: For each machine the next task scheduled is the one that minimizes this evaluation. This hill-climbing strategy depends mightily on the order in which we consider the machines. We will apply the strategy to a number of randomly chosen orderings of the machines and select the best schedule found.

First we have some symbolic constants and declarations of data arrays:

```
#define M 6              // Number of machines
#define JOBS M           // Number of jobs
#define TASKS JOBS       // Number of tasks

int duration[JOBS][TASKS]; // Time to run task t of job j
int task_time[M][JOBS];    // Time to run job j's task on m
int machine[JOBS][TASKS];  // Machine for task t of job j

#define ITERATIONS 15  // Number of randomized runs
```

To represent the finishing time of the task from job j which runs on machine m, we introduce a continuous variable `FinTime[m][j]`; we also introduce a continuous variable `MakeSpan` for the objective function:

```
continuous FinTime[M][JOBS];  // Finish of j's task on m
continuous MakeSpan;          // Finish of last task
```

6.3 Randomization and local search

TABLE 6.3 Duration of Task/Machine for Task

	Task 0	Task 1	Task 2	Task 3	Task 4	Task 5
Job 0	1/2	3/0	6/1	7/3	3/5	6/4
Job 1	8/1	5/2	10/4	10/5	10/0	4/3
Job 2	5/2	4/3	8/5	9/0	1/1	7/4
Job 3	5/1	5/0	5/2	3/3	8/4	9/5
Job 4	9/2	3/1	5/4	4/5	3/0	1/3
Job 5	3/1	3/3	9/5	10/0	4/4	1/2

Entering the data of Table 6.3 is straightforward. We will also need to have the amount of time that job j requires on each machine m. To that end, we will store the amount of time job j's task runs on machine m in the slot `task_time[m][j]`:

```
data()      // 6 by 6
{
    duration = {
        1,3,6,7,3,6,
        ...
        3,3,9,10,4,1
    };
    machine = {
        2,0,1,3,5,4,
        ...
        1,3,5,0,4,2
    };
    and(int j=0;j<JOBS;j++)
        and(int t=0;t<TASKS;t++)
            task_time[machine[j][t]][j] = duration[j][t];
}
```

Constraints can be used to enforce the ordering relations on the tasks of each job. To express that task k+1 follows task k, we stipulate that the finishing time of task k+1 is at least the time it takes to perform task k+1 later than the finishing time of task k. Finally, the continuous variable `Makespan` must majorize the finishing times of the last task of each job. The `setup` routine follows:

```
setup()
{
    and(int jb=0;jb<JOBS;jb++)
        and(int tsk=0;tsk<TASKS;tsk++)
            duration[jb][tsk]
                <= FinTime[machine[jb][tsk]][jb];
```

```
    and(int jb=0;jb<JOBS;jb++)
        and(int tsk=0;tsk<TASKS-1;tsk++)
            FinTime[machine[jb][tsk]][jb]
            + duration[jb][tsk+1]
            <= FinTime[machine[jb][tsk+1]][jb];

    and(int jb=0;jb<JOBS;jb++)
        MakeSpan >= FinTime[machine[jb][TASKS-1]][jb];
}
```

The heuristic will proceed machine by machine, scheduling the tasks that must run on that machine. When the schedule is being built, it is not critical to determine the actual time that a task will be launched; it is sufficient to determine the ordering among the tasks. We will schedule the tasks in the order based on the heuristic evaluation function. To schedule machine m, a routine `schedule` with signature

```
schedule(continuous Ft[],int time[])
```

will be sent the slices `FinTime[m]` and `task_time[m]`. A local integer array `p[JOBS]` will be used to keep track of the ordering of the jobs. Initially the candidate ordering will be 0,...,JOBS-1. Among all the jobs, the one with the best evaluation will be selected to go first, and `p[0]` will be set to this job number. The remaining tasks will be from jobs `p[1],....,p[N-1]`. Then `p[1]` will be determined. To ensure that the task from job `p[0]` is run before that from job `p[1]`, we add the constraint

```
    Ft[p[0]] + time[p[1]] <= Ft[p[1]];
```

and so on. The code for the `schedule` routine can be written as follows:

```
schedule(continuous Ft[],int time[])
{
int p[JOBS];

    and(int j=0;j<JOBS;j++)
        p[j] = j;

    and(int i=0;i<JOBS;i++){
        pick_next_job(i,p,Ft,time);
        if i > 0;
        then Ft[p[i-1]] + time[p[i]] <= Ft[p[i]];
    }
}
```

Let us turn to the task of writing the `pick_next_job` routine. The idea is to choose the job whose task yields the best evaluation. To eval-

6.3 Randomization and local search

uate a job i, we use negation-as-failure to look into the future. We consider the minimum value of the objective function MakeSpan when all the remaining jobs tasks are required to follow this job's task on this machine:

```
pick_next_job(int i, int p[], continuous Ft[], int time[])
{
double prize;
    prize = 1e6;
    and(int j=i;j<JOBS;j++) {
        if not {     // Try for improvement
            if i > 0;    // p[0],...,p[i-1] already scheduled
            then Ft[p[i-1]] + time[p[j]] <= Ft[p[j]];
            and(int k=i;k<JOBS;k++)  // Try job j's task first
                if p[k] != p[j];
                then Ft[p[k]] >= Ft[p[j]] + time[p[k]];
            min: MakeSpan;
            wp(MakeSpan) < prize;   // Better evaluation ?
            switch(i,j,p,prize);    // Then take first place
        }
        then ;       // If no improvement, do nothing
    }
}

switch(int i,j, p[], double & prize)
{
int winner;

    winner = p[j];
    p[j] = p[i];
    p[i] = winner;
    prize = wp(MakeSpan);
}
```

A call to `pick_next_job(i,p,Ft,time)` returns with the selected candidate in `p[i]`.

A random run of this heuristic will start with a permutation `q[M]` of the machines, and we apply the heuristic to each machine one after the other in this permuted order. We will keep track of the best makespan found in a variable `best`. When a new random run is launched, we can insist that it achieve a finishing time of no worse than `best` by stipulating `MakeSpan <= best`. If the randomized run cannot achieve this bound, it can be terminated as soon as this is detected. What is more, since the data are all integral, a better schedule will reduce the finishing time by a least 1. Therefore we can up the ante and ask that the next run achieve a schedule with `MakeSpan <= best-1`.

```
random_run(int k, double & best)
{
int q[M];

    printf("Entering iteration %d\n",k);

    find_next_random_sequence(q,M);     // Order the machines
    MakeSpan <= best-1;                 // Force the issue
    and(int i=0;i<M;i++)                // Heuristic
        schedule(FinTime[q[i]],task_time[q[i]]);
    min: MakeSpan;      // To store best possible value
    best = wp(MakeSpan);                // Record result
}
```

Finally, the main routine is

```
2lp_main()
{
double best;

    data();
    setup();

    seed(4684);     // Seed random number generator
    best = 1e6;     // Initialize to very high value

    and(int k=0;k<ITERATIONS;k++)
        if not random_run(k,best);
        then printf("Iteration %d failed\n",k);

    printf("The best result was %.0f\n",best);
}
```

The code finds a solution with a finishing time of 59. The result is quite good compared to what a truly naive approach can yield. In fact, with the methods of Section 9.2, we will see that the optimal solution is 55. So our heuristic has brought us within 8% of optimality.

End of Model 6.3

Job shop problems are notoriously difficult. However, with contemporary computer platforms, a 6-by-6 example is no longer formidable. This example and a famous 10-by-10 example appeared in a paper of Fischer and Thompson in the volume edited by Muth and Thompson (1963). The problems are usually referred to as MT6 and MT10. Determining the optimal solution to MT10 remained an open problem for over 25 years until it was solved by Carlier and Pinson (1989).

6.3 Randomization and local search

(a) Greedy tour   (b) Better tour

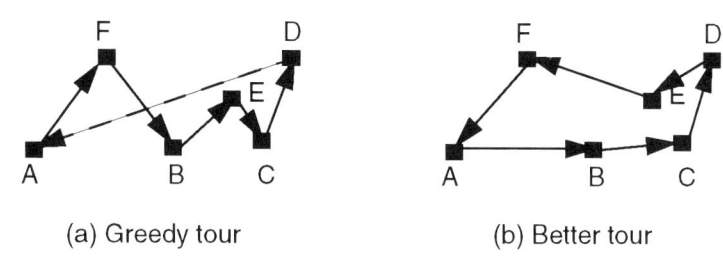

FIGURE 6.1 A greedy tour and a better tour

Let us point out a distinguishing feature of the job shop problem among scheduling problems: There are no given ready times and no given due times for the individual tasks that make up a job. On the other hand, for each job the sequence in which the tasks are to be processed is given. In the next section we will discuss a special case of the job shop problem.

Scheduling was one of the very first areas to which randomized algorithms were applied. In fact, our next model is an early classic. The application is known as the *traveling salesperson problem* or *TSP*. The method of solution is that of Lin (1965). It is a discrete model. The code could as well be written in a traditional programming language, but we persist with 2LP so as not to change syntax in mid-stream.

Model 6.4 We Open in Venice

A small circus troupe, equipped with tent, has been contracted by the Ministero del Turismo, Cultura e Spettacolo to put on performances in 17 northern Italian cities in the month of June. The tour will start in Venice, and an additional closing performance will be given in Venice at the completion of the tour. Each of the cities has a seldom used fairgrounds, and so the troupe can decide the order in which it will play them. To save wear and tear on its personnel, its vehicles, and its gasoline bills, the troupe would like to follow a route that keeps the total distance traveled down as much as possible. The distances between pairs of cities are given in Table 6.4

A route comes down to a permutation of the cities, starting with Venice, city 0. The number of possible routes is astronomical, and so there is no possibility of generating them all to choose the best among them. The first thing that comes to mind is a greedy strategy: From each city, go to the unvisited city closest to it. However, this strategy is too simplistic for this kind of application. In Figure 6. the kind of difficulty that the simple greedy strategy runs into is illustrated.

But given a tour of the cities, the next thing to do is to try to improve the tour by making changes. One simple way to perturb a tour is to find successive cities i,i+1 and successive cities j,j+1 on the tour

TABLE 6.4 Distances between Pairs of Cities

	0	1	2	3	4	5	6	7	8
	9	10	11	12	13	14	15	16	
City 0	0	633	257	91	412	150	80	134	259
	505	353	324	70	211	268	246	121	
City 1	633	0	390	661	227	488	572	530	555
	289	282	638	567	466	420	745	518	
City 2	257	390	0	228	169	112	196	154	372
	262	110	437	191	74	53	472	142	
City 3	91	661	228	0	383	120	77	105	175
	476	324	240	27	182	239	237	84	
City 4	412	227	169	383	0	267	351	309	338
	196	61	421	346	243	199	528	297	
City 5	150	488	112	120	267	0	63	34	264
	360	208	329	83	105	123	364	35	
City 6	80	572	196	77	351	63	0	29	232
	444	292	297	47	150	207	332	29	
City 7	134	530	154	105	309	34	29	0	249
	402	250	314	68	108	165	349	36	
City 8	259	555	372	175	338	264	232	249	0
	495	352	95	189	326	383	202	236	
City 9	505	289	262	476	196	360	444	402	495
	0	154	578	439	336	240	685	390	
City 10	353	282	110	324	61	208	292	250	352
	154	0	435	287	184	140	542	238	
City 11	324	638	437	240	421	329	297	314	95
	578	435	0	254	391	448	157	301	
City 12	70	567	191	27	346	83	47	68	189
	439	287	254	0	145	202	289	55	
City 13	211	466	74	182	243	105	150	108	326
	336	184	391	145	0	57	426	96	
City 14	268	420	53	239	199	123	207	165	383
	240	140	448	202	57	0	483	153	
City 15	246	745	472	237	528	364	332	349	202
	685	542	157	289	426	483	0	336	
City 16	121	518	142	84	297	35	29	36	236
	390	238	301	55	96	153	336	0	

6.3 Randomization and local search

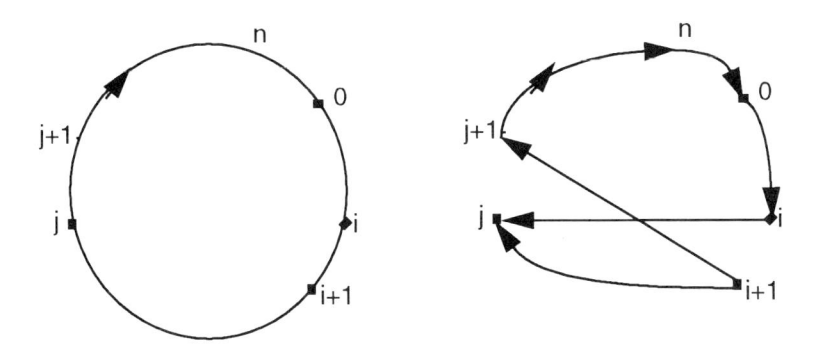

FIGURE 6.2 Switching two cities.

and to switch so as to go from i to j and from i+1 to j+1. If we represent the first tour as a circle, this kind of switch changes it to a figure eight as in Figure 6.2. Let us go one step further and consider making three switches at a time; this yields the kind of change pictured in Figure 6.3. In the figure the first new step is from i to j; two other first steps are to go from i to j+1 or to go from i to k. We are most interested in making switches in the tour that decrease the total distance traveled.

A natural hill-climbing strategy emerges: Starting with an initial tour, continually improve the tour by making switches of two and three cities until a tour is reached that cannot be improved this way. A tour with this property is called a *local optimum*. A local optimum will always be reached because each step in the hill climb strictly improves the objective function. The traditional picture of local and global optima is that from calculus, Figure 6.4.

The insight of Lin (1965) is that the way to get a rich sampling of local optima is to run the hill-climbing algorithm on a sequence of randomly generated initial tours. The hope is that by spreading out the

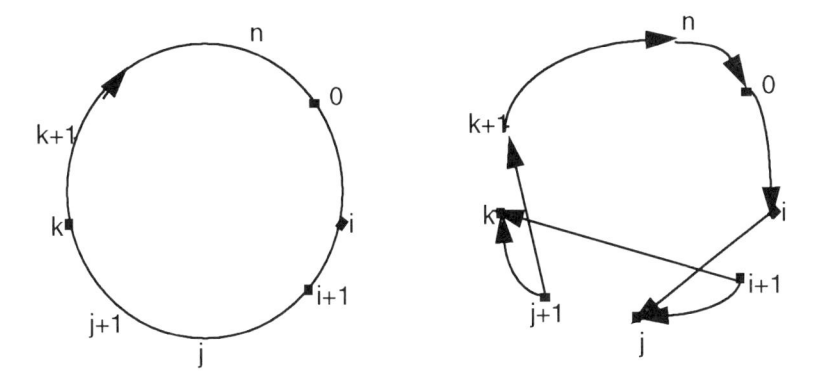

FIGURE 6.3 Switching three cities.

initial tours, one can reach very different local optima, some of which might be equal to or reasonably close to a global optimum.

Let us get down to work and write code for the problem at hand. For the data we will need an integer array to record the distance between each pair of cities:

```
#define T 17
int d[T][T];
```

and this can be initialized in a procedure `data`.

The program can be organized in a top-down fashion starting with the main procedure:

```
#define ITERATIONS 20     // Number of runs
21p_main()
{
int tour[T];
int best_tour[T];
int best_distance;

    data();
    seed(2981);            // Seed random number generator
    best_distance = 1e6;   // Initialize to high value

    and(int i=0;i<ITERATIONS;i++) {
        get_next_random_tour(tour);// Generate random tour
        apply_hillclimb(tour);     // Reach local optimum
        compare(tour,best_distance,best_tour);//Record best
    }
    printf("The best we found was %d\n",best_distance);
}
```

The routine to generate a random tour can make use of the routine `find_next_random_sequence` used above. We will send this routine

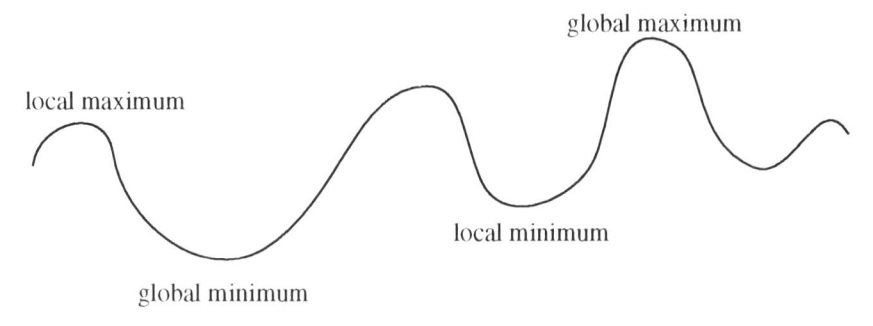

FIGURE 6.4 Local and global optima.

6.3 Randomization and local search

the slice of q[T] consisting of q[1],..., q[T-1] and fill this subvector with a random permutation of the integers 0, ..., T-2. Then we increment all these entries by one to make room for Venice:

```
#define VENICE 0
get_next_random_tour(int q[])
{
    find_next_random_sequence(q[1],T-1);// From above
    and(int c=1;c<T;c++)    // Add one to eliminate 0
        q[c] = q[c] + 1;    // ... Mantua and Padua ...
    q[0] = VENICE;          // We open in Venice
}
```

We need routines for making three switches and two switches. For the latter case, we can write

```
#define fail 0==1
two_switch(int path[])
{
int temp1,temp2;

    and(int i=0;i<T-3;i++)
        and(int j=i+2;j<T-1;j++){
            temp1=d[path[i]][path[i+1]]+
                  d[path[j]][path[j+1]];
            temp2=d[path[i]][path[j]]+
                  d[path[i+1]][path[j+1]];
            if temp1 > temp2;
            then {
                switch_2(i,j,path);
                return;
            }
        }
    fail;
}

switch_2(int i, j, path[])// Switch between cities i and j
{
int m,temp;
    m = j;
    and(int n=i+1;n<m;n++){
        temp=path[n];
        path[n]=path[m];
        path[m]=temp;
        m = m-1;
    }
}
```

The case where three cities are switched is similar and is left as an exercise. The hill-climbing code is straightforward:

```
apply_hillclimb(int tour[])
{
    and(;;)
        c_either three_switch(tour);
        or two_switch(tour);
        or break;        // Local optimum reached
}
```

Comparing tours and storing the best is routine enough:

```
compare(int tour[], & best_distance, best_tour[])
{
int dist;

    dist = 0;
    and(int c=1;c<T;c++)
            dist = dist + d[tour[c-1]][tour[c]];
    dist = dist + d[T-1][0];
    if dist < best_distance;
    then {
        best_distance = dist;
        and(int c=0;c<T;c++)
            best_tour[c] = tour[c];
    }
}
```

The results are interesting. We found a solution at 2095. In fact, the data are from TSPLIB and we know the best possible solution is 2085. Lin's algorithm therefore provides a solution within 0.5% of optimality. One can access TSPLIB by means of anonymous `ftp` to `softlib.cs.rice.edu`.

End of Model 6.4

For the TSP, Lin's algorithm is considerably strengthened in Lin and Kernighan (1973); in particular, simple two and three switches are replaced by more complex switches that are determined dynamically as the hill climb proceeds; the hill climb itself is modified to include more complex mechanisms.

Two weak points in local search algorithms are that they can converge to poor local optima and that the history of how a state is reached is not taken into account when making the next step. Two very important responses to this are *simulated annealing* and *tabu search*. These

6.3 Randomization and local search

methods are called *metaheuristics* because they can be applied to entire families of heuristic search algorithms.

In brief, simulated annealing introduces a time element and a probabilistic element into local search. With a certain probability, transitions that do not improve the objective function are allowed; this mitigates the tendency of local search to "rush" to a local optimum. Over time this probability is reduced to zero, and in the final stage the algorithm reduces to local search. The gradual reduction of the probability of taking a step that does not improve the objective function value is called the *annealing schedule* and the probability itself is called the *temperature*. The intuition behind this is to simulate the annealing method used to bring metals to a proper state of tension. This is another interesting example of how computational power places a new problem-solving technique in one's hands. The literature is rich and varied. A now classic paper is Kirkpatrick, Gelatt, and Vecchi (1983); a more recent contribution is van Laarhoven, Aarts, and Lenstra (1992).

Tabu search makes use of the history that led to the current state in a local search; it was introduced in Glover (1986) and in Hansen (1986). For example, when a local optimum is reached, the algorithm will begin to make moves that worsen the objective function. Then steps that improve the objective function will be resumed; to prevent the search from going back to the last local optimum, a list of interdicted moves, the *tabu list*, is maintained. To prevent the search from missing a move to a better new local optimum, counterbalancing techniques known as *aspiration criteria* are used. In fact, a whole panoply of methods can be gainfully added in. The literature in this field is growing rapidly; a place to start is Glover, Taillard, and de Werra (1993). For further developments in metaheuristics in general, see Osman and Kelly (1996).

In Chapter 8 and succeeding chapters, we discuss *backtracking* and *enumeration*, methods that can be used to avoid the problem of being trapped in local optima. Naturally there will be trade-offs of varying kinds, and different techniques and hybrids will be needed for different applications.

In the next section, we treat *genetic algorithms*, another member of the family of metaheuristics.

Exercises

6.3.1. Write the code for the three switch in Lin's algorithm.

6.3.2. Randomize the Drop/Add heuristic approach to the set covering problem.

6.3.3. Randomize the progressive roundoff heuristic approach to the problem of Model 2.3, Call 911.

TABLE 6.5 Duration of Task/Machine for Task

	0	1	2	3	4	5	6	7	8	9
0	29/0	78/1	9/2	36/3	49/4	11/5	62/6	56/7	44/8	21/9
1	43/0	90/2	75/4	11/9	69/3	28/1	46/6	46/5	72/7	30/8
2	91/1	85/0	39/3	74/2	90/8	10/5	12/7	89/6	45/9	33/4
3	81/1	95/2	71/0	99/4	9/6	52/8	85/7	98/3	22/9	43/5
4	14/2	6/0	22/1	61/5	26/3	69/4	21/8	49/7	72/9	53/6
5	84/2	2/1	52/5	95/3	48/8	72/9	47/0	65/6	6/4	25/7
6	46/1	37/0	61/3	13/2	32/6	21/5	32/9	89/8	30/7	55/4
7	31/2	86/0	46/1	74/5	32/4	88/6	19/8	48/9	36/7	79/3
8	76/0	69/1	76/3	51/5	85/2	11/9	40/6	89/7	26/4	74/8
9	85/1	13/0	61/2	7/6	64/8	76/9	47/5	52/3	90/4	45/7

6.3.1. Run Model 6.3 with different seeds and with different values for ITERATIONS. Comment on the results.

6.3.2. Table 6.5 holds the data for MT10, the 10-by-10 job shop scheduling problem. The optimal solution is 930. Find as good a solution as you can.

6.4 Genetic algorithms

We have another class of algorithms that are based on the ability of the computer to imitate natural processes, this time the workings of natural selection. Let us describe the method informally. Suppose that a set P_n of solutions to an application has been developed. A copy of an element of P_n is called a *reproduction* or *clone*. Two additional kinds of solutions can be generated by applying two other operations to elements of P_n: *mutation* and *crossover*. The mutation operation imitates the random mutations of genetics. Its effect is to diversify the search. The crossover operation combines two or more members of the population to produce a new member, called an *offspring*. A by-product of crossover is that if two solutions in the current population have each solved disjoint subproblems effectively, then their offspring hopefully combines both. From the clones, mutations and crossovers generated from the population P_n, a new population Q_n of solutions is formed. Then the *fittest* of this group will constitute the new pool P_{n+1}, where fitness is measured by an evaluation function. Each population P_n is called a *generation*. When the process is halted, typically the most fit individual in the last generation is the solution chosen. Let us illustrate these ideas with a classic scheduling problem known as the *flow shop problem*.

6.4 Genetic algorithms

TABLE 6.6 Duration of tasks

Task	0	1	2	3	4	5	Task	0	1	2	3	4	5
Job 0	10	3	6	7	23	6	Job 1	8	5	10	10	10	4
Job 2	5	24	8	9	1	7	Job 3	5	5	25	3	8	9
Job 4	9	3	35	4	3	1	Job 5	23	3	9	10	4	1
Job 6	1	32	6	7	3	6	Job 7	18	5	10	10	10	4
Job 8	5	14	8	9	11	7	Job 9	25	5	5	3	8	9
Job 10	9	3	5	24	3	1	Job 11	3	13	19	10	4	1

Model 6.5 The Workstations on the Assembly Line

An assembly line of 12 workstations has been set up to process 6 kinds of jobs. A job will enter the line and will be processed at the workstations in sequence. Each job must go through the stations in order and the stations must treat the jobs in the order of their arrival. The goal is to find a schedule that keeps the makespan low.

Clearly this is a simple form of the job shop problem. The workstations correspond to the machines of the job shop problem, and the tasks for each job must be performed without preemption on the machines in a fixed order. The order in which the jobs are processed at the first workstation determines the order in which they will be processed on all subsequent workstations. The data are in Table 6.6.

As in Model 6.3, we begin with the following symbolic constants and declarations:

```
#define JOBS 12
#define TASKS 6

int duration[JOBS][TASKS];
continuous Makespan;      // The objective function
```

However, because of the simpler structure of this model, we introduce continuous variables for the finishing time of the task t of job j:

```
continuous Ft[JOBS][TASKS];// Finish time of task t of job j
```

This time we will try a genetic algorithm. Every permutation of the jobs is a valid schedule, and conversely. Therefore an initial population can be generated by taking a collection of random permutations of 0,...,JOBS-1.

A mutation will simply be a new permutation obtained from a given one by interchanging the positions of two elements. A crossover occurs when a new solution π_3 is formed from two existing solutions π_1 and π_2. In this application we can define a crossover scheme as follows: π_3 agrees with π_1 on the first JOBS/2 elements; then on the remaining elements, it follows the order of π_2. The choice of mate for a solution will be made randomly.

In this model, fitness will be measured by makespan: The shorter the makespan, the more fit the individual is. To develop the machinery for the genetic algorithm, let us introduce some definitions and work arrays:

```
#define POP 20       // Number of individuals in population
#define M 3
#define CURRENT 0
#define MUTATIONS 1
#define OFFSPRING 2

int population[POP][JOBS];   // The current population
int variations[POP][JOBS];   // The mutations
int crossover[POP][JOBS];    // The offspring
double fitness[M][POP];      // Evaluation of fitness

#define ITERATIONS 10        // Number of generations
```

From the top down, the structure of the model is quite simple. First, following Model 6.4, The Body Shop, the constraints that prioritize the tasks of each job can be laid down. Then an initial population can be generated and evaluated for fitness. At this point a number of iterations are made; at each iteration two sets of new individuals are created by means of mutation and reproduction. Then the next generation is produced by culling the fittest individuals from the three sets available.

```
2lp_main()
{
    data();
    setup();

    initialize(population,fitness[CURRENT]);

    and(int it=0;it<ITERATIONS;it++) {
        generate_pool();     // New individuals
        survival_of_the_fittest(); // Natural selection
    }

    and(int p=0;p<POP;p++)
        printf("fitness of %d is %f\n",
                                p,fitness[CURRENT][p]);
}
```

To obtain an initial population, we can send the array population and the slice fitness[CURRENT] to the following routine:

6.4 Genetic algorithms

```
initialize(int pop[][JOBS],double fitness[])
{
    seed(1968);      // Prime pseudorandom number generator
    and(int k=0;k<POP;k++) {
        find_next_random_sequence(pop[k],JOBS);
        compute_fitness(pop[k],fitness[k]);
    }
}
```

To compute the fitness of an individual solution, we will enforce the constraints corresponding to that solution and minimize the makespan. Negation-as-failure is needed to undo this solution.

```
compute_fitness(int p[],double & fit)
{
    if not {
        and(int jb=0;jb<JOBS-1;jb++)
            and(int tsk=0;tsk<TASKS;tsk++)
                Ft[p[jb]][tsk] + duration[p[jb+1]][tsk]
                <= Ft[p[jb+1]][tsk];
        min: Makespan;
        fit = wp(Makespan);
    }
    then {
        printf("Error\n");  // Every permutation is a
        exit();              // valid solution
    }
}
```

To produce a pool of new individuals, we resort to mutation and crossover. The mutations will be stored in the array `variations`, and the offspring will be stored in the array `crossover`:

```
generate_pool()
{
    mutations(variations,fitness[MUTATION]);
    reproductions(crossover,fitness[OFFSPRING]);
}
```

For simplicity, we will produce a collection of POP new individual through each process. To produce a set of mutations, we have

```
mutations(int pool[][JOBS], double fitness[])
{
int s,t;
int temp;
```

```
    and(int p=0;p<POP;p++) {
        and(int k=0;k<JOBS;k++)
            pool[p][k] = population[p][k];
        s = random()*JOBS;      // random elements
        t = random()*JOBS;      // in [0,...,JOBS-1]
        temp = pool[p][s];
        pool[p][s] = pool[p][t];
        pool[p][t] = temp;
        compute_fitness(pool[p],fitness[p]);
    }
}
```

The offspring will be produced by mating each individual of the current population with a randomly chosen mate from the same population:

```
reproductions(int pool[][JOBS], double fitness[])
{
int s,top;

    and(int p=0;p<POP;p++) {
        s = random()*POP;  // random from [0,...,POP-1]
        top = JOBS/2;
        and(int j=0;j<JOBS;j++)
            pool[p][j] = population[p][j];
        and(int j=0;j<JOBS;j++) {
            if and(int i=0;i<top;i++)
            pool[p][i] != population[s][j];
            then {
                pool[p][top] = population[s][j];
                top = top+1;
            }
            if top == JOBS;
            then break;
        }
        compute_fitness(pool[p],fitness[p]);
    }
}
```

To select the fittest from among the three population groups, we can loop through the new populations, replacing less fit individuals from the current population:

```
survival_of_the_fittest()
{
    select_from(MUTATION,variations);
    select_from(OFFSPRING,crossover);
}
```

6.4 Genetic algorithms

```
select_from(int type, int pool[][JOBS])
{
double gauge;
int temp;

    and(int k=0;k<POP;k++) {   // Individual from new group
        gauge = fitness[type][k];  // Record its fitness
        temp = POP;
        and(int s=0;s<POP;s++) // Individual from old group
            if gauge < fitness[CURRENT][s];// Compare
            then {  // Less fit individual found
                temp = s;   // Least fit individual so far
                gauge = fitness[CURRENT][s];// Its fitness
            };
        if temp !=POP;  // Less fit individual has been found
        then {  //Place new individual in current population
            and(int t=0;t<JOBS;t++)
                population[temp][t] = pool[k][t];
            fitness[CURRENT][temp] = fitness[type][k];
        }
    }
}
```

For completeness, let us list the setup routine, which is somewhat different from that of Model 6.3, The Body Shop, because here the mapping from tasks to machines is the identity function:

```
setup()
{
    and(int jb=0;jb<JOBS;jb++)
        and(int tsk=0;tsk<TASKS;tsk++)
            sigma(int s=0;s<=tsk;s++) duration[jb][s]
            <= Ft[jb][tsk];

    and(int jb=0;jb<JOBS;jb++)
        and(int tsk=0;tsk<TASKS-1;tsk++)
            Ft[jb][tsk] + duration[jb][tsk+1]
            <= Ft[jb][tsk+1];

    and(int jb=0;jb<JOBS;jb++)
        Makespan >= Ft[jb][TASKS-1];
}
```

The best solution found with this code has a makespan of 174. This is not the optimal solution; using techniques of Chapters 9 and 11, somewhat better solutions can be found. However, this can require an inordinate amount of time, unless quite sophisticated techniques are introduced. This, then, is a great strength of genetic algorithms, the ability to reach solutions of excellent quality quickly.

It can be instructive to print out the fitness of the generations as they are formed to see how the algorithm is converging. It is also instructive to study the effect of turning off mutation or crossover. The range of experiments like these that one can make with genetic algorithms is almost limitless.

End of Model 6.5

In the above model, as in Model 6.3, continuous variables are only used to make the programming relatively simple. Indeed, this is a consummation devoutly to be wished. However, for reasons of performance, on large problems or constantly re-run applications, it is instructive and useful to code the problem as a discrete application. A situation where discrete code is more natural is encountered in the exercises.

In the above example the coding was made easier by the fact that any permutation of the jobs provides a feasible solution. In applications where things are not so simple, one can work with partial solutions and assign their fitness evaluation a penalty based on "how infeasible" the partial solution is. This is one possibility; there are many other things to do.

The importance of genetic algorithms was signaled by the work of Holland in the 1970s. For further reading on genetic algorithms, see Goldberg (1989), Holland (1992) and Mitchell (1996).

Exercises

6.4.1. Do the flow shop problem with Lin's algorithm. Do not use constraints; write code in a traditional programming language.

6.4.2. Refine the code of Model 6.5, The Workstations on the Assembly Line, so that each entry in the pool is a local optimum for the operation of two-switches. That is, after the mutations and crossovers are made, each of these solutions should be transformed into a local optimum. (It is here that simple discrete code will prove useful, since evaluating the effect of innumerable switches could prove a bottleneck with respect to time if continuous variables and incessant not nots are used.)

6.4.3. The following is known as the *quadratic assignment problem*: Given costs c_{mn} and quantities q_{ij} with $m,n = 0,...,N-1$ and $i,j = 0,...,N-1$, find a permutation π which minimizes (or comes close to minimizing)

$$\sum_{i,j} c_{\pi(i)\pi(j)}\, q_{i,j}$$

6.4 Genetic algorithms

TABLE 6.7 Quantities that must be shipped from i to j

i/j	0	1	2	3	4
0	0	1	1	2	3
1	1	0	2	1	2
2	1	2	0	1	2
3	2	1	1	0	1
4	3	2	2	1	0

TABLE 6.8 The cost of shipping from m to n

m/n	0	1	2	3	4
0	0	5	2	4	1
1	5	0	3	0	2
2	2	3	0	0	0
3	4	0	0	0	5
4	1	2	0	5	0

For the data sets of Tables 6.7 and 6.8, the optimal solution is 50 and is given by $\pi = \{3,0,4,1,2\}$. Develop search algorithms for this family of problems. A collection of data sets for this problem is maintained at the QAPLIB whose web address is

```
http://www.diku.dk/~karisch/qaplib
```

7

Sensitivity Analysis

In a linear programming application the constant terms and coefficients in the constraints are fixed, and the optimization process returns a single witness point where the optimal value of the objective function is realized. A change in the value of a coefficient or constant term is called a *perturbation* of the linear program. When data are perturbed, it is instructive to analyze the corresponding changes in the optimal value of the objective function and in the vertex at which the optimum is realized. In fact, it is a often a good idea to perturb the data precisely for this reason.

In the present chapter we discuss two cases in detail: the case where the bound on a continuous variable is perturbed, and the case where a coefficient in the objective function is perturbed. Examples are

```
// b is a bound on a variable
X <= b;

// c[i_0] is a coefficient in the objective function
max: sigma(int i=0;i<N;i++) c[i]*X[i];
```

For these two situations, there are the powerful tools of *sensitivity analysis*: The optimal value of the objective function will vary continuously as a function of the perturbed data, and its graph will be concave up or down and piecewise linear.

To study the effect of perturbations in other situations, one can always resort to "what-if" analysis by rerunning the program with new data and graphing the results. For example, in Model 4.1, Risk vs Reward, we carried out a "what-if" analysis in order to graph the trade-off between the optimal value of the objective function and the fickle factor. In that model we kept changing the multiplier r, which modifies the coefficients in the ratio constraint:

```
sigma(int i=0;i<N;i++) ff[i]*X[i]      //(ff[i]-r) is the
    <=  r*sigma(int i=0;i<N;i++) X[i]; //coefficient of X[i]
```

In Figure 4.1 the graph is not concave. Even worse, in a situation like this, where coefficients on continuous variables in constraints are varied, the optimum value of the objective function does not necessarily vary continuously with the perturbation. An example is to be found in the exercises.

7.1 Shadow prices and reduced costs

In this section we follow what happens to the optimal value of the objective function when the bound on a variable is made to vary across a wide range. After that, we consider the analogous question for a coefficient on a continuous variable in the objective function.

Let us start our analysis with the simplest kind of product mix application, one where there is one resource and one product. Suppose that a resource X is transformed into a product P and that each barrel of P requires 3 barrels of X. An unlimited number of barrels of P can be sold at a price of $10 each. However, there is a limit b on the amount of resource X that is available. Turning this into a code segment for computing the optimal revenue, we have

```
3*P == X;
X <= b;
max: 10*P;
```

Viewed this way, the optimal value of the objective function is now a function of the parameter b. We will call this function the *optimum function*; it is determined by the objective function and the resource X, and it depends on the bound b. Therefore we will denote the optimum function by $z^*_x(b)$ or, simply, z^*_x. In the example under consideration, the optimum function $z^*_x(b)$ depends linearly on the upper bound b for x, as is sketched in Figure 7.1.

If there is a limit, say 90, to how may barrels of product P can be sold, the code segment becomes

```
3*P == X;
X <= b;
```

FIGURE 7.1 z^* versus b.

7.1 Shadow prices and reduced costs

```
P <= 90;
max: 10*P;
```

With the bound on P, the graph of the optimum function flattens out when b reaches the usable limit of 3*90 = 270, as in Figure 7.2 This time, for values of b below 270, the slope of z^*_x is 3.333333; for values of b greater than 270, the slope of z^*_x is 0.

Now let us suppose that two more products Q and R can also be made by transforming resource X, with Q requiring 5 barrels of X and R requiring 8. Suppose also that up to 60 barrels of Q and up to 70 barrels of R can be sold at the prices of $11 and $9, respectively. Now the code becomes

```
3*P + 5*Q + 8*R == X;
X <= b;
P <= 90;
Q <= 60;
R <= 70;
max: 10*P + 11*Q + 9*R;
```

This is still a simple situation. Since there is only one multivariable constraint, it is a knapsack problem, and the greedy algorithm provides the optimal solution. In order of profitability the products are P, Q, and R. So the solution of the knapsack as a function of b will be to produce only P until b reaches the limit of resource X that can be used on P, namely b=3*90=270. At that point Q will enter the picture, and P and Q will be produced until the limit of production of both P and Q is reached at b = 270 + 5*60 = 570. At this point R will enter the production picture, joining P and Q; then all three products will be produced until b reaches 570 + 70*80 = 1130. After b = 1130, no further sales are possible, and the optimum curve flattens out. The graph of the optimum as a function of b is given is Figure 7.3.

The key thing to note is that the optimum function is continuous, concave down, piecewise linear and nondecreasing. Moreover the function only has a finite number of points where it bends; such points are

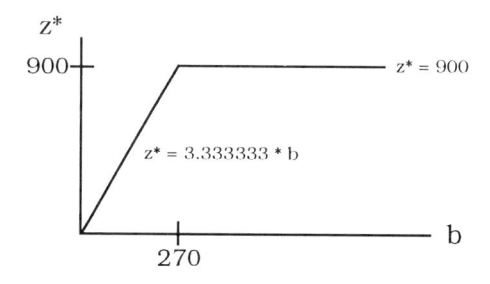

FIGURE 7.2 z* versus b with limit.

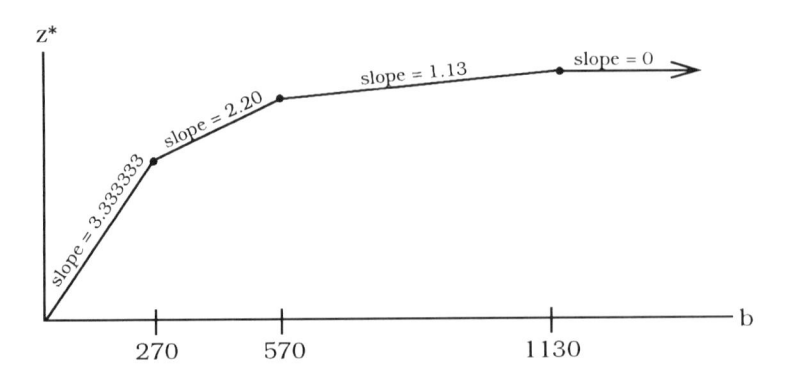

FIGURE 7.3 The optimum function.

called *singular* because the function is not differentiable there. Eventually the function has constant slope zero. The points where the slope changes correspond to values of b where the next most profitable product can begin to contribute to the optimal solution. For example, when b reaches 270, the variable Q leaves its lower bound of 0 and the set of products in the optimal mix changes. This is called a *structural change* in the solution. The function flattens out when the limits on the amounts of the three products that can be sold are reached and further structural changes are impossible.

When several resources enter into the production scheme, we no longer have a linear knapsack problem, and graphing the dependency of the optimal value of the objective function on a variable bound becomes more challenging. At first glance, since each value of b determines a different linear program, there could be a continuum of linear programs to solve! However, things turn out to be quite tractable. To start, we have the following beautiful theorem which affirms that the graph of the optimum function z^*_x will always have the same general shape as that in Figure 7.3:

Theorem (Parametric Analysis Theorem) *Suppose that F is a bounded feasible region defined by a consistent set of constraints $\mathbf{Ax} \leq \mathbf{b}$. Let the variable x and the bound $b_0 \geq 0$ be such that the constraints $\mathbf{Ax} \leq \mathbf{b}$, $x \leq b_0$ are consistent. For $b \geq b_0$, let $z^*_x(b)$ be the value of the optimal solution to the linear programming problem*

$$\mathbf{Ax} \leq \mathbf{b}$$
$$x \leq b$$
$$\max: \mathbf{cx}$$

Then z^*_x is a continuous, piecewise linear, concave down, nondecreasing function on the interval $[b_0, \infty)$.

7.1 Shadow prices and reduced costs

The proof of this theorem is given in Chapter 15; it comes out of the analysis of the duality that is present in linear programming. This theorem and Figure 7.3 must be contrasted with Figure 4.1; in the latter the graph obtained by connecting points is not concave down. The key difference is that in Model 4.1, the datum that is varied is not a bound on a variable but a coefficient:

```
sigma(int i=0;i<N;i++) ff[i]*X[i]
<=  r*sigma(int i=0;i<N;i++) X[i];
```

A continuous piecewise linear function is smooth or differentiable except at the singular points where the function bends and changes slope. At those points the function still has both left- and right-sided derivatives. For a maximization problem and for the analysis of the effect of increasing the upper bound, the *shadow price* of x at b is formally defined to be the right-sided derivative of the function $z^*_x(b)$ at b; the shadow price of x at b is denoted by $sp_x(b)$.

For all points except the singular points of z^*_x we have

$$sp_x = \frac{dz^*_x}{db}$$

and for all points we have

$$sp_x = \frac{d^+ z^*_x}{db}$$

where d^+ indicates that it's the right-sided derivative. Therefore the optimum function z^*_x is an antiderivative of sp_x, the shadow price of x. Since we are assuming F to be bounded, the optimum function z^*_x eventually becomes constant, and the shadow price eventually becomes 0.

The shadow price function is a step function whose discontinuities correspond to the singular points of the optimum function. In Figure 7.4, the shadow price function sp_x corresponding to the optimum function z^*_x of Figure 7.3 is illustrated.

The term shadow price comes from its use in economics; see Dorfman, Samuelson and Solow (1958). The simplex algorithm stores a

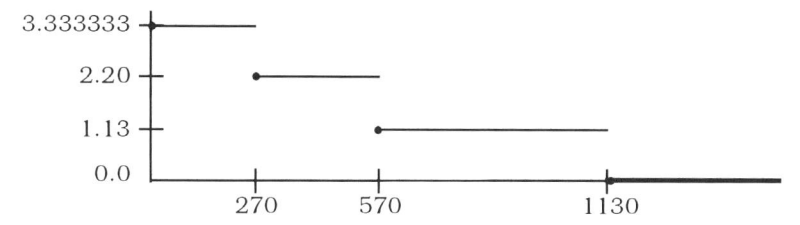

FIGURE 7.4 The shadow price function.

value known as the *reduced cost* of the variable and a second value called the *range*. Over this range, the reduced cost is equal to the shadow price. After a call to optimize an objective function in 2LP, the reduced cost is available for each continuous variable and is returned by the built-in function `rc`. This function's signature is

```
double rc( continuous )
```

To distinguish the case where a variable is at its upper bound from the case where it is at its lower bound, the function `rc` returns the reduced cost when the variable is at its upper bound and -1 times the reduced cost when the variable is at its lower bound. If the variable is in between bounds, the reduced cost is 0 and this is the value returned by `rc`. So the reduced cost of the variable X is always given by `fabs(rc(X))`.

In 2LP there is a built-in function `range` that returns an estimate of the amount the bound on a variable can be altered without changing its reduced cost. This estimate will be exact or conservative; over this range the reduced cost is equal to the shadow price. The signature of this function is

```
double range( continuous )
```

Let us look at an example in order to illustrate the notion of shadow price and its relation to reduced costs and ranges.

Model 7.1 The Price of Resource X

A company makes two products A and B. Each measure of A requires 1 hour of resource X and .33 tons of resource Y; it brings in a selling price of 100. Each measure of B requires 1 hour of X and 1 ton of resource Y; it brings in a selling price of 225. However, there are upper limits on the amount of product for which there is a market at these prices; at the present time these upper limits are 65 and 70 for A and B, respectively. Any amount of Y that is not used in this production cycle can be sold at the price of 65 per ton, and up to 85 tons can be resold. However, for each ton of Y resold this way, .40 hours of resource X are needed.

Further we are told that right now the company has 110 tons of Y on hand and that 90 hours of X have been allocated for this operation. The thing for the company to do at this point in time for its production plan is to maximize revenue, since the resources X and Y are already in place. The company also has the opportunity of adding further amounts of these two resources and would like to have an idea of how much it could pay for additional resources and profitably add to its revenue.

To represent this information, we can use continuous variables A, B for the amounts of products A and B to produce, continuous variables X,

7.1 Shadow prices and reduced costs

Y to represent the amounts of resources *X* and *Y* available, and continuous variables Yused and Ysold for the amounts of resource *Y* that are used in production and resold, respectively. Adding a continuous variable Revenue for the objective function, we have the declarations

```
continuous A,B,X,Y,Yused,Ysold,Revenue;
```

We can summarize the foregoing information in a procedure:

```
setup()
{
    A <= 65;
    B <= 70;
    Ysold <= 85;
    Yused + Ysold <= Y;
    1.0/3*A + 1*B == Yused;
    1*A + 1*B + .40*Ysold <= X;
    Revenue == 140*A + 225*B + 65*Ysold;
}
```

So the code for this analysis is

```
2lp_main()
{
    setup();
    analyze_current_situation();
}
```

The current situation consists of the given bounds on X and Y:

```
analyze_current_situation()
{
    X <= 90;
    Y <= 110;
    max: Revenue;
    printf("To maximize Revenue:\n");
    printf("Revenue will be %.2f\n",Revenue);
    output();
}
```

The built-in functions now do all the work:

```
output()
{
    printf("Produce %.2f of A and %.2f of B\n",A,B);
    printf("Use %.2f units X, use %.2f units Y,
             sell %.2f units Y\n",X,Yused,Ysold);
```

```
        printf("\n");
        printf("Reduced cost of X is %.2f with range %.2f\n",
                                        rc(X),range(X));
        printf("Reduced cost of Y is %.2f with range %.2f\n",
                                        rc(Y),range(Y));
}
```

The output will be

To maximize Revenue:
Revenue will be 18896.15
Produce 4.62 of A and 70.00 of B
Use 90.00 units X, use 71.54 units Y, sell 38.46 units Y

The reduced cost of X is 136.54 with range 52.33
The reduced cost of Y is 10.38 with range 10.00

The outcome is that right now in order to optimize its use of the resources that have already been paid for, the company should produce 4.62 measures of A and 70 of B. This plan calls for full utilization of resource X but also calls for a substantial sell-off of resource Y. Moreover, if the company can obtain more of resource X, as long as it doesn't pay more than 136.54 per hour of X, revenue will be enhanced. This analysis is good for purchases of additional X up to 52.33 hours at least. After that point, further analysis would have to be made. On the other hand, if it is Y that can be obtained, so long as the price per ton does not exceed 10.38, revenue will increase for a purchase of up to 10 more tons of Y at least.

In this analysis the costs of obtaining the current supply of resources X and Y have not been taken into account. Rather, they have been treated as "sunk costs." The part played by these resources in producing revenue is a contribution to profit. Their effect on net profit is a different matter.

It is important to emphasize that the reduced cost is determined both by the variable in question and the objective function of the linear program. To help make this point and to address the question of net profit, let us introduce the fact that the price of resource X is 50 dollars per hour and the fact that the price of resource Y is 100 dollars per ton, information that we have not needed up to now. If we want to study the sensitivity of net profit rather than gross revenue to increases in the availability of the resources X and Y, we suppose that X and Y have not been acquired yet but that they are available up to the limits of 90 and 110, respectively, at these prices. The objective function will be

```
        Revenue - 50*X - 100*Y
```

and we would use the routine

7.1 Shadow prices and reduced costs

```
analyze_new_situation()
{
    X <= 90;
    Y <= 110;

    max: Revenue - 50*X - 100*Y;
    printf("To maximize Net:\n");
    printf("Net will be %.2f\n",Revenue - 50*X - 100*Y);
    output();
}
```

The output from this routine is quite different, of course, from that of the analysis of Revenue.

To maximize Net:
Net will be 6383.33
Produce 20.00 of A and 70.00 of B
Use 90.00 measures X, use 76.67 measures Y, sell 0.00 measures Y

The reduced cost of X is 56.67 with range 45.00
The reduced cost of Y is 0.00 with range Inf

Now the reduced costs are telling us what we can pay above the current price of 50 for each additional hour of X, namely 56.67. This already takes into account the fact that to use additional X, additional Y must also be purchased at the current rate of 100. This analysis will hold for up to 45 additional hours of X. What happens at 90+45 = 135 and why things change there are addressed with the methods of the next section, Parametric Analysis. As for Y, the 0 reduced cost is telling us not to purchase further amounts of Y beyond 76.67 tons, since it will not add to net profit at all. The reason for this is that the constraint X <= 90 is blocking the utility of any further amount of resource Y. The range is telling us that 0 will remain the reduced cost of Y no matter how the upper bound of Y is increased.

End of Model 7.1

All along we have heralded the duality relation between the left-hand sides of the constraints and the continuous variables in a linear program. Indeed, it is this duality that underlies the analysis discussed in this chapter. These points are taken up in Chapter 15.

Sensitivity analysis also applies to the case where the right-hand side term in a multivariable constraint is perturbed:

```
// b is a constant term
sigma(int i=0;i<N;i++) a[i]*X[i] <= b;
```

To apply sensitivity analysis to a constraint like this, one introduces a slack or surplus variable:

```
sigma(int i=0;i<N;i++) a[i]*X[i] + S == b;
```

and the reduced cost of the slack variable S gives the rate of change in the optimum function at b. For resource constraints and other constraints where the left-hand side `sigma(int i=0;i<N;i++) a[i]*X[i]` stays nonnegative, one can introduce a new continuous variable, say, R, and write

```
sigma(int i=0;i<N;i++) a[i]*X[i] == R;
R <= b;
```

Now the analysis of the original resource constraint is made in terms of R.

Thus far we have been emphasizing the role of resources in product mix problems. Now let us turn to the dual question, the analysis of products. A very important aspect of linear programming is the effect of changes in the value of coefficients in the objective function. For example, if optimization leads to a plan in which a product does not get produced, one can ask how does the selling price of the product influence this result. In particular, one can ask to what value must the selling price of such a product be increased in order for it to be included in the optimal scheme of production. More generally, we can ask how much the objective function coefficient of the variable must be increased in order for it to figure more prominently in the optimal solution.

As with reduced cost there are built-in functions in linear programming systems that provide range information on how much the objective function coefficient on a continuous variable X can change before the value wp(X) changes in the optimal solution. In 2LP the function `objcup` gives an estimate of how far the coefficient can be increased; its signature is

```
double objcup( continuous )
```

This function returns an exact or conservative estimate of how much the coefficient of the continuous variable can be increased without requiring a change in the witness point that yields the optimal solution. In the example of a product, say, P, whose production is zero in the optimal solution, the value `objcup(P)` will estimate how much the selling price of P must be increased in order for its production to be warranted. This is in fact a case that builds on the analysis made in terms of reduced costs. If P is at its lower bound, its reduced cost is measuring how much production of a unit of P will hurt the value of the objective function. Therefore, the objective function must be compensated by the addition of `-rc(P)` to the current coefficient on P, and this is the value returned

7.1 Shadow prices and reduced costs

by objcup(P). However, objcup(P) will also return the appropriate value if P is between its lower and upper bounds, and it will return infinity if P is already at its upper bound.

For decreasing the objective function coefficient, the corresponding function is objcdn, and it has similar signature:

```
double objcdn( continuous )
```

As an example of the use of this kind of function, let us consider not parts in the example of Model 3.3, Motorcycles Inc., but the motorcycles themselves. Certain models are not planned to be produced at all in the optimal solution to maximize income, and others are to be produced at a level below the allowed limit. The following procedure will use the function objcup to find the selling price that a motorcycle will need for it to be included in the optimal plan at a level higher than its current level and the function objcdn to find how low the selling price can go before its level will go down:

```
21p_main()
{
    data();
    setup();
    max: Z;
    and(int m=0;m<MODS;m++)
        obj_analyze(m);
}

obj_analyze(int m)
{
    if {
        wp(Bike[m]) > 0.0;
        wp(Bike[m]) < hi[m] ;
    }
    then {
        printf("Model %d is in between\n",m);
        printf("Increase price by %f\n",objcup(Bike[m]));
        printf("Decrease price by %f\n",objcdn(Bike[m]));
    }
    else
        if  wp(Bike[m]) > 0.0;
        then {
            printf("Model %d is at upper limit\n",m);
            printf("Decrease price by %f\n",
                                        objcdn(Bike[m]));
        }
        else{
            printf("Model %d is not being made\n",m);
```

```
            printf("Increase price by %f\n",
                                        objcup(Bike[m]));
        }
}
```

The output is

Model 0 is not being made
Increase by 188.182553
Model 1 is in between
Increase price by 20.627010
Decrease price by 74.909968

...

Model 10 is at upper limit
Decrease price by 92.290792
Model 11 is in between
Increase price by 470.576606
Decrease price by 75.000000

We see that for Model 0 to be brought into the optimal product mix, it would take a substantial increase in selling price. Model 1 will stay in the optimal mix so long as its price is not decreased by more than 74.91. If its price is increased by as much as 20.62, the mix will not change. However, an increase beyond that will cause other items to change their role in the optimal mix, and so on.

With this example, we see why it is useful to use continuous variables to model a situation where the items involved are discrete, such as motorcycles, but where the scale of the application means that fractional solutions are accurate enough. Continuous methods allow us to use the linear programming analog of derivative to study the model as it undergoes various transformations.

Let us end this section with something of a cautionary remark. Sensitivity analysis through reduced costs applies to linear programs. If there are integrality requirements as in Models 2.3 and 3.4, "what-if" analysis must be done by running the model with perturbed data. In the same way goal-programming models with prioritized goals must also be treated by "what-if" analysis, and the entire program must be rerun with new data to gauge the effect of a perturbation on the goal program as a whole. If the goal program is formulated as a single linear program with weighted deviation variables in the objective function, then sensitivity analysis could be applied, though its interpretation might well prove difficult.

Exercises

7.1.1. For $a \geq 0$, consider the minimization problem: $(1/(1+a))x - (1-a)y = 1$, $x - y = 1$, $x \geq 0$, $y \geq 0$, min: x. At $a = 0$, the objective function has optimal value 1. Show that as $a \to 0$, the optimum function tends to infin-

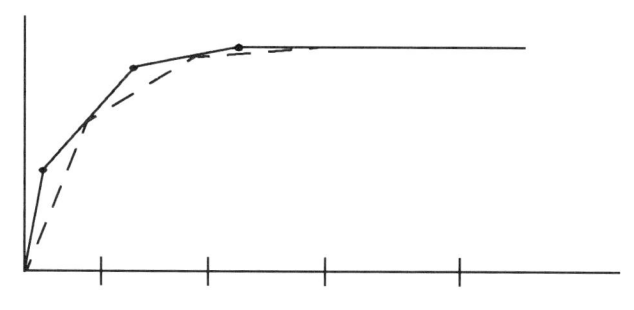

FIGURE 7.5 The interpolated function.

ity. Conclude that the optimum function is not necessarily a continuous function when it is a coefficient on a continuous variable in a constraint that is varied.

7.1.2. Apply sensitivity analysis to Model 3.1, The Pacific Rim.

7.2 Parametric analysis

Finding or approximating the shadow price and optimum functions is often an important part of the analysis of a linear programming application. These functions tell how a variable is contributing to the optimal value of the objective function, and as we will see, the discontinuities in the shadow price are pointers to structural changes in the optimal solution to the linear program.

The Parametric Analysis Theorem provides a qualitative description of the shape of the graph of the optimum function; the result strongly suggests that an effective quantitative analysis of the optimum function can be made by plotting points. If we sample the value of $z^*_x(b)$ at points $b_1, b_2, \ldots, b_n$ and connect the points by linear interpolation, we determine an approximation to z^*_x which we will call the *interpolated function*. This situation is illustrated in Figure 7.5. One can see that the linear interpolation of a continuous, concave down, nondecreasing function is itself continuous, piecewise linear, concave down, and nondecreasing, and so the interpolated function has a shape very similar to that of z^*_x.

The interpolated function has the property that the slope leaving a sample point b_i is always less than or equal to the slope of the function z^*_x leaving b_i. On the other hand, the slope of the interpolated function approaching a sample point b_i from below is always greater than or equal to that of z^*_x. Naturally, at each sample point b_i, the two function values coincide. We also know that at all points b, the optimum function value $z^*_x(b)$ is greater than or equal to the value given by the interpolated function. It follows from all this that if the interpolated function has the same slope from b_1 to b_2 as from b_2 to b_3, then the interpolated function and the optimum function z^*_x must agree on the interval $[b_1, b_3]$. Therefore, on such an interval, the slope of the interpolated function is equal to the shadow price.

If the slope of the interpolated function on $[b_1, b_2]$ is greater than that on $[b_2, b_3]$, it is possible, though not necessarily the case, that at some or even all intermediate points $b \ne b_2$ in the interval (b_1, b_3), the value $z^*_x(b)$ is strictly greater than the value of the interpolated function at b.

By taking sample points $b_1, b_2, ..., b_n$ at distances commensurate with the scale of the data for a linear programming problem, one can obtain a good approximation to the function z^*_x and even be able to determine regions where the approximation is exact. In all cases the interpolated function gives exact information as to how the objective function will change if the upper bound on the variable x is increased all the way from b_i to b_{i+1}.

Let us return to the programming situation where the feasible region F is given by the set of constraints of a 2LP program, Z is a continuous variable that defines the objective function, and X is a continuous variable of the program that represents a resource or some other aspect of the application. We want to analyze the effect as the value of b changes in the bounding constraint X <= b. In this case we will write Z wrt X to refer to the optimum function z^*_x obtained by varying the upper bound on X. In this notation "wrt" is shorthand for "with respect to."

Returning to Model 7.1, The Price of Resource X, let us analyze the relation between revenue and the resource by studying the behavior of the optimum function Revenue wrt X. To fix ideas, let us have the bound on X go through 11 sample points in increments of 10, from 30 less than the current bound to 70 beyond the current bound. For this we will need to change the upper bound on X repeatedly and re-optimize the objective function. Here we have a choice. Following the method used in Model 4.1, Risk vs Reward, we can run a loop from b = 70 + 90 down to b = 90 - 30, making the feasible region progressively smaller. To run the loop in the forward direction from b = 90 - 30 up to b = 70 + 90, we must repeatedly relax the upper bound on X from X <= b to X <= b + 10; this can be done by wrapping the body of the loop in the if/not combination, for negation-as-failure will undo the constraint X <= b after it is used. Since it will be important a bit further on to run the loop in the forward direction, let us do it that way here. These considerations lead to code such as

```
analyze_Revenue_wrt_X()
{
    and(int b=90-30;b<=90+70;b=b+10)
        if not {
            Y <= 110;    // Kept the same
```

7.2 Parametric analysis

```
        X <= b;    // Allowed to vary
        max: Revenue;
        printf("At b=%d, Revenue: %.2f, X: %.2f\n",
                        b,Revenue,X);
    }
    then printf("Problem infeasible at b=%d\n",b);
}
```

Calling this procedure instead of `analyze_current_situation`, we find that the output is

At b=60, Revenue: 13500.00, X: 60.00
At b=70, Revenue: 15750.00, X: 70.00
At b=80, Revenue: 17375.00, X: 80.00
At b=90, Revenue: 18896.15, X: 90.00
At b=100, Revenue: 20261.54, X: 100.00
At b=110, Revenue: 21626.92, X: 110.00
At b=120, Revenue: 22992.31, X: 120.00
At b=130, Revenue: 24357.69, X: 130.00
At b=140, Revenue: 25723.08, X: 140.00
At b=150, Revenue: 26041.67, X: 142.33
At b=160, Revenue: 26041.67, X: 142.33

The results give rise to an interpolation of the optimum function Revenue wrt X graphed in Figure 7.6. This graph is telling us how Revenue will grow if more of resource X can be had. The slope of the piecewise linear curve as it leaves b = 60 is 225.00. Therefore Revenue will be increased by at least 225.00 dollars per hour of X up to the next sample point. At sample point b = 70, the slope is equal to 162.50. The slope of the piecewise linear curve as it leaves b = 80 is somewhat less, 152.16. Therefore Revenue will be increased by at least 152.16 dollars per hour of X from b = 80 up to the next sample point at b = 90. In the interval from 90 to 140, the slope of the interpolated function is con-

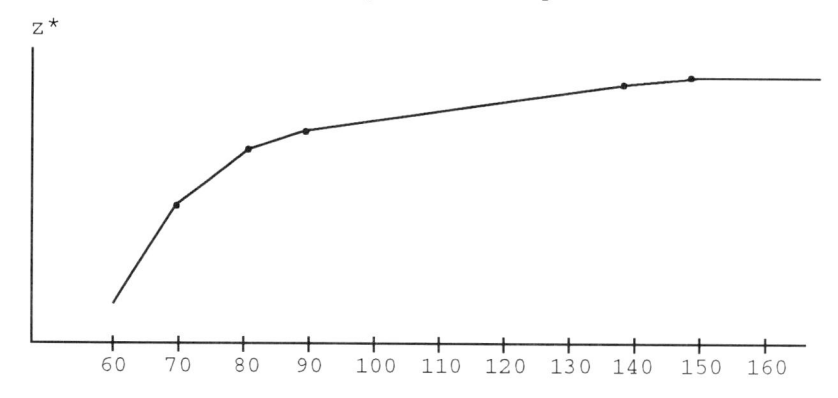

FIGURE 7.6 The interpolated optimum function.

stantly 136.54. Therefore we know that this is exactly the shadow price in the interval [90, 140). After b = 140, the curve flattens out. From the output we can surmise that 142.33 represents some kind of structural limit on the amount of resource X that can be gainfully utilized; the curve has slope 0.0 from b = 150 on. The graph is also telling us that the optimum function must bend somewhere in the interval [70,80] and again somewhere in [80, 90]. Printing more output would show, for example, that in the interval [80, 90] the underlying plan recommended by the linear program undergoes a structural change, since production of product A is 0.0 at b = 80 but this product has entered the picture by b = 90.

The objective function Revenue does not bring the cost of X into it. Hence the slope 136.54 is an upper bound on the price per hour of X that one would pay now when one has 90 hours of X already committed, up to 140 hours without decreasing revenue; at any lower cost, revenue would be increased. Since 136.54 is the value of the derivative of the optimum function on the interval (90,140), the resource X contributes to the optimal value of revenue at this point at the instantaneous rate of $136.54 per hour.

If we want to improve the fit between the interpolated function and the true optimum function, it is enough to redo the interpolation with more sample points. Since the number of singular points that z^*_x has is finite (in fact bounded by "m choose n," where m is the number of constraints including bounding constraints and n is the number of continuous variables), it follows that for a fine enough partition the two functions coincide except possibly on a set of very small total length.

From the points sampled in the discussion so far, we were able to conclude that we had the exact shadow price on the intervals [90,140) and [150,∞). However, it would be interesting to pin down what happens in the intervals [70,80], [80,90], and [140,150], where the shadow price jumps and structural changes take place.

The discontinuities of the shadow price are interesting because they typically represent a moment at which the plan determined by the linear program goes through a structural change. One idea is to localize these points by shrinking the interval around them at which samples are taken for the interpolated function. Another idea is to use the functions rc and range as analytical tools for pinning down the points of discontinuity of the shadow price. Let us see how to apply this approach to our example of the function Revenue wrt X of Model 7.1. In this example we want to follow events as the bound on X is increased. There are many ways in which this can be done. We will follow a simple, but reasonable plan. At each sample point the reduced cost and its range will be computed. If the reduced cost is 0, the analysis can be ended. Otherwise, if the range is positive, the next sample point is set to be the current one plus this value; if the range is 0, the next sample point is chosen delta units further along, where delta is a pre-chosen increment. Since this strategy is perfectly general, we put together some

7.2 Parametric analysis

routines that can be reused later. For the current example we have to sample points in the direction of increasing upper bounds. On the other hand, to allow for development along the lower bound on a variable, the routines will carry a parameter dir that will indicate the direction of the loop.

We will set things up for an objective function given by a single variable. Optimization can call for this variable to be minimized or maximized. For the maximization case we use the routine

```
optimize(continuous Z)
{
    max: Z;
}
```

Next let us develop a procedure to set up the bounding constraint, optimize the objective function, and record the reduced cost. This procedure will be passed the resource R, the objective function Z, the bound b on R, and parameters to store the reduced cost and range of R. To differentiate between the case of analysis along a lower bound from that along an upper bound, a parameter dir can be used; this parameter will be equal to +1 for the upper bound and -1 for the lower bound situation. This, plus some printing of results, can be encapsulated in a procedure:

```
sample(double & r, & c, b, continuous R, Z, int dir)
{
    dir*R <= dir*b;    // The bounding constraint
    optimize(Z);
    r = range(R);
    c = fabs(rc(R));

    printf("At b=%.2f,",b);
    printf("Obj: %.2f, Res: %.2f, Rd Cst: %.2f, Rg: %.2f\n",
                            Z,R,c,r);

    local_output();    // Application specific
}

local_output() // For Model 7.1
{
    printf("Produce %.2f of A and %.2f of B\n",A,B);
    printf("Use %.2f measures X, use %.2f measures Y,
            sell %.2f measures Y\n",X,Yused,Ysold);
}
```

The routine we will build to carry out the parametric analysis will take several parameters. Continuous variables X and Z for the resource under analysis and the objective function will be passed, as will be the increment `delta`, the starting bound `b`, and a restriction `limit` on the range of the analysis. Also to be passed is the directional flag `dir` to indicate the direction in which to carry out the analysis.

The parameter `limit` will give an upper or lower limit on the value that the bound `b` will be allowed to take. The setup constraints or other parts of the program might well have stipulated a previous upper bound on X, such as X <= bb, which will nullify the effect of X <= b when b > bb and confuse the analysis. To avoid such situations, one can use the built-in functions ub and lb which return the operative upper and lower bounds on continuous variables:

```
double ub( continuous )
double lb( continuous )
```

To verify that the value of the parameter `limit` does not exceed already existing limits, a check can be made before launching the parametric analysis:

```
check_bound(continuous R, int dir, double limit)
{
    if dir == 1;
    then ub(R) >= limit;
    else lb(R) <= limit;
}
```

The routine to carry out the parametric analysis will have local storage for recording the reduced costs and ranges found. The code will run a loop that will record the results of sampling the optimum function at a sequence of values of b. If the reduced costs reaches 0, or if the value of b goes beyond the range of `limit`, the code breaks out of the loop and returns:

```
parametric_analysis(continuous X,Z, double delta, b, limit,
                                                    int dir)
{
double c,r;

    if not check_bound(X,dir,limit);
    then {
        printf("Limit exceeds known bound\n");
        return;
    }

    and(;;) {   // Loop until break
        if not
            sample(r,c,b,X,Z,dir);//Do and undo constraints
        then {      // Adjustment for infeasibility
```

7.2 Parametric analysis

```
            printf("Infeasibility detected at b=%f\n",b);
            b = b + dir*limit/10;   // Move b along
        }
        else {
            if c == 0.0;    // If reduced cost is 0.0
            then break;     // end analysis
            update(b,r,delta,dir);      // Revise b
        }
        if dir*b > dir*limit;  // If no room left
        then break;     // exit the loop and end analysis
    }
}

update(double & b, r, delta,int dir)
{
    if r <= delta;
    then b = b + dir*delta;
    else b = b + dir*r;
}
```

It all can be put together in the main procedure:

```
2lp_main()
{
    setup();
    Y <= 110;
    parametric_analysis(X,Revenue,2.0,60.0,400.0,1);
}
```

With this code the analysis will be made on the function `Revenue` wrt X by increasing the upper bound on X starting at 60; for situations where the range is too small the increment in b will be equal to `2.0`. The parametric analysis will break off if the bound reaches `400.0`, or if a reduced cost of `0` is encountered. For this call, with our current data, the output is

```
Analyzing X vs Revenue
At b=60.00, Obj: 13500.00, Res: 60.00,Rd Cst: 225.00, Rg: 10.00
Produce 0.00 of A and 60.00 of B
Use 60.00 measures X, use 60.00 measures Y, sell 0.00 measures Y
At b=70.00, Obj: 15750.00, Res: 70.00,Rd Cst: 225.00, Rg: 0.00
Produce 0.00 of A and 70.00 of B
Use 70.00 measures X, use 70.00 measures Y, sell 0.00 measures Y
At b=72.00, Obj: 16075.00, Res: 72.00,Rd Cst: 162.50, Rg: 14.00
Produce 0.00 of A and 70.00 of B
Use 72.00 measures X, use 70.00 measures Y, sell 5.00 measures Y
At b=86.00, Obj: 18350.00, Res: 86.00,Rd Cst: 136.54, Rg: 56.33
Produce 0.00 of A and 70.00 of B
```

Use 86.00 measures X, use 70.00 measures Y, sell 40.00 measures Y
At b=142.33, Obj: 26041.67, Res: 142.33,Rd Cst: 0.00, Rg: Inf
Produce 65.00 of A and 70.00 of B
Use 142.33 measures X, use 91.67 measures Y, sell 18.33 measures Y

Note that b = 70 appears as a sample point but the range of the reduced cost is 0.0. The information that the range is 0.0 indicates that this point is a discontinuity in the shadow price function. So we step past this point in a gingerly manner and restart the analysis at b = b + delta. At the point b = 72, a nonzero amount of excess Y is sold off. By comparing the reduced cost at b = 72 with the slope of the interpolated function on the interval [72, 86], we know that this reduced cost is the true shadow price. At the next sample point, something similar happens. Although there is no sample taken at 86 + 2, production of A enters the picture in the interval [86, 142.33]. As we can tell by looking ahead to the breakpoint b = 142.33, from b = 86 to b = 142.33, production of A increases to 65 and sale of excess Y dwindles to 18.33. From this point on, the reduced cost is equal to the shadow price of 0, and nothing can change.

Thus far we have concentrated on the upper bound of a variable; as we increase this bound and analyze an objective function that is being maximized, the shadow price and reduced cost start off positive and eventually stabilize at 0 because we have assumed that the initial feasible region is bounded. When this point is reached, the variable will remain at a value lower than the upper bound that is being allowed.

It is also possible, when the objective function is optimized, for a variable to be sent to its lower bound. In fact, if a product is "costing us money," then, when revenue is optimized, the variable P representing this product will likely be sent to its lowest possible value b. If a variable is at its lower bound, then its reduced cost is obtained by multiplying the value returned by the function rc by -1. In this case the reduced cost measures the rate at which this product is currently detracting from revenue. Once again, when rc returns a negative value, it simply indicates that the variable is currently at its lower bound, and it is the absolute value that is equal to the reduced cost.

The upshot is this. When rc returns a positive value, the lower bound on the variable can be relaxed indefinitely without changing the reduced cost or the value of the optimum function; when rc returns a negative value, the upper bound can be increased indefinitely to similar effect. When rc returns 0, the variable can be at either or neither bound. In this case the upper bound can be increased, and the lower bound can be decreased indefinitely without changing the reduced cost or the value of the optimum function.

If we analyze the behavior of the maximum value of the objective function as the lower bound on a variable is reduced, the function z^*_x we obtain will have the shape illustrated in Figure 7.7. In this situation the lower bound on a variable is being decreased, and so we define the

7.2 Parametric analysis

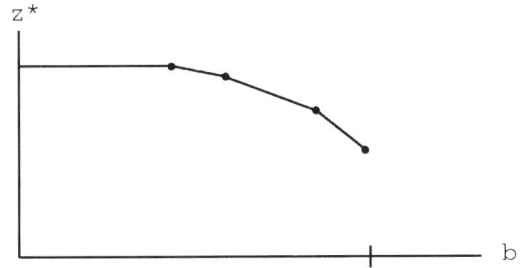

FIGURE 7.7 Varying the lower bound for a maximization problem.

shadow price to be the left-sided derivative of the optimum function z^*_x. When the continuous variable being analyzed is at its lower bound, the value returned by rc is a lower bound on the shadow price. The shadow price is the directional derivative of the optimum function in the direction in which the bound *b* is moving.

Let us turn to minimization problems. When minimizing an objective function and increasing the upper bound on a variable, the graph of the optimum function has the shape illustrated in Figure 7.8. Since we are increasing the upper bound on the variable, the shadow price is defined to be the right sided derivative of the optimum function. The reduced cost for a variable at its upper bound is nonnegative, but multiplying the reduced cost by -1 will now give a lower bound on the shadow price. The range function will estimate how far the upper bound can be increased without changing the reduced cost; over this range the reduced cost is the shadow price multiplied by -1.

Finally, if we develop an optimum function for a minimization problem by lowering the lower bound on a variable, we find a function whose graph is shown in Figure 7.9 In this case the shadow price is defined to be the left-sided derivative of the optimum function. If the variable is at its lower bound, multiplying the reduced cost by -1 gives a lower bound on the shadow price; this is of course the value returned by the function rc. As before, the range function estimates how far the lower bound can be decreased without changing the reduced cost. Perhaps a more

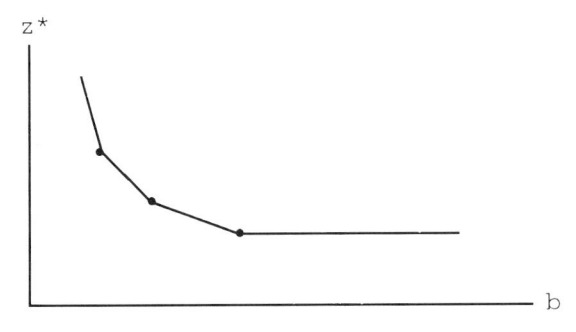

FIGURE 7.8 Varying the upper bound for a minimization problem.

intuitive way of looking at this case is that the negative value returned by rc gives the rate at which the optimum function is going down. If the function rc returns a nonnegative value, the shadow price is zero, and the lower bound can be decreased indefinitely without changing the optimal value of the objective function.

Let us apply some of these tools to refine the analysis of Model 4.4, The Grid. This will be a minimization problem, and we will analyze the effect of lowering bounds on continuous variables.

Model 7.2 The Grid Analyzed

The results of Model 4.4 show that the utility is delivering somewhat more power than consumers need to the substations, presumably because of its policy of maintaining a flow of at least 110 megawatts through each intermediate node on the network. Engineers would be interested to know if any of these nodes are working against the goal of minimizing the amount of power generated at the utility's plant. The question is whether such nodes exist and, if so, how badly are these nodes affecting things.

The objective function in Model 4.6 is to minimize Power[0], the amount of energy produced at the central plant. For the analysis we want to make now, we can record the values returned by the function rc for the nonterminal nodes and proceed to analyze those that have negative values. These are nodes that are at their lower bound of 110. Moreover the nonzero reduced cost indicates that lowering the bound will lower the amount of power that must be generated.

We begin with some additional declarations.

```
#define NT 12        // Number of nonterminal nodes
double signed_reduced_cost[NT];// To record reduced costs
```

We modify the setup routine of Model 4.6 by commenting out the lower bounds on nodes:

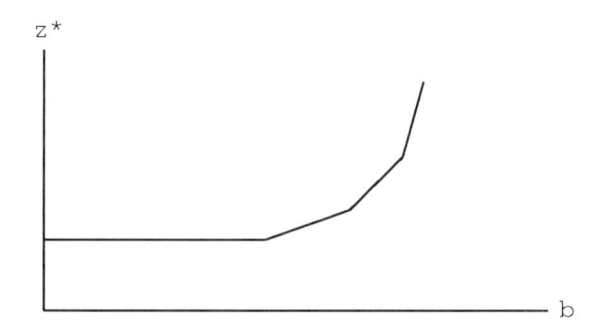

FIGURE 7.9 Varying the lower bound for a minimization problem.

7.2 Parametric analysis

```
new_setup()
{
    and(int n=0;n<NODES;n++) {
        Power[n] <= node_cap[n];
        //Power[n] >= node_demand[n]; FREE UP LOWER BOUNDS
    }
    and(int arc=0;arc<ARCS;arc++)
        Flow[arc] <= arc_cap[arc];
    flows_in();
    flows_out();
}
```

In a first pass we restore the lower bounds on the nodes of the network, optimize the objective function, and record the reduced costs on the nonterminal nodes:

```
first_pass()
{
    and(int n=0;n<NODES;n++)
        Power[n] >= node_demand[n];
    min: Power[0];
    and(int i=1;i<NT;i++)
        signed_reduced_cost[i] = rc(Power[i]);
}
```

For each nonterminal node that is to be analyzed, we will reinstate the lower bounds on all the other nodes:

```
keep_the_very_same(int p)
{
    and(int n=0;n<NODES;n++)
        if n != p;
        then Power[n] >= node_demand[n];
}
```

We revise the optimize routine for minimization:

```
optimize(continuous Z)
{
    min: Z;
}
```

and turn off the routine for further output

```
local_output()
{
    ;          // Null statement
}
```

Let us introduce an error-handling routine, that will invoke the `exit` statement to leave the program after printing an appropriate message:

```
error_hndlr(int error)
{
    printf("Program halted: error condition #%d\n",error);
    exit();
}
```

Finally, the main routine will call the `analyze` procedure with a `delta` of .5, an initial value of b of 110, a limit of 0.0, and with the directional flag `dir` set to -1:

```
21p_main()
{
    data();
    tables(1,in,nnz_in);
    tables(0,out,nnz_out);
    new_setup();
    if not first_pass();
    then error_hndlr(-1);           // Report and exit
    and(int p=1;p<NT;p++)
        if signed_reduced_cost[p] < 0.0;
        then
            if not {// Negation will restore feasible region
                printf("Analyzing node %d\n",p);
                keep_the_very_same(p);
                parametric_analysis(
                    Power[p],Power[0],.5,110,0.0,-1);
            }
            then error_hndlr(p);    // Report and exit
}
```

The analysis shows that nodes 6, 7 and 10 can be profitably reduced: In fact node 10 can go all the way down to 8.32. However, none of these nodes can be eliminated entirely if demand is to be met. For example, the output for node 10 is

```
Analyzing node 10
At b=110.00, Obj: 1117.75, Res: 110.00, Rd Cst: 1.20, Rg: 11.64
At b=98.36, Obj: 1103.80, Res: 98.36, Rd Cst: 1.16, Rg: 72.13
At b=26.22, Obj: 1020.32, Res: 26.22, Rd Cst: 1.16, Rg: 0.00
At b=25.72, Obj: 1020.24, Res: 25.72, Rd Cst: 0.16, Rg: 0.75
At b=24.98, Obj: 1020.12, Res: 24.98, Rd Cst: 0.16, Rg: 0.00
At b=24.48, Obj: 1020.08, Res: 24.48, Rd Cst: 0.09, Rg: 13.03
At b=11.44, Obj: 1018.87, Res: 11.44, Rd Cst: 0.09, Rg: 0.00
At b=10.94, Obj: 1018.84, Res: 10.94, Rd Cst: 0.05, Rg: 2.62
At b=8.32, Obj: 1018.72, Res: 8.32, Rd Cst: 0.00, Rg: Inf
```

7.2 Parametric analysis

Note that the initial reduced costs are positive while the variable is following its lower bound. When the reduced cost reaches 0, the range is listed as infinite even though the continuous variable has a default lower bound of 0; this means that things would not change at this point even if negative lower bounds for this variable were entertained.

End of Model 7.2

There are some other interesting views of a linear programming application that can be explored by parametric analysis. By way of illustration, in the example studied in Model 7.1 with products A,B and resources X,Y we considered the objective functions Revenue and Revenue - 50*X - 100*Y. With the first objective function, the purchase of resource Y was treated as a sunken cost; with the second, the amount of Y to be purchased was determined by the optimal solution. It is also interesting to study yet another revenue function that treats the purchase of Y as a sunken cost but takes into account the entire amount of resource Y that is purchased. As we vary the upper bound b on Y, this objective function is

```
max: Revenue - 50*90 - 100*b;
```

But something has changed here; we are now also perturbing a constant term in the objective function as opposed to a coefficient of a continuous variable. This means that the Parametric Analysis Theorem, given in Section 7.1, does not directly apply. However, if we add a variable Yupper and the constraint Y <= Yupper, then by enforcing

```
Yupper == b
```

while varying b and optimizing

```
max: Revenue - 50*90 - 100*Yupper;
```

we obtain the same optimum function. For this case the appropriate theorem is a variant of the Parametric Analysis Theorem in which the half spaces $x \le b$ are replaced by hyperplanes $x = b$; then the conclusion is that the optimum function is concave down, continuous, and piecewise linear, but not necessarily nondecreasing. The shape of the optimum function in these cases means that parametric analysis can be carried out effectively by plotting points. An alternative is to replace the constraint Yupper == b with Yupper <= b and carry out parametric analysis increasing b until Yupper is no longer at the upper bound b in the optimal solution.

Another area where parametric analysis can be pursued is the effect on the optimum function of changes in a coefficient in the objective function. For example, as a positive coefficient is increased in a maximization problem, the optimum function will be concave down, continuous, piecewise linear, and nondecreasing. For this kind of parametric analysis, the `objcup` and `objcdn` functions are used instead of the `rc` and `range` functions.

We have the analog of the Parametric Analysis Theorem.

Theorem (Parametric Objective Coefficient Analysis Theorem) *Suppose that F is a bounded feasible region defined by a set of constraints* $\mathbf{Ax} \leq \mathbf{b}$. *For $c \geq 0$, let $z^*_x(c)$ be the value of the optimal solution to the linear programming problem*

$$\mathbf{Ax} \leq \mathbf{b}$$
$$\max: \mathbf{cx} + c x$$

Then z^*_x is a continuous, piecewise linear, concave up, and nondecreasing function on the interval $[0, \infty)$.

This theorem is also discussed in Chapter 15. As an example of the use of this theorem, let us again consider Model 3.3, Motorcycles Inc., and do a parametric version of the analysis of Section 7.1. Certain models are not planned for in the optimal solution to maximize income. The following procedure will use the function `objcup` to search how the selling price of a model of motorcycle will change the optimal plan. The parameters passed to the routine are the number of the model and an upper bound on how far the selling price can be increased.

```
coeff_analysis(int i,double limit)
{
double c,now;
double delta;
    delta = 1;          // Initialize the increment
    now = wp(Bike[i]);  // Record current value

    and(;;) {           // Work until done
        if objcup(Bike[i]) < delta;
        then c = c + delta;     // Minimal increment
        else c = c + objcup(Bike[i]);  // Computed increment
        if c > limit;   // Reality test
        then {
            printf("Next selling price would pass limit\n");
            break;
        }

        max: Z + c*Bike[i];
        if wp(Bike[i]) != now;   // Solution changes
        then {
            printf("For model %d\t",i);
```

7.2 Parametric analysis

```
            printf("At price %.2f, produce %.2f\n",
                            income[i]+c,Bike[i]);
            now = wp(Bike[i]);
        }
    }
}
```

The following program flags models that are not to be produced and searches for the selling price required to make them profitable relative to the other models using the routine `coeff_analysis`:

```
21p_main()
{
    data();
    setup();
    and(int m=0;m<MODS;m++) {
        max: Z;
        if wp(Bike[m]) == 0.0;
        then coeff_analysis(m,500.0);
    }
}
```

The output starts with

For model 0 At price 249.18, produce 95.25
For model 0 At price 301.00, produce 120.00
Next selling price would pass limit

Let us note from the output that when a model enters the production picture, it does so at a substantial level at once. This is to be contrasted with what we saw previously when we varied the upper bound on a variable.

A final remark is in order. Parametric analysis along the upper or lower bound on a variable continually alters the feasible region because the bound on the variable has to be readjusted. For parametric analysis on an objective function coefficient, the feasible region doesn't change; it stays the same while different objective functions are optimized.

Exercises

7.2.1. Suppose that with X <= b, we have rc(X) = r > 0 and that, for delta > 0 with X <= b + delta, we have rc(X) = r. Prove that the shadow price of X on the interval [b, b+delta) is equal to r.

7.2.2. A company makes 10 products. Each product uses a certain amount of labor and other resources. There are 12 resources each of which is measured in an appropriate metric. Table 7.6 lists the amount

TABLE 7.9 Resource Requirements

Resource Number	Number of Products Requiring Resource	Products Requiring Resource	Quantity of Resource Required by Product	Amount of Resource Available
0	3	0,2,4	22,23,12	2500
1	4	0,1,4,5	30,14,44,9	3300
2	4	1,2,6,11	21,24,11,22	1650
3	3	3,4,5	33,10,21	2722
4	4	4,6,7,8	11,14,16,32	3411
5	3	6,7,8	42,52,12	1820
6	5	3,5,6,7,8	12,26,9,8,5	2250
7	5	3,4,9,10,11	5,29,9,7,25	2710
8	6	1,4,6,7,8,9	4,30,6,4,7,9	3130
9	4	3,4,10,11	24,14,19,31	2930

of each resource consumed in the production of a unit of each product. Table 7.7 gives the revenue generated by a unit of each product. Write a linear program to find a production strategy which optimizes revenue. Perform parametric analysis on the resources and on the products.

TABLE 7.10 Revenue Data

Product	0	1	2	3	4	5	6	7	8	9	10	11
Revenue	40	80	73	80	40	88	90	110	220	145	66	55

8

Backtracking

We return to the interaction between constraints and disjunction and develop programming support for an approximation to the way disjunction is used in classical logic. In particular we will use disjunctive constructs to develop search techniques that can be used when an application has integrality requirements or disjunctive logical requirements.

8.1 Persistent disjunction

As remarked in Section 5.1, in ordinary mathematics a branch of a disjunction can be explored and other branches held in abeyance so that they can be returned to later if necessary. A version of this kind of disjunction is supported in 2LP by means of the either/or construct, whose syntax is

```
either <statement>
or <statement>
    ...
or <statement>
```

The either/or construct blends the mathematical and the programming versions of disjunction: To proceed past a disjunction, one of the disjuncts must be committed to, but the alternative of coming back and considering further disjuncts is maintained. We call this interpretation of disjunction *persistent disjunction*.

Let us look at the following disjunction of constraints:

```
either X <= 3;
or X >= 9;
```

The declarative meaning of this is that from this point in the program, X is restricted to vary over the union of the intervals $[0,3]$ and $[9,+\infty)$. To capture this meaning, the operational semantics of 2LP treat this disjunction as a statement with two alternative parts. When encountering the call to either/or, the system will put this call on its program stack and record the fact that this statement has a second alternative. We say that a *choice point* has been created. The program first tries the constraint X <= 3; if it is not feasible, the alternative X >= 9 is tried straightaway. If X <= 3 is feasible, the program will commit to it and proceed to the next statement. The existence of the choice point will enable the program to return later, undo the constraint X <= 3, and try the alternative. Returning to choice points is called *backtracking*.

Let us look at some other simple examples. The code

```
2lp_main()
{
continuous X;
    X >= 5;
    either X <= 3;
    or X >= 9;
    printf("The witness point is now %.0f\n",wp(X));
}
```

will output

The witness point is now 9

as will the code

```
2lp_main()
{
continuous X;
    either X <= 3;
    or X >= 9;
    X >= 5;
    printf("The witness point is now %.0f\n",wp(X));
}
```

since both pieces of code terminate with the feasible region $[9,+\infty)$. However, in the first case the inconsistency of the constraint X <= 3 with X >= 5 is detected straightaway, while in the second case many more things take place:

1. The feasible region is changed to $[0,3]$.
2. The inconsistency with X >= 5 is detected.
3. Backtracking occurs.
4. X <= 3 is discarded.
5. The feasible region becomes $[9,+\infty)$.
6. The constraint X >= 5 is added but without changing the feasible region.

8.1 Persistent disjunction

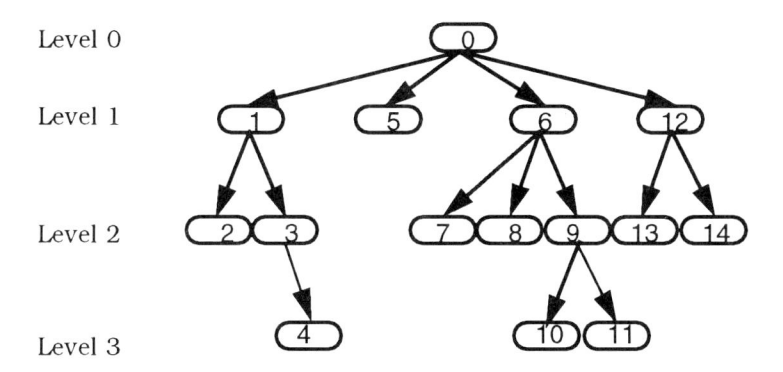

FIGURE 8.1 A depth-first search.

The process of branching on alternatives is naturally represented by a tree. In Figure 8.1 we have a diagram of a tree that consists of nodes and directed arcs connecting the nodes; the nodes of the tree are labeled 0-14. In the literature of the mathematical sciences, trees grow downward, somewhat *contra naturam*.

The tree must be searched in order to find a node that provides a solution to the problem; such a node is called a *goal node*. Successors of a node are also called its *children*, and the node itself is called the *parent* node. Nodes with the same parent are called *siblings*. Nodes of the tree without any successors are called *leaves, leaf nodes*, or *terminal nodes*. The task of finding a right combination of alternatives generates a search through the tree, and a successful solution yields a path through the tree from its root to a goal node. Another term for the tree associated with a search process is *search space*. The process of avoiding nodes in the search that cannot lead to a solution is called *pruning* the search tree or *pruning* the search space.

In Figure 8.1 the nodes are numbered in the order in which they would be searched in a depth-first search. A depth-first search proceeds down the leftmost branch of the tree until it reaches a leaf. The next node to be searched is found by going back up the tree until a node with an unexplored branch is found; the search then again proceeds down the leftmost available branch. This process is repeated until a goal node is found. If there is no goal node and the tree is finite, the entire tree is searched. If the tree is infinite, the search will terminate when a goal node is found during the search; otherwise, the search will continue indefinitely. This last possibility is illustrated in Figure 8.2. This eventuality must be guarded against by the programmer.

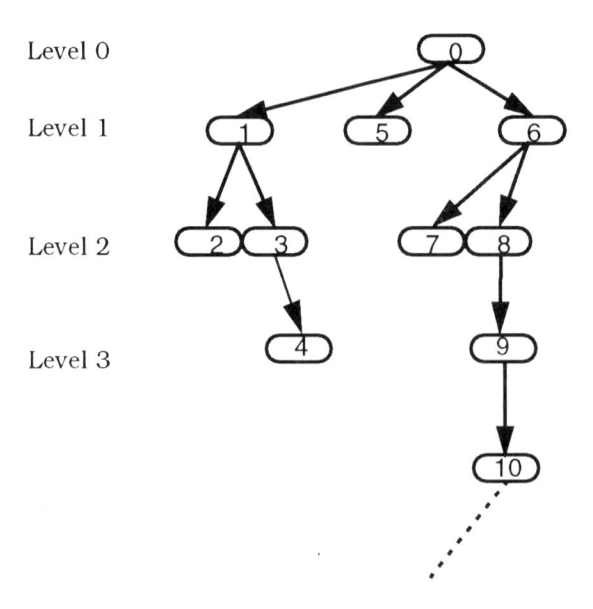

FIGURE 8.2 A depth-first search with an infinite path.

In several models such as Model 4.3, Fairer Co-op Tax, we have encountered situations where there were competing objectives. In the next model we consider another situation of this sort. We return to Model 3.1, where the company has to plan its production strategy subject to resource constraints on available components.

Model 8.1 The Pacific Rim Revisited

The strategy found in Model 3.1 calls for the company to produce only the higher-priced Huang and Kosciusko models. The sales department is not happy with this development, since salespeople need a more variegated product line in order to satisfy customer tastes. However, the more competitively priced Fuji and Hanla models are now thought to be too similar in style and function. So it is considered best that only one of them be included in the product line, and the same is true of the pair of higher-priced models. The sales department has requested that at least 750 units of whichever lower-priced model is decided upon be produced and that one of the higher-priced models could be manufactured. The accounting office has asked that the projected revenue on the new product line be within 12.5% of that which was originally projected before these changes. Is there a feasible solution that meets all these demands?

The data, the basic production constraints, and the objective function can be set up as in Model 3.1. So the first three calls in the main procedure will again be

8.1 Persistent disjunction

```
    data();
    constraints();
    objective();
```

The objective `Revenue` is maximized in the procedure `objective`; hence, to guarantee that `Revenue` is within 12.5% of what it would have been without the additional conditions, we can then call

```
    Revenue >= .875*wp(Revenue);
```

before addressing the logical requirements on the workstations.

One way to express the requirement that at least 750 units of one model be produced and 0 units of another model is by means of a routine such as

```
one_but_not_the_other(int a,b)
{
    Workstations[a] >= 750;
    Workstations[b] == 0.0;
}
```

To express the logical requirement that it is either the Fuji model or the Hanla model that is to be produced, we employ persistent disjunction:

```
    either one_but_not_the_other(FU,HA);
    or    one_but_not_the_other(HA,FU);
```

Similarly, to require that one of the two higher-priced models be eliminated, we can write

```
    either Workstations[HU] == 0.0;
    or     Workstations[KO] == 0.0;
```

So we only need to modify Model 3.1 by changing the main procedure as follows:

```
2lp_main()
{
    data();
    constraints();
    objective();    // Revenue is maximized here

    new_code();

    max: Revenue;
    output();
}
```

```
new_code()
{
    Revenue >= .875*wp(Revenue);
    either one_but_not_the_other(FU,HA);
    or one_but_not_the_other(HA,FU);
    either Workstations[HU] == 0.0;
    or Workstations[KO] == 0.0;
}
```

The output will now be

The plan calls for production of
    0.00 Fuji models
 5150.00 Hanla models
    0.00 Huang models
  350.00 Kosciusko models
Revenue will be 484775.00

The solution calls for the production of the Hanla and Kosciusko lines.

As can be easily checked by commenting out the second `either/or`, had it not been required that only one of the Fuji and Hanla models be carried in the product line, then producing the Fuji, Huang, and Kosciusko models would have been the solution.

From this observation and the actual output, we can draw some conclusions about the flow of the program. After the constraint

```
Revenue >= .875*wp(Revenue);
```

the sequence of events is

1. `one_but_not_the_other(FU,HA)` is called successfully.
2. `Workstations[HU] == 0.0` fails.
3. `Workstations[KO] == 0.0` also fails.
4. the constraints generated by `one_but_not_the_other(FU,HA)` are undone.
5. `one_but_not_the_other(HA,FU)` is called successfully.
6. `Workstations[HU] == 0.0` fails again.
7. `Workstations[KO] == 0.0` succeeds.
8. `max: Revenue` is called.

To emphasize the difference between persistent disjunction and conditional disjunction, consider what would happen if we were to change the first disjunction to

```
c_either one_but_not_the_other(FU,HA);
or one_but_not_the_other(HA,FU);
```

8.2 Generate-and-test

Now the program will fail, because the first alternative succeeds initially but leads to failure later, and there is now no way to backtrack to the second alternative.

End of Model 8.1

Model 8.3 below, Salt and Mustard, provides an example where conditional disjunction and persistent disjunction are used together in a complementary fashion.

Exercises

8.1.1. In Exercise 4.3.4, the objective function measured the average time it took a scout to do the job assigned. Let us change the objective to minimizing the longest time it takes any one scout to do the job assigned. (This is the makespan). Is there a solution that makes this number no greater than 32? What is the best possible solution to this version of the problem?

8.2 Generate-and-test

Backtracking is commonly used in artificial intelligence, operations research and other fields to find a solution to a challenging problem. In the simplest situation, a straightforward search is made through candidate solutions until one that satisfies all the conditions of the problem is found. This problem solving paradigm is called *generate-and-test*. If the problem size is small, this method can be effective. However, in general, techniques must be devised to reduce the complexity of the search process. Many examples will be seen where strategies and heuristic techniques are used to make the search process more efficient. In this section we consider some simple examples where generate-and-test is the appropriate paradigm.

The term *0-1* is used to refer to problems where all the variables in the solution must be either 0 or 1. Such applications arise with surprising frequency and can prove daunting. The following example, however, is very simple. The model also illustrates the way the loop variables are restored when backtracking occurs. When a persistent disjunction occurs in the body of a loop and backtracking takes place, the loop control variable is restored to the value it had when the first alternative of the disjunction was encountered. It is for this reason that the restriction has been placed on the loop control variables that they can only be assigned in the loop header.

Model 8.2 The 0-1 Solution I

The linear equations

$$x_0 + 4x_1 + x_2 = 2$$
$$2x_0 + 2x_1 - x_2 = 1$$

define a line in three-dimensional space. Is there a 0-1 solution to these equations?

Now there are eight possible ways of assigning the values 0 and 1 to the three variables. The following code runs through these combinations and tests each one until one of them yields a solution to the equations. This problem in three variables is of course trivial, but the example is an illustration of how the backtracking mechanism works to solve the problem by means of generate-and-test.

```
21p_main()
{
int x[3];

    // Generate
    and(int i=0;i<3;i++)
        either x[i] = 0;
        or x[i] = 1;

    // Test
    2*x[0] + 2*x[1] - x[2] == 1;
    x[0] + 4*x[1] + x[2] == 2;

    // Print results
    printf("There is a 0-1 solution:\n");
    and(int i=0;i<3;i++)
        printf("x[%d] is %d\n",i,x[i]);
}
```

Let us step through what happens. Here the and loop with the either/or generates a value for each x[0], x[1], and x[2] before proceeding to test the equations. When the and loop is exited, all the x[i] are 0, but the first equation fails to hold, and the program backtracks to try x[2] = 1 with the assignments x[0] = 0 and x[1] = 0 still in force. Since the first test fails again, and there are no other choices for x[2], the program backtracks again and tries the assignment x[1] = 1. After testing x[0] = 0, x[1] = 1, x[2] = 0, which fails to satisfy the first equation, it tries x[0] = 0, x[1] = 1, and x[2] = 1. This satisfies the first equation but not the second. At this point the program has exhausted all the alternatives resulting from the assignment x[0] = 0. So the program backtracks into the and loop to i = 0 but now tries the

8.2 Generate-and-test

assignment x[0] = 1. Once again there are two alternatives for x[1]. First the assignment x[1] = 0 is made, then the alternatives for x[2] are considered. The choice x[2] = 0 violates the first test equation, but the second choice x[2] = 1 leads to a solution of both test equations. Thus the solution x[0] = 1, x[1] = 0, x[2] = 1 has been found, and the program prints the results and halts.

This example illustrates an important point about choice points. The either/or statement

```
either x[i] = 0;
or x[i] = 1;
```

occurs within the scope of the and loop and this and loop is exited successfully five different times. However, in the first four instances the test equations fail, and control returns to the last either/or in the loop that has an alternative left. If a choice point is created within an and loop, that choice point remains active even if the loop is exited successfully; when the choice point is returned to, the loop control variable is restored to its proper value. Thus, when control returns to an either/or statement, both the constraints on the continuous variables and the values of the loop variables are restored to their state when the either/or statement was first encountered. In other words, the code

```
and(int i=0;i<3;i++)
    either x[i] = 0;
    or x[i] = 1;
```

is equivalent to the three disjunctions

```
either x[0] = 0;
or x[0] = 1;
either x[1] = 0;
or x[1] = 1;
either x[2] = 0;
or x[2] = 1;
```

In Figure 8.3 the search tree for this 0-1 problem is illustrated. The nodes in all white are those that are "visited" in the search but discarded later; the nodes in white with a heavy black border are those that form part of the successful search path, while those with stripes are nodes that are not "visited" in the search. Note that depth in the search tree corresponds to the conjuncts generated by the and loop and that breadth in the tree corresponds to the disjuncts generated by the either/or statement.

In this example the search strategy is to enumerate all nodes of the search tree until a goal node is found. This is called *enumeration* or *complete enumeration*. The tandem of backtracking and enumeration traces

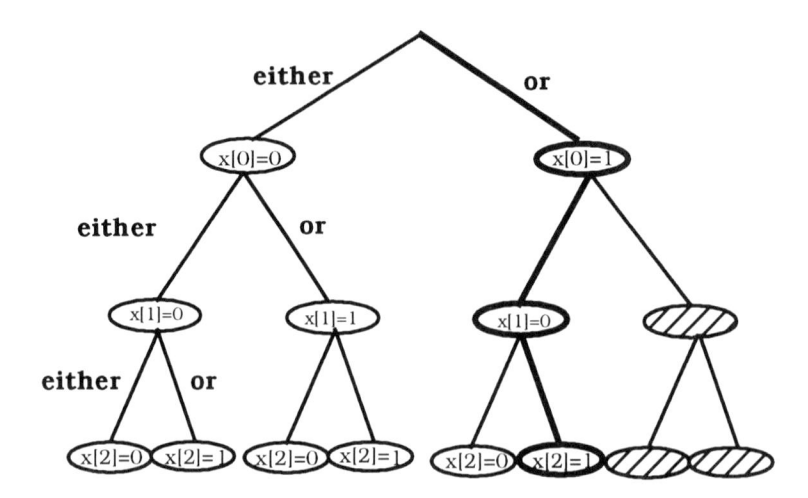

FIGURE 8.3 The search tree using "generate and test".

back to Lucas (1882) and a discussion of the problem of navigating through a labyrinth.

End of Model 8.2

In the above example the *generate* phase of the generate-and-test process was encoded by means of an and loop whose body consisted of an either/or statement, in other words, by means of a conjunction of disjunctions. This logical structure is characteristic of an important family of combinatorial problems, as we will see in many applications to come. In the above example this combination of 3 conjuncts and 2 disjuncts led to a search tree with 2^3 leaf nodes. More generally, M conjuncts of N disjuncts each yields N^M possible combinations. This phenomenon is known as *combinatorial explosion*. When confronted with a search problem with this potential, it is important to find ways of guiding the search efficaciously. This is a fundamental theme in logic and optimization work.

In 2LP, conjunction can be expressed both by juxtaposition and by the and loop; conditional disjunction has the c_either/or construct and the c_or loop. Persistent disjunction can also be expressed in two ways: by means of the either/or construct and by means of the or loop. As an example of the or loop notation, the disjunction

```
either x[i] = 0;
or x[i] = 1;
```

of the above model can be written

8.2 Generate-and-test

```
or(int k=0;k<2;k++)
    x[i] = k;
```

The syntax of the `or` loop is perfectly analogous to that of the `and` loop and `c_or` loop:

    `or` <*loop header*> <*statement*>

Once again, the control variable cannot be assigned in the body of the loop or accessed after the loop is exited. Also, if the control variable is passed as a parameter to a procedure in the body of the loop, this must be a pass by value.

The generate-and-test method is naturally coded by using the combination of the `and` loop and the `or` loop.

```
and(int i=0;i<M;i++)
    or(int j=0;j<N;j++)
        { ... }
```

The nested and/or loop combination expresses the structure of the search tree as having depth M and branching factor N. The tree structure generated by this nested loop is illustrated in Figure 8.4. We will return to the discussion of the intrinsic role of the and/or combination in solving difficult combinatorial problems in Section 9.3.

Nodes are generated in the search process by the `either/or` construct and the `or` loop. One can therefore count nodes by placing a counter in the disjunctive code. For example, to count the number of nodes visited in Model 8.2, we would introduce an identifier `count` of type `int` which is initially set to 0 and rewrite the and loop as

```
and(int i=0;i<3;i++)
    either {
        count = count + 1;
        x[i] = 0;
    }
    or {
        count = count + 1;
        x[i] = 1;
    }
```

At the end of the program `count` will be equal to 11.

We have seen that a choice point that is created in the body of an and loop can remain open, even after the loop itself is exited. What is more, when a choice point is created within a procedure, it stays open even after the procedure is exited; if backtracking is required, the procedure is then re-entered and processing begins again with the next untried branch of the choice point. For these reasons we say that choice

```
and(int i=0;i<M;i++)
    or(int j=0;j<N;j++) {    }
```

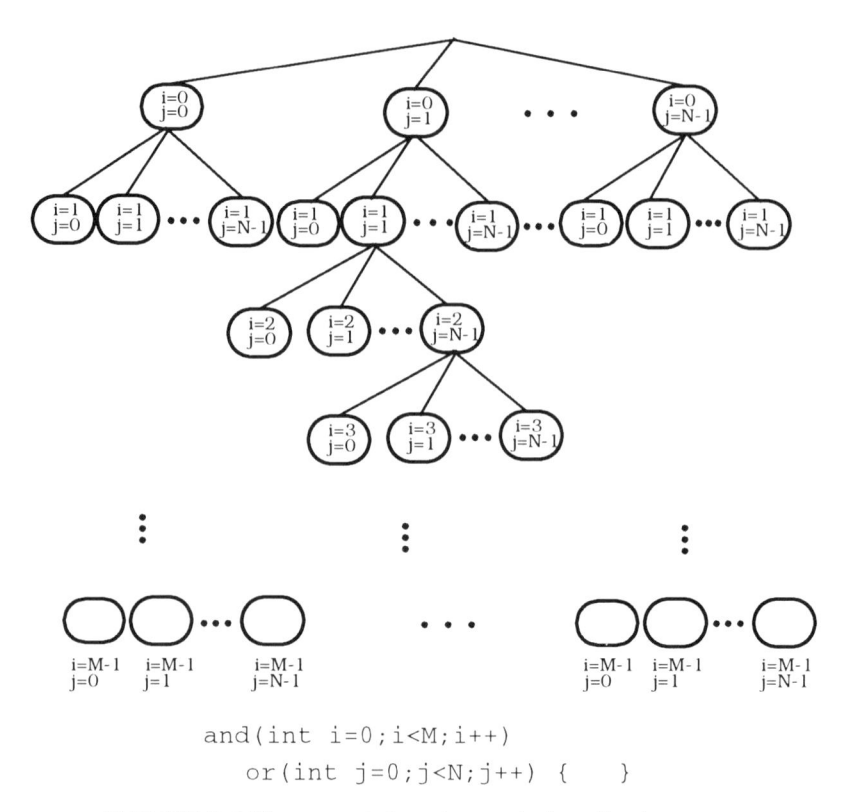

FIGURE 8.4 The search tree for nested and/or loops.

points are *robust*. This property of choice point is illustrated in the following model, a puzzle from Lewis Carroll that is based on his life as an Oxford don.

Model 8.3 Salt and Mustard

The following narrative describes the situation:

Five friends, Barry, Cole, Dix, Lang, and Mill, agreed to meet every day at a certain *table d'hôte*. They devised the following rules, to be observed whenever beef appeared on the table.

If Barry takes salt, then either Cole or Lang takes one only of the two condiments, salt and mustard; if he takes mustard, then either Dix takes neither condiment, or Mill takes both.

If Cole takes salt, then either Barry takes only one condiment, or Mill takes neither; if he takes mustard, then either Dix or Lang takes both.

If Dix takes salt, then either Barry takes neither condiment or Cole takes both; if he takes mustard, then either Lang or Mill takes neither.

8.2 Generate-and-test

If Lang takes salt, then either Barry or Dix takes only one condiment; if he takes mustard, then either Cole or Mill takes neither.

If Mill takes salt, then either Barry or Lang takes both condiments; if he takes mustard, then either Cole or Dix takes only one.

The problem is to discover whether these rules are compatible....

In this problem it is assumed that the phrase 'If Barry takes salt' allows two possible cases, viz. (1) He takes salt only; (2) He takes both condiments. And so with all similar phrases.

It is also assumed that the phrase, 'Either Cole or Lang takes one only of the two condiments' allows of two possible cases, viz. (1) Cole takes one only, Lang takes both or neither; (2) Cole takes both or neither, Lang takes one only.

It is also assumed that every rule is to be understood as implying the words 'and *vice versa.*' Thus the first rule implies that if either Cole or Lang takes only one condiment, then Barry takes salt.

The simplest way (and pretty much the only way) to handle this puzzle is by means of a generate-and-test approach. At the top of the program we will generate the choices that each Oxford don makes by means of an assignment statement and then test to see if the logical requirements are met. The way the author has set this puzzle up, it is not possible to interleave the generating and testing phases of the program, since each block of tests involves all of the participants.

To start, we can introduce symbolic constants NEITHER, SALT, MUSTARD, BOTH to represent the four possible choices that each don can make:

```
#define NEITHER 1
#define SALT 2
#define MUSTARD 3
#define BOTH 4
```

Each don will be represented by a global identifier of type int

```
int barry,cole,lang,dix,mill;
```

and the choices for the dons can be created by a call to a routine choices that loops through all the possibilities, assigning each to the don in turn.

```
choices(int & don)      // Pass by reference
{
    or(int i=NEITHER; i<=BOTH; i++)    // Backtracking loop
        don = i;
}
```

This generation phase of the program needs persistent disjunction because these assignments are tentative and will need to be changed repeatedly until a consistent combination is found. Since choice points are robust, when backtracking is required, the `choices` procedure will be re-entered and a new choice for the don will be tried.

The tests also make heavy use of disjunction, for example,

> ... either Cole or Lang takes one only of the two condiments, salt and mustard

but the tests only need conditional disjunction, since no re-assignments of choices for the dons will be made in the test phase. So these disjunctions will be coded by means of the `c_either/or` construct. Thus the above prose turns into the code

```
c_either salt_or_mustard(cole);
or      salt_or_mustard(lang);
```

where the `salt_or_mustard` routine itself also uses the conditional disjunction:

```
salt_or_mustard(int don)
{
    c_either don == SALT;
    or       don == MUSTARD;
}
```

The other tests are straightforward:

```
takes(int don, condiment)
{
    c_either don == condiment;
    or       don == BOTH;
}

neither(int don)
{
    don == NEITHER;
}

both(int don)
{
    don == BOTH;
}
```

The puzzle is such that each don will require his own routine to test if the conditions set by the puzzle are met. Thus for Barry the code will be

8.2 Generate-and-test

```
do_barry()
{
    if takes(barry,SALT);
    then
        c_either salt_or_mustard(cole);
        or salt_or_mustard(lang);
    if
        c_either salt_or_mustard(cole);
        or salt_or_mustard(lang);
    then takes(barry,SALT);
    if takes(barry,MUSTARD);
    then
        c_either neither(dix);
        or both(mill);
    if
        c_either neither(dix);
        or both(mill);
    then takes(barry,MUSTARD)
}
```

To solve the puzzle, we write the code for the other Oxonians and call a procedure such as

```
solve_salt_and_mustard()
{
    choices(barry);
    choices(cole);
    choices(lang);
    choices(dix);
    choices(mill);
    do_barry();
    do_cole();
    do_lang();
    do_dix();
    do_mill();
}
```

To obtain a more pleasant message than "No feasible solution" in the case the puzzle is unsolvable, let us write the main procedure so as to guarantee that the 2LP program at least will be successful:

```
2lp_main()
{
    if solve_salt_and_mustard();
    then output(barry,cole,dix,lang,mill);
    else printf("The puzzle has no solution\n");
}
```

The `output` routine is left to the reader. The program reveals that the dons can enjoy beef at lunch, since the conditions they have set for themselves are indeed consistent. This model also has an and/or structure, but here the role of the and loop is played by the five juxtaposed calls to `choices`.

To count the nodes generated in pursuit of a solution to this Lewis Carroll puzzle, one would change the or loop in the `choices` routine to read

```
or(int i=NEITHER;i<=BOTH; i++) {
    count = count + 1;
    don = i;
}
```

At the end of the program, `count` will contain the number of alternatives that had to be generated before the solution was found.

End of Model 8.3

In the generate-and-test examples in this section, backtracking has been triggered by the failure of a test. In the next section we will introduce a related paradigm where the backtracking will be triggered by the failure of constraints on continuous variables.

Exercises

8.2.1. Two friends Michael and John played a tennis match a while back. Michael won the match 6 games to 3. In tennis each player serves alternate games. They remember that the person who was serving won 5 of the games. Michael thinks that he was the one who served first. Can John prove Michael wrong?

8.2.2. Show that there is a way to place the numbers 1,...13 in 3 boxes so that if i and j are in the same box, then $i+j$ is in a different box. What about 1,...,14? (This puzzle is due to W. Schur.)

8.2.3. If the goal of a generate-and-test application is to produce a permutation of the integers from 0,...N-1, it can be useful to enumerate the permutations in lexicographic order, for this way tests can be run on partial solutions and if a test fails, all extensions of the current partial permutation can be eliminated. Write code that will print a permutation, and then fail and go back and produce the next permutation in lexicographic order, print that one and so on. (See McAloon and Tretkoff 1995).

8.2.4. Show that for every re-ordering of the digits 0,...,9 either there a subsequence of length 4 that is in the natural order or there is a subsequence of length 4 that is in the reverse of the natural order. Use the machinery of the previous exercise. Show this is not true for 0,...,8. (This puzzle is due to P. Erdös.)

8.3 Constrain-and-generate

In applications requiring search, the trees are often very large, and some method of pruning together with cleverness in the search must be used if the problem is to be solved. In the *constrain-and-generate* paradigm, constraints are set forth at the beginning and are in force as search is made to find a solution that meets all the requirements of the application. Continuous variables and linear constraints provide an important tool for dealing with this kind of situation. In fact, we have already seen how the progressive roundoff heuristic uses the linear relaxation to get a practical solution to difficult applications. In an application requiring search, continuous variables can provide a linear relaxation before and during the process of progressing down the tree and backtracking. The constrain-and-generate paradigm can often automatically reduce the number of nodes in the search tree, and hence reduce the execution time. As we will see, it can also make for easier coding and thus speed up the total development time considerably.

Let us redo the 0-1 problem of Model 8.2 using continuous variables and the constrain-and-generate paradigm.

Model 8.4 The 0-1 Solution, II

This time, to solve the problem posed in Model 8.2, we will set up the equations as constraints before we begin to fix values for the variables. Thus we declare the variables X[3] as continuous and stipulate that

```
and(int i=0;i<3;i++)
    X[i] <= 1.0;
2*X[0] + 2*X[1] - X[2] == 1;
X[0] + 4*X[1] + X[2] == 2;
```

at the outset. After setting up these constraints, the code will then try to fix the continuous variables at 0-1 values:

```
and(int i=0;i<3;i++)
    either X[i] == 0;
    or X[i] == 1;
```

Note that we have interchanged the order of things. The test equations of Model 8.2 have been replaced by equality constraints, but the constraints are set up before values for the X[i] are generated. Hence, this paradigm is given the name *constrain-and-generate*. So the code becomes

```
// A solution to the 0-1 problem using
// constrain-and-generate

21p_main()
{
continuous X[3];

    // Constrain
    and(int i=0;i<3;i++)
        X[i] <= 1;

    2*X[0] + 2*X[1] - X[2] == 1;
    X[0] + 4*X[1] + X[2] == 2;


    // Generate
    and(int i=0;i<3;i++)
        either X[i] == 0;
        or X[i] ==1;

    //Print results
    and(int i=0;i<3;i++)
        printf("X[%d] is %.0f\n",i,X[i]);
}
```

Let us step through what happens. After bounding the continuous variables by 1 and setting up the equations, the and loop and either/or statement set up the two possibilities for each variable X[i]. The first alternative is to try the constraint X[0] == 0 which has the effect of reducing the equality constraints to

```
    2*X[1] - X[2] == 1;
    4*X[1] + X[2] == 2;
```

Since X[0] == 0 succeeds, the alternative X[0] == 1 is held in reserve. Next the first alternative for X[1] is tried, namely X[1] == 0. However, when X[0] and X[1] are both equal to 0, the first equality constraint reduces to X[2] == -1, which is infeasible since X[2] is constrained to be greater than or equal to 0. So the either statement fails for X[1], and the or part of the statement is executed, namely X[1] == 1. But this reduces the second equality constraint to X[2] == -2, which is impossible. Therefore the program backtracks to the alternative for X[0] that has been held in abeyance. So X[0] == 1 is tried and succeeds, reducing the equality constraints to

8.3 Constrain-and-generate

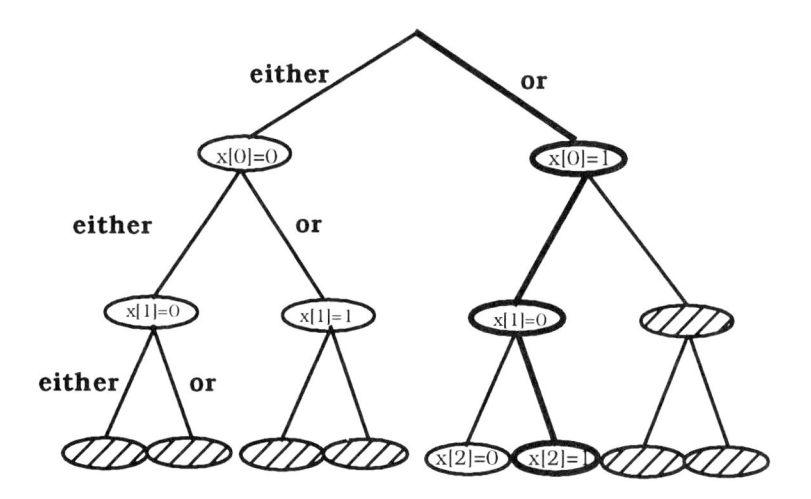

FIGURE 8.5 The search tree using "constrain and generate".

```
2*X[1] - X[2] == -1;
4*X[1] + X[2] == 1;
```

At the next level in the search tree, the first alternative for X[1] is tried again. This time X[1] == 0 reduces the equality constraints to

```
- X[2] == -1;
  X[2] == 1;
```

which is feasible. Then X[2] == 0 is tried, which proves infeasible. Finally X[2] == 1 is added to the current set of constraints with success and the solution is printed.

The search tree for this solution to the 0-1 problem is drawn in Figure 8.5. In contrast with Model 8.2, this time the four leaf nodes under the alternative X[0] == 0 are not visited. It is determined one level up that the constraints are already infeasible. So the strategy is an improvement over complete enumeration and is called *implicit enumeration*. Backtracking and enumeration strategies like this trace back to Lucas (1882). A device for detecting infeasibility in the search tree by anticipating failure that lies ahead is an example of a lookahead. Often a lookahead mechanism requires carefully crafted code. As we see in this model, however, the constrain-and-generate paradigm can automatically provide for a lookahead that prunes the search space.

End of Model 8.4

The next model is a classic puzzle. It has been formulated in myriad AI systems. The pets that are involved not at all realistic but are part of the tradition of the puzzle.

Model 8.5 Who Has the Zebra?

A small street has five differently colored houses on it. Five men of different nationalities live in these five houses. Each man has a different profession, each man likes a different drink, and each has a different pet animal. We have the following information:

1. The Englishman lives in the red house.
2. The Spaniard has a dog.
3. The Japanese is a painter.
4. The Italian drinks tea.
5. The Norwegian lives in the first house on the left.
6. The owner of the green house drinks coffee.
7. The green house is on the right of the white house.
8. The sculptor breeds snails.
9. The diplomat lives in the yellow house.
10. They drink milk in the middle house.
11. The Norwegian lives next door to the blue house.
12. The violinist drinks fruit juice.
13. The fox is in the house next to the doctor's.
14. The horse is in the house next to the diplomat's.

The question is who has the zebra and who drinks water?

Since this is a puzzle, we can reasonably suppose that the questions asked are fair and that each has only one possible answer. So the challenge is to find a consistent way of placing the men with their drinks, pets, and professions in houses of different colors.

Let us number the houses 0, 1, 2, 3, 4 going from the leftmost house on the block to the rightmost. Then the idea of the puzzle is to associate every man with a house, every drink with a house, every pet with a house, every profession with a house and every color with a house so as to fulfill requirements 1-14. In more mathematical jargon, the idea is to map each man to a house and the mapping must be such that no two men can be assigned to the same house. Since the number of men is equal to the number of houses, if we also represent the men by the numbers 0, 1, 2, 3, 4, the mapping becomes a permutation. The same is true of course for the other categories: drink, pet, profession, and color.

The simplest approach to the problem would be to generate mappings for each category and then to test to see if the logical conditions are satisfied. This generate-and-test approach could be costly, since there are 5! or 120 ways of permuting five elements, and this means that as many as 120^5 combinations can be generated. Things can certainly

8.3 Constrain-and-generate

be improved by running each test as soon as enough information is available to make it valid. By moving to a constrain-and-generate approach, we can have the constraints do this kind of thing automatically.

To get started, let us introduce some symbolic constants, the first one being for the dimension of the problem:

```
#define N 5
```

Then for each man, we define a symbolic constant between 0 and 4 corresponding to the man's appearance in the puzzle narrative:

```
#define ENG 0
#define SPA 1
#define JAP 2
#define ITA 3
#define NOR 4
```

Similarly for the professions:

```
#define PAI 0
#define SCU 1
#define DIP 2
#define VIO 3
#define DOC 4
```

And so on, for the house colors, drinks, and pets. Finally, let us have constants for the categories man, profession, drink, pet, and house color:

```
#define MAN 0
#define PRO 1
#define DRI 2
#define PET 3
#define COL 4
```

Now we can declare an array of continuous variables:

```
continuous House[N][N];
```

The following routine will take a one-dimensional array of continuous variables and run through all possible permutations of 0,...,4 trying to fix the continuous variables:

```
permutation(continuous X[])
{
    and(int i=0;i<N;i++)              // Each individual
        or(int h=0;h<N;h++){          // each possible value
            and(int k=0;k<i;k++)
                nint(X[k]) != h;      // if it's unused
            X[i] == h;                // try to fix the
        }                             // variable at that value
}
```

We will be able to send this routine the vectors House[i].

At this point we can simply write down conditions 1-14 as constraints and persistent disjunctions of constraints and then try to generate a solution. Let us encapsulate the information given in the puzzle in a procedure to set things up:

```
puzzle()
{
    bounds();  // It's good practice to state known bounds
               // on continuous variables
    House[MAN][ENG] == House[COL][RED];           // 1
    House[MAN][SPA] == House[PET][DOG];           // 2
    House[MAN][JAP] == House[PRO][PAI];           // 3
    House[MAN][ITA] == House[DRI][TEA];           // 4
    House[MAN][NOR] == 0;                         // 5
    House[COL][GRE] == House[DRI][COF];           // 6
    House[COL][GRE] == House[COL][WHI] + 1;       // 7
    House[PRO][SCU] == House[PET][SNA];           // 8
    House[PRO][DIP] == House[COL][YEL];           // 9
    House[DRI][MIL] == 2;                         // 10

    either House[COL][BLU] == House[MAN][NOR] + 1;
    or House[COL][BLU] == House[MAN][NOR] - 1;    // 11
    House[PRO][VIO] == House[DRI][FRU];           // 12
    either House[PET][FOX] == House[PRO][DOC] + 1;  // 13
    or House[PET][FOX] == House[PRO][DOC] - 1;
    either House[PET][HOR] == House[PRO][DIP] + 1;  // 14
    or House[PET][HOR] == House[PRO][DIP] - 1;
}

bounds()
{
    and(int i=0;i<N;i++)
        and(int j=0;j<N;j++)
            House[i][j] <= N-1;
}
```

8.3 Constrain-and-generate

The main procedure can be

```
21p_main()
{
    puzzle();

    and(int i=0;i<N;i++)
        permutation(House[i]);

    output();
}
```

A very simple way to determine the results of the program is with an output routine such as

```
output()
{
    printf("The number of nodes visited is %d\n",cnt);
    and(int i=0;i<N;i++)
        if House[MAN][i] == House[PET][ZEB];
        then printf("Man Number %d has the zebra\n",i);
    and(int i=0;i<N;i++)
        if House[MAN][i] == House[DRI][WAT];
        then printf("Man Number %d drinks water\n",i);
}
```

The code determines that the Japanese has the zebra and that the Norwegian drinks water. As is expected of puzzles, this one has exactly one solution; we treat the question of proving this in Section 11.1.

There are many situations where it can be useful to add constraints to the constrain phase of the application in order to help prune the search space. The problem in this model is to map each man to a different house, each profession to a different house, and so on, in order to meet the requirements of the puzzle. Since there are five houses and five men, it was not necessary to express the implicit fact that every house was to have someone in it. In other words, by mapping five men to five houses, it had to follow that each house would be occupied. However, the fact that each house had to be occupied by one of the five men could be stipulated in a fuzzy logic way at the outset by means of the constraint

```
    sigma(int j=0;j<N;j++) House[MEN][j] ==
        sigma(int house=0;house<N;house++) house;
```

In fact, since the same remarks apply to the other categories as well, we could add the following five constraints to the setup routine:

```
and(int i=0;i<N;i++)
    sigma(int j=0;j<N;j++) House[i][j] ==
        sigma(int h=0;h<N;h++) h;
```

These constraints are all unnecessary in that they must be satisfied by any solution to the puzzle. Although their presence does not change the set of solutions, they provide fuzzy approximations to the implicit requirements that each house be associated with a man, with a profession, with a pet, and with a drink. The role of these fuzzy approximations is to act as a catalyst on the search for solutions to the puzzle, providing a declarative assist to the procedural search process. Adding them has a dramatic effect on the number of nodes visited and on the solution time, improving both more than fourfold as is easily checked.

End of Model 8.5

What we are doing with the constrain-and-generate paradigm is dividing the knowledge needed to meet the requirements of the application into two parts. The constraints provide the part that can be encapsulated declaratively; the generate part, which is the search process, codes up the knowledge that cannot be reduced to simple constraints. For example, in the last model, the fact that the solution had to provide permutations of the different categories could not be stated in terms of linear constraints on continuous variables but rather was captured by the way progress could be made down the search tree. We also saw that adding a fuzzy approximation to the permutation requirement in the constraints significantly improved the search process. The declarative constraints and the procedural process are complementary to one another. Both encode parts of the knowledge that goes into defining a solution. The search process is the nonlinear part of the application, typically the computationally expensive part. The closer the linear constraints are to modeling the full problem, the easier the search process will be. This is a theme we will return to often.

When working in the purely discrete context, in order to develop a constrain-and-generate application, programmers must build their own system of constraint checking or use a language that supports the appropriate type of constraint. A pioneering constraint language of this sort was ALICE of Laurière (1978). The language approach was developed further in CHIP by Dincbas, van Hentenryck, Simonis and Aggoun (1988). Another approach has been to integrate the discrete constraint solvers into a C++ library; this was done in Puget (1994) for the ILOG Solver. Background on languages which support continuous constraints is given in the next section.

8.3 Constrain-and-generate

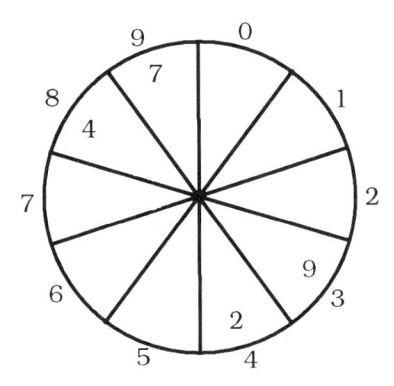

FIGURE 8.6 The Pie Puzzle.

Exercises

8.3.1. This exercise is a brain teaser that has earned students extra credit in schools around the world. Figure 8.6 represents a pie with 10 slices numbered from 0 to 9. The goal is to place a number from 0 to 9 in each slice of the pie so that no two slices get the same digit but that the sum of the digits in two consecutive slices is always equal to the sum of the digits in the two opposite slices. By way of illustration, if slices 3 and 4 are assigned 9 and 2 respectively, then the sum of the digits in slices 8 and 9 must be 11, and conversely. Solve this using both the generate-and-test and constrain-and-generate paradigms. Compute the number of nodes generated in each case.

8.3.1. Find a solution to the following equation where all the variables are required to be digits:

$$\sum_{i=0}^{150} ix_i = 101925$$

9

Classical Disjunction and Combinatorially Hard Problems

In this chapter we treat some topics from theoretical computer science. First we discuss the relationship between classical logic and programming languages. Then we introduce a very important class of constrain-and-generate applications, known as *disjunctive linear programs*. These applications provide an example where the programming picture and the mathematical picture converge nicely. Finally, we discuss some of the concepts that are used to classify combinatorial problems and algorithms.

9.1 Classical disjunction

In classical logic it is possible to establish the disjunction $p \vee q$ without confirming which of the two do hold. Indeed, the cornerstone of classical logic is the Law of the Excluded Middle, which is expressed as the tautology $p \vee \sim p$. This law is critically important in mathematics, although it has been challenged in the mathematical world itself, notably by Brouwer and the intuitionist school. The classical interpretation of disjunction was defended by Hilbert and Russell among others and is the one used in current mathematical practice. Persistent disjunction in 2LP and other logic based languages provides a programming approximation to classical disjunction.

To give some idea of the power of classical disjunction, let us look at a popular mathematical argument. The challenge is to prove that there exist two irrational numbers a, b (not necessarily distinct) such that a^b is rational. A classical proof can be made as follows. Either $\sqrt{2}^{\sqrt{2}}$ is rational or $\sqrt{2}^{\sqrt{2}}$ is not rational. This is an example of Law of the Excluded Middle. We now show that either branch of the disjunction leads to a solution of the problem. Indeed, first suppose that $\sqrt{2}^{\sqrt{2}}$ is rational; since $\sqrt{2}$ is irrational, we can set $a = b = \sqrt{2}$ to solve the problem.

9 Classical Disjunction and Combinatorially Hard Problems

Backtracking to the other alternative, we suppose that $\sqrt{2}^{\sqrt{2}}$ is irrational, but now we can take $a = \sqrt{2}^{\sqrt{2}}$ and $b = \sqrt{2}$. So we have proved the existence of a remarkable pair of irrational numbers a, b without having to determine the true values of a and b! And we did not have to work hard; the Law of the Excluded Middle did it for us. However, even the most uninquisitive classical logician will want to know which of the two alternatives holds. Happily one can invoke a deep theorem of number theory that shows that $a = \sqrt{2}^{\sqrt{2}}$, $b = \sqrt{2}$ is indeed the solution. This problem was considered daunting by Hilbert himself and was solved on the way to the solution of Hilbert's 7th Problem; see Tijdeman (1974). In the exercises, the power of the laws of classical logic is illustrated in another mathematical tale, but one which does not have quite the same kind of happy ending.

The point to note is that classical disjunction is a powerful logical concept. It is pervasive in mathematics, and its use goes far beyond number theoretic points, serving, for example, in proofs that algorithms converge. Support for an approximation to classical disjunction in programming languages is designed to make problem solving with programming more akin to ordinary mathematics, albeit in a modest way.

In 2LP persistent disjunction is supported by means of the `either/or` construct and the `or` loop. Persistent disjunction can serve as an approximation to the working problem solver's periodic need to start all over again. The connection between the way persistent disjunction restores the state of the continuous variables and the loop control variables and the way one solves problems can be understood informally: the mathematician reaches an environment in the argument where a certain number of lemmas have been established, where bounds on variables have been determined, and so on. At that point, there might be several ways to continue the argument. One way is chosen. The ensuing argument tightens bounds and further narrows things until a conclusion is reached or the line of reasoning fails. In case of failure, the line of reasoning jumps back to the environment that was in force when the path was taken discarding all intermediate results. This can be emulated by persistent disjunction: upon backtracking, the feasible region and the loop control variables are restored and the programs starts again where it left off.

Persistent disjunction can also serve as an approximation to classical disjunction. In programming as in ordinary mathematical problem solving, the Law of the Excluded Middle is not usually invoked in stark $p \vee \sim p$ terms. Indeed in 2LP programming, the interpretation of negation as negation-as-failure makes this unworkable. Instead, what is done is simply to break the problem into two or more all inclusive natural alternatives. For example, if X is a continuous variable in a 0-1 application, the dichotomy that X must be fixed at 0 or X must not be fixed at 0 is refined to the dichotomy that X must be fixed at 0 or X must be fixed at 1. In code, this becomes

9.1 Classical disjunction

```
either X == 0;
or X == 1;
```

Suppose our goal is to show that no solution can be reached at this point. Persistent disjunction enables the code to choose each branch in turn and to verify in a robust manner that no solution can be found on either branch. Then the previous environment is restored. Note that we have switched to the *no solution* side of things; the reason for this will developed in Section 9.3.

Moreover, in some traditional programming languages, there is a mechanism akin to the either/or construct. In C, it is the setjmp/longjmp combination, in Scheme it is Call/cc, in LISP it is catch/throw. With these constructs the program can jump back to the program stack environment that existed at the point of creation of the jump. In C, for example, after a longjmp call, the program stack is restored to the state it had when the corresponding setjmp was created. Register variables are restored, but global and local variables are not; of course in C there is no type such as the continuous variable to restore. Moreover longjmp is not robust in that the setjmp point cannot be returned to after the procedure in which it was created is exited. However, there is a profound link between the setjmp/longjmp construct and classical logic. In fact, setjmp/longjmp is a means of capturing an analog of Proof by Contradiction in programming. Proof by Contradiction is expressed by the proposition $\sim\sim p \to p$. This proposition is a variant of the Law of the Excluded Middle. In fact, the formula $\sim\sim p \to p$ is obtained from $\sim p \vee p$ via the equivalence of $\sim a \to b$ with $a \vee b$. Proof by Contradiction provides an alternative to the Law of the Excluded Middle for axiomatizing classical propositional logic.

The intuitive connection between setjmp/longjmp and Proof by Contradiction can be also be understood informally: When a contradiction is reached, the proof jumps back to the environment that was in force when the argument by contradiction was begun. For research on connections between classical logic and procedural programming mechanisms; two places to start are Griffin (1990) and Krivine (1994). For work on the unity of logic and computation, see (Girard 1993).

The key features that provide the programming support for persistent disjunction in 2LP are choice points and the backtracking mechanism. When a choice point is returned to in a program, backtracking restores the feasible region and the loop control variables to their original state. As noted, choice points are robust: When a choice point is created within a procedure, it stays open even after the procedure has been exited. If backtracking takes place, the procedure is reentered and processing begins again with the next untried branch of the choice point.

The 2LP interpreter, which supports both linear constraints and persistent disjunction, is based on two fundamental technologies. The type continuous is maintained by means of an incremental linear programming solver based on the *revised simplex method*. This algorithm is discussed in Chapter 15. The choice point machinery is based on the *Warren Abstract Machine* architecture which is discussed in Warren (1983) and Ait-Kaci (1991). This architecture is commonly called the WAM. The 2LP interpreter employs an appropriately engineered version of the WAM, which is adapted to support the types int, double and continuous; it is a virtual machine that is emulated in C. A 2LP program is compiled into code for this virtual machine as in McAloon and Tretkoff (1990). The theoretical analysis that underlies the 2LP implementation can be found in Cox, McAloon, and Tretkoff (1992). Languages that support logic programming constructs and constraints are called *constraint logic programming languages* or *CLP languages*, (Jaffar and Lassez 1987). The first to support constraints on continuous variables were Prolog III, Colmerauer (1987), and CLP(R), Jaffar, Michaylov, Stuckey and Yap (1992). For work on CLP languages which support constraints over discrete domains, see van Hentenryck (1989). For a full survey, see Jaffar and Maher (1994).

Let us distinguish between ways in which backtracking can occur. Consider code such as

```
p()
{
    either <statement1>;
    or <statement2>;
    q();
}

q()
{
    <statement3>;
}
```

When the procedure p() is entered, the call to <statement1> can fail and the persistent disjunction immediately will switch over to call <statement2>. This form of backtracking is called *sidetracking*. The other possibility is that <statement1> succeeds, but <statement3> fails when called from the procedure q(). Control then comes back to the either/or statement in procedure p() and <statement2> is called. This form of coming back anew to a persistent disjunction is called *deep backtracking*. Therefore conditional disjunction supports sidetracking only while persistent disjunction supports both sidetracking and deep backtracking. In Section 10.3 we will see that a nonrobust form of persistent disjunction can be simulated by means of conditional disjunction and recursion.

It is quite easy to construct situations where the either/or construct fails to satisfy commutativity or idempotence. For the former we have the same example as with conditional disjunction:

9.1 Classical disjunction

```
    a = 0;
    either {
        a = 1;
        a == 0;
    }
    or a == 0;
```

With similar ingenuity, one can construct a statement to illustrate that idempotence can also fail for the either/or construct.

When a program backtracks to a choice point, the feasible region is restored to the same polyhedral set as when the choice point was first encountered. However, while the constraints define the same polyhedral set as before, the witness point is not necessarily the same vertex as before. This is due to the fact that the witness point is maintained by the underlying procedural apparatus of the programming language interpreter. Something similar can happen with conjunctions of constraints: as we have seen in Exercise 1.2, changing the order of two constraints does not change the feasible region generated but can lead to different witness points. It is also important to note that identifiers of types double and int, with the exception of loop control variables, are not reset to their previous values upon backtracking. The following code illustrates these points:

```
#define fail 0==1   // A macro definition

2lp_main()
{
continuous X,Y;
int a;

    Y <= 9.9;
    Y + X == 9.9;

    printf("wp(Y) is %f\n",wp(Y));    // wp(Y) is 0.000000
    printf("wp(X) is %f\n",wp(X));    // wp(X) is 9.900000

    either {
        Y == 9.9;
        fail;              // Force sidetracking
    }
    or ;                   // NULL statement

    printf("wp(Y) is %f\n",wp(Y));    // wp(Y) is 9.900000
    printf("wp(X) is %f\n",wp(X));    // wp(X) is 0.000000

    a = 0;
    printf("a is %d\n",a);     // a is 0
```

```
    either {
        a = 1;
        fail;        // Force sidetracking
    }
    or ;             // NULL statement
    printf("a is %d\n",a);    // a is still 1
}
```

A key fact about the 2LP programming constructs implication, conditional disjunction and negation-as-failure is that they do not leave choice points behind. Thus the three code segments

```
// Segment 1
    if
        either printf("A\n");
        or printf("B\n");
    then fail;  // Macro for 0==1 as above

// Segment 2
    c_either printf("A\n");
    or printf ("B\n");
    fail;

// Segment 3
    not
        not
            either printf("A\n");
            or printf("B\n");
    fail;
```

will all print the message A but never B. In effect, for Segment 1, before executing the then statement, the if discards the second alternative once the first has succeeded. Similar remarks apply to conditional disjunction, negation-as-failure, and the other two segments. So the critical difference is that conjunction and persistent disjunction keep alternatives alive, while the other constructs discard all but the first sequence of alternatives to succeed. In other words, these other constructs succeed or fail and then exit without leaving any new choice points that can be re-entered. On a positive note, this functionality can be used if one wants to seal off a section of code from re-entry.

The procedural way in which if/then is used in 2LP and other programming languages means that the programmer must guard against ascribing declarative meaning to this construct. For example, in Model 8.1, The Pacific Rim Revisited, we had to choose between the Fuji and Hanla lines of workstation. The choice was formulated as

```
    either one_but_not_the_other(FU,HA);
    or one_but_not_the_other(HA,FU);
```

9.1 Classical disjunction

If we formulate this requirement in the terms "if there are Fuji models, then there are no Hanla models, and *vice versa*," we would be inclined to code this as

```
if Workstations[FU] >= 750;
then Workstations[HA] == 0.0;
else {
    Workstations[HA] >= 750;
    Workstations[FU] == 0.0;
}
```

But this would lead to an infeasible situation, since the `if` part of the conditional will succeed and commit the program to the constraints

```
Workstations[FU] >= 750;
Workstations[HA] == 0.0;
```

As discussed in Model 8.1, this choice is incompatible with the disjunctive requirements on the higher-priced workstations.

The message here is simply that one must be aware that expressions such as `if/then` have one meaning in natural language, another meaning in mathematics, and yet another meaning in programming. Keeping them apart is not particularly hard, but one must be aware of the distinctions. It can be said, however, that the programming operators conjunction and persistent disjunction, when applied to constraints, stay closer to their meaning in ordinary mathematics than do the other logic-based constructs.

Mathematics and computer science are not the only fields where the role of the logical connectives must be analyzed carefully. In quantum mechanics, for example, a theoretician cannot categorically assert that either the cat in the box is alive or the cat is not alive, until the box is opened and an observer peers in. In fact, the analysis of the logic of quantum mechanics has a rich history, starting with Birkhoff and von Neumann (1936); see Hughes (1989).

Exercises

9.1.1. Prove that the square root of p is irrational for p prime. Does your proof rely on the Law of the Excluded Middle or on Proof by Contradiction?

9.1.2. For an alternative proof of the existence of irrationals a, b such that a^b is rational, try the following line of argument: First prove that e is transcendental; then show that $\ln(2)$ is irrational and set $a = e$, $b = \ln(2)$. (This line of reasoning was pointed out to us by the author of Ginsberg (1993), who credits an impatient undergraduate with the idea.) This proof is not as elementary as the one in the text in that it requires a proof or belief that e is not algebraic.

9.1.3. As is well known, both e and π are transcendental numbers. Prove that either the sum $e + \pi$ is transcendental or the product $e * \pi$ is transcendental. Hint: Denote the first proposition by *sum* and the second by *product*. By the Law of the Excluded Middle, we have

$$(sum \vee product) \vee \sim (sum \vee product)$$

Then show the second alternative is impossible. This time we cannot say whether it is the sum or it is the product that is transcendental. Deciding which of the two is transcendental or if both are is an open problem, see Siegel (1949). (This example was pointed out to us by Harvey Friedman.)

9.1.4. Show by example that the either/or construct is not always idempotent.

9.2 Disjunctive programs

In Model 8.4, The 0-1 Solution II, an initial feasible region in three dimensional space is defined by the setup constraints

```
// Constrain
and(int i=0;i<3;i++)
    X[i] <= 1;
2*X[0] + 2*X[1] - X[2] == 1;
X[0] + 4*X[1] + X[2] == 2;
```

Let us call this feasible region F.
The central part of the code was

```
// Generate
and(int i=0;i<3;i++)
    either X[i] == 0;
    or X[i] == 1;
```

For each i, the constraint X[i] == 0 and the constraint X[i] == 1 both define polyhedral sets, hyperplanes that fix the value of the X[i] coordinate of the witness point. Let us call these sets H_{i0} and H_{i1}. Geometrically the disjunction

```
either X[i] == 0;
or X[i] == 1;
```

9.2 Disjunctive programs

defines the union of polyhedral sets $H_{i0} \cup H_{i1}$. So a solution to the problem is a point in the set

$$F \cap \bigcap_{i<3} \bigcup_{j<2} H_{ij}$$

This can also be written as

$$\bigcap_{i<3} \bigcup_{j<2} (F \cap H_{ij})$$

What we have here is an intersection of unions of feasible regions. An important class of combinatorial problems can be expressed this way, and so we introduce some terminology.

A union of polyhedral sets is called a *disjunctive set*. The problem of determining whether the intersection of a family of disjunctive sets is nonempty is called the *disjunctive linear programming problem* or *disjunctive programming problem*. In mathematical notation the problem is to determine if a set of the following form is nonempty:

$$\bigcap_{i<M} \bigcup_{j<N_i} F_{ij}$$

This set is called the *solution set* of the disjunctive programming problem. In many situations all the N_i will be equal, and so the problem can be denoted

$$\bigcap_{i<M} \bigcup_{j<N} F_{ij}$$

In fact, by taking N large enough and setting unneeded F_{ij} to be the empty set, all disjunctive programming problems can be written in this form. Oftentimes this standardization will serve to simplify the discussion.

Typically in an application that gives rise to a disjunctive programming problem, each $i < M$ will index a logical requirement L_i and each union of polyhedra $\bigcup F_{ij}$ will arise as a set of alternative ways of meeting that requirement. The question in a disjunctive programming problem is whether there are solutions that satisfy all these requirements L_i simultaneously. When an objective function is present, the question becomes that of both determining whether a solution exists and that if so, finding the best solution one can.

Disjunctive constraint is a term often used to denote a condition that is expressed as a disjunctive set; to avoid confusion with the usual sense of constraint, we will use the expression *disjunctive requirement* or *logical requirement*.

As an example of a disjunctive programming problem that we can represent graphically, consider the two-dimensional situation with axes labeled x_0 and x_1 and with the sets F_{ij} given by

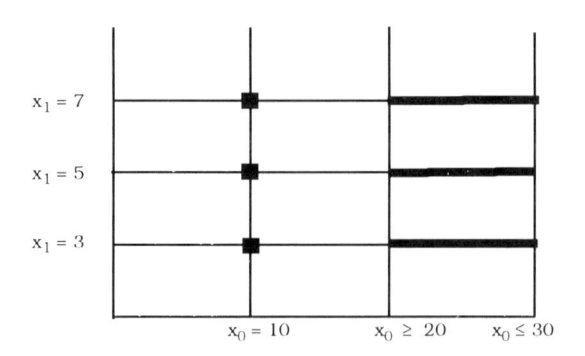

9 Classical Disjunction and Combinatorially Hard Problems

FIGURE 9.1 A disjunctive program

F_{00}: $x_0 = 10$
F_{01}: $x_0 \geq 20$, $x_0 \leq 30$
F_{10}: $x_1 = 3$
F_{11}: $x_1 = 5$
F_{12}: $x_1 = 7$

Then the solution set is the set of heavily shaded points and line segments of Figure 9.1

In practice, the feasible regions in a disjunctive programming problem will usually be subsets of a given initial region. So the disjunctive programming problem will take the form

$$F \cap \bigcap_{i<M} \bigcup_{j<N} F_{ij}$$

which of course is equivalent to

$$\bigcap_{i<M} \bigcup_{j<N} (F \cap F_{ij})$$

The effect of restricting to an initial feasible region is illustrated in Figure 9.2.

The code to solve a disjunctive programming problem is called a *disjunctive program*. By abuse of language, the disjunctive programming problem itself is often simply called a disjunctive program. The problem of finding a point in the intersection of disjunctive sets is coded by

```
and(int i=0;i<M;i++)         // ∩
    or(int j=0;j<N;j++)      // ∪
        cc(i,j);             // F_ij
```

where `cc(i,j)` generates the conjunction of constraints that defines the feasible region F_{ij}. Thus, in practice, disjunctive programs take the form

9.2 Disjunctive programs

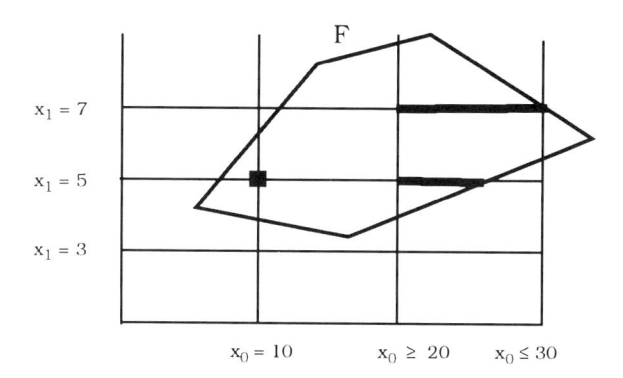

FIGURE 9.2 A disjunctive program with initial feasible region

of a constrain-and-generate application, with a setup routine restricting the program to an initial feasible region:

```
setup();      // Constraints for initial feasible region
and(int i=0;i<M;i++)      // ∩
    or(int j=0;j<N;j++)   // ∪
        cc(i,j);          // F_ij
```

With a disjunctive programming problem, the computational task is to determine whether a solution exists and, if so, to produce a feasible region that provides a solution. More precisely, finding a solution means determining a sequence $j_0,\ldots,j_{N-1}$ such that the polyhedron

$$F \cap F_{0j_0} \cap \ldots \cap F_{N-1\,j_{N-1}}$$

is nonempty. A vertex on this polyhedron will then provide a numerical solution to the problem. In programming terms this means that the sequence of calls $cc(0,j_0),\ldots,cc(N-1,j_{N-1})$ will succeed and the witness point will furnish a concrete solution. Let us note that, although there are only M*N feasible regions F_{ij} in the formulation of the disjunctive program, the number of sequences $j_0,\ldots,j_{N-1}$ that are candidates to generate a solution is N^M. So, once again, we encounter the specter of combinatorial explosion.

Disjunctive programs have a nice mathematical property that will turn out to be very important from a computational point of view. Since set intersection is a commutative operation, permuting the indices i < M does not change the solution set. Similarly set union is commutative, and for given i, nothing is changed if the indices j < N are permuted. Thus, for example, the above program can also be written

```
setup();        // Constraints for initial feasible region
and(int i=M-1;i>=0;i--)        // Permute the i
   or(int j=N-1;j>=0;j--)      // Permute the j
      cc(i,j);
```

Mathematically, permuting indices does not change the underlying set of solutions. In terms of programming, permuting the indices can mean that a different solution will be found if there is one. It can also mean a difference in the time required to find a solution or in the time to determine that there is no solution. In the sequel we will exploit this flexibility in many applications.

Among the examples we have seen so far, Model 8.1, The Pacific Rim Revisited, and Model 8.4, The 0-1 Solution II, are disjunctive programs. However, not all constrain-and-generate applications are disjunctive programs. For example, Model 8.5, Who Has the Zebra, is not; the permutation conditions that each of the categories, man, profession, and so on, be mapped in a one-to-one way onto the five houses is coded by means of the permutation routine. The body of the or loop in this routine does not consist of a conjunction of constraints.

In this section we will develop models that illustrate three perennial sources of disjunctive requirements. The first kind is known as a *threshold requirement*: If a product A is to be produced, then at least a units must be produced. If the continuous variable A represents the amount of A to be produced and a represents the threshold number, we can express this requirement in code as

```
either A >= a;     // Produce A
or A == 0;         // Do not produce A
```

The other two examples of recurring disjunctive conditions that we will see a bit further on are *integrality requirements* and *temporal requirements*.

We have seen several resource allocation problems, such as Model 3.1, The Pacific Rim, and Model 3.3, Motorcycles Inc. These are straightforward linear programming models. However, adding disjunctive conditions can mean that the model requires search and backtracking to solve. This was the case in the passage from Model 3.1 to Model 8.1. In a similar vein, let us return to Model 3.3, pick up the narrative, and deal with some threshold requirements that have sprung up.

Model 9.1 Motorcycles Inc. Reconsidered

Until now, production planning has based on the considerations summarized by Tables 3.5 and 3.6. However, because of the difficulty involved in setting up production and problems with quality control, labor has asked management to agree that if a model is to be produced at all, at least a substantial minimum number of units of it must be

9.2 Disjunctive programs

made during the week's production cycle. Local management consulted with corporate management and were told that the key thing was to maintain a level of production of at least 900 motorcycles per period. So local management has come back to the workers with the demand that at least at 900 motorcycles be made. The first question is whether all these requirements can be met. Moreover, while reaching production of at least 900 units is the critical goal, local management is also interested in a production plan that provides corporate management with a good revenue prediction from the factory's production. So we want not only to meet the various requirements but also to do this in a way that leads to good revenue figures.

The relevant new data are listed in Table 9.1

TABLE 9.1 Additional Bike Model Data

Bike	0	1	2	3	4	5	6	7	8	9	10	11
Lower Limit	120	100	200	50	50	100	100	125	150	150	250	330

Notation can be set up as in Model 3.3, with the addition of an array `lo[MODS]` of type `double`. The data can again be entered in the appropriate arrays by a call to a routine `data` and the constraints set up by a call to the routine `setup`. The most salient difference with Model 3.3 is the threshold requirement that the level of production of model `i` either be 0 or at least `lo[i]`. Now the task of determining a good production planning strategy has been made much more complex.

Let us address management's requirement, adding the following demand constraint to the resource constraints:

```
management_constraint()
{
        sigma(int i=0;i<MODS;i++) Bike[i] >= 900;
}
```

Now we turn to labor's requirement. Two naive attempts to adjust to labor's demand present themselves at once: one could be very timid and try to set

```
        and(int i=0;i<MODS;i++)
           Bike[i] == 0;
```

or very greedy and set

```
        and(int i=0;i<MODS;i++)
           lo[i] <= Bike[i];
```

The first attempt is doomed because of management's demand. The second is probably infeasible and, even were it feasible, is unlikely to produce a particularly profitable plan because it can lead to overproduction of low-profit items. Let us explore better ways of deciding the fate of each `Bike[i]`.

For each `i`, the two choices correspond to two different constraints: `Bike[i] == 0.0` which defines a hyperplane F_{i0} and `Bike[i] >= lo[i]` which defines a halfspace F_{i1}. Each alternative is described by a single constraint. To code these alternatives we have the code

```
cc(int i,j)
{
    if j == 0;
    then Bike[i] == 0.0;        // F_i0
    else Bike[i] >= lo[i];      // F_i1
}
```

and the central and/or loop can be written as

```
    and(int i=0;i<MODS;i++)
       or(int j=0;j<2;j++)
           cc(i,j);
```

Since the alternatives are simple and few in number, we can also formulate the alternatives as follows:

```
decide(continuous Bike, double low)    // Version I
{
    either Bike == 0;
    or Bike >= low;
}
```

and write the central loop as

```
    and(int i=0;i<MODS;i++)
        decide(Bike[i],lo[i]);
```

Commuting things, we could also write the `decide` routine as

```
decide(continuous Bike, double low)    // Version II
{
    either Bike >= low;
    or Bike == 0;
}
```

For this example, the first version of `decide` favors fixing `Bike[i]` at 0, which amounts to a timid strategy. The second version is a greedy strategy. We could also write the central loop as

9.2 Disjunctive programs

```
    and(int i=MODS-1;i>=0;i--)
        decide(Bike[i],lo[i]);
```

which favors the bikes with later indices.

In any case these approaches are guaranteed to find a feasible solution if one exists and lead to a main routine such as the following:

```
21p_main()
{
    data();
    setup();
    management_constraint();

    and(int i=0;i<MODS;i++)         // From 0 to MODS-1
        decide(Bike[i],lo[i]);      // Version II

    max: Z;     // There is still a linear program left

    printf("The projected revenue is %0.f\n",Z);

    detailed_output();    // Left to reader
}
```

This program determines that a feasible plan does exist. Note that we call

```
        max: Z;
```

after the loop decides which products are to be made. At this point we still have a linear programming problem to solve. The projected revenue with this program is 99450.

There is, however, a slight lacuna in our solution of this problem. Although it does not happen with these data, it is possible for the sum of floor(Bike[i]) to be less than the threshold of 900 set by management! If this should happen, the shortfall would be at most 6, the number of models divided by 2, and presumably management could live with this. However, in a situation where this kind of point is important, other methods such as those developed in the next and later models have to be applied.

In this example, we could randomize the order in which the different motorcycles are considered in the search for a solution. We could also randomize the treatment of the threshold requirements. This will lead to a richer sampling of solutions.

<u>End of Model 9.1</u>

A second form of disjunctive requirement that comes up again and again is the situation where one or more of the continuous variables in a model must take an integer value in the solution. These are examples of *integrality requirements*. We have already seen numerous examples, including Model 2.3, Call 911, and Model 8.4, The 0-1 Solution II. In the first case, we were able to exploit the special structure of the model and a powerful heuristic to obtain a very good solution. In the second case, we resorted to enumeration and backtracking to force each continuous variable to an appropriate integer witness point value. The next model will also address the issue of integer valued solutions by means of a disjunctive programming formulation. Here the integrality requirements will still be simple 0-1 conditions; in the following chapters, we will deal with more complex integrality requirements. The model will also help explain why the disjunctive program and the and/or loop are so pervasive in combinatorial problems. First, some definitions are in order.

In mathematical logic both a propositional variable and its negation are called *literals*. A disjunction of literals is called a *clause*. By way of example, $p \vee \sim q \vee r$ is a clause, while $p \wedge q$ is not and $\sim (p \vee q)$ is not. A formula that is a conjunction of clauses is said to be in *Conjunctive Normal Form* or CNF. The formula $(q \vee \sim r) \wedge (p \vee \sim q \vee r)$ is in CNF, while the formulas $(q \vee (\sim r \wedge p)) \wedge (p \vee \sim q \vee r)$ and $(p \vee q) \rightarrow r$ are not. However, every formula is equivalent to one in CNF. Thus $p \wedge q$ is already in CNF, $\sim (p \vee q)$ is equivalent to $\sim p \wedge \sim q$, and $p \vee (q \wedge r)$ is equivalent to $(p \vee q) \wedge (p \vee r)$. An assignment of truthvalues to the propositional variables of a formula that makes the formula true is called a *satisfying assignment*. In terms of truthtables, a satisfying assignment is a row that makes the formula true.

CNF plays a central role in computational logic. The thing to note is that CNF formulas are conjunctions of disjunctions, while disjunctive programs have the form and/or.

There is a duality in CNF between the propositional variables and the clauses. The propositional variables are elements, and the clauses are sets of signed elements. We will see this duality arise in the next model.

Model 9.2 Silicon Logic

Engineers working on the logic of a new computer chip have a gate that computes a logical function of a set of input bits and outputs a single bit. The gate is in fact the result of taking the AND of a set of component gates, each of which is a disjunction of inputs and inversions of inputs, as in Figure 9.3. An input can be identified with a propositional variable, and an inversion can be identified with a negation. So the component gates are all disjunctions of literals, and the entire gate is a conjunction of clauses; that is, a formula in CNF.

Thus far all the test data that have been used on this gate output a 0, so the engineers suspect that the gate reduces to a constant value of

9.2 Disjunctive programs

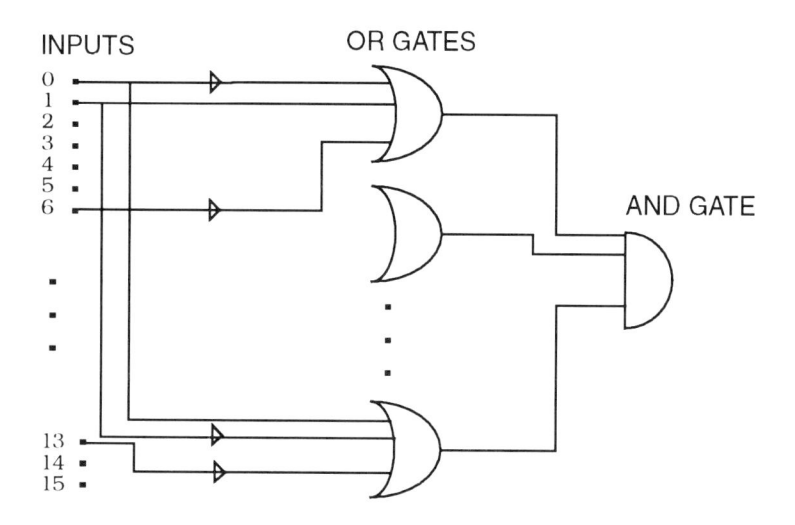

FIGURE 9.3 The layout of the OR gates and AND gate

0. The gate and its components are described in sparse fashion in Table 9.2. The task is to determine if there is a set of inputs so that the gate does output a 1. Such a set of inputs will be a satisfying assignment for the CNF formula. In this table an entry 0 in the Parities column means that the corresponding propositional variable is negated in the clause.

Let us introduce symbolic constants C for the number of clauses, N for the number of inputs or variables, and TOT for the total number of literals that occur in the clauses. Thus C is the number of rows in the table and TOT is the total number of entries in the Variables column. Let us introduce arrays of type int to hold the data. An array vars[TOT] can hold the entries in the second column: An array parity[TOT] can hold the entries in the Parities column. To record the Begin column, let us use an array begin[C+1], where begin[C] will be set equal to TOT. This use of an extra slot will obviate the need for a special case for the last clause.

We introduce continuous variables X[N] to represent the propositional variables. These variables will be fixed at 0 or 1 by a search process in order to find a satisfying assignment. Making all variables global for simplicity, we can begin the program with the definitions and declarations:

```
#define N 16
#define C 15
#define TOT 50

int vars[TOT],parity[TOT],begin[C+1];
continuous X[N];
```

9 Classical Disjunction and Combinatorially Hard Problems

TABLE 9.2 Data for Silicon Logic

Clause	Variables	Parities	Begin
0	1,6,0	1,0,0	0
1	0,15,3,2	0,1,0,1	3
2	4,15,14,2,7	0,0,1,1,1	7
3	7,6,12,4	0,1,0,1	12
4	7,3	1,1	16
5	13,6,7,8	1,0,1,1	18
6	1,9,5	0,1,0	22
7	2,6	1,0	25
8	11,0,3	0,1,1	27
9	7,9,2	0,0,1	30
10	11,4,13,14	0,1,0,0	33
11	0,2,3,7	1,1,1,1	37
12	11,12,13	1,1,1	41
13	11,14,15	0,0,0	44
14	1,0,13	0,1,0	47

The input data can be assembled in a routine:

```
data()
{
    vars = {
            1, 6, 0,
                ...
            1, 0, 13
    };

    parity = {
            1, 0, 0
                ...
            0, 1, 0
    };

    begin =  { 0,...,47,TOT };
}
```

We can code whether there is a satisfying assignment by simply following the specification of the problem. This leads us to loop over clauses in an attempt to satisfy each clause by making one of its literals take the value 1:

9.2 Disjunctive programs

```
21p_main()
{
    data();
    and(int i=0;i<C;i++)                    // For each or gate
        or(int k=begin[i];k<begin[i+1];k++)// Find an input
            X[vars[k]] == parity[k]; // To make gate true
}
```

The program succeeds and shows that the gate is not a constant function after all.

Two facts about continuous variables underlie this code. First, a continuous variable cannot be fixed at both 0 and 1. Thus, if X[vars[k]] is bound to parity[vars[k]], any later attempt to fix the same variable must bind it to the same parity for it to be successful. Second, upon backtracking to clause i, only the bindings that occurred before clause i was reached are still in force, all others having been undone. Since the only constraints here are the ones that fix the continuous variables to 0 or 1, the continuous variables are really only used for bookkeeping purposes to prevent a variable from being made both true and false. In effect, the programs codes up a generate-and-test algorithm where the tests are made clause by clause.

Though the above code is wonderfully compact, its performance suffers from the fact that it is entirely sequential in its treatment of the clauses. Thus, for example, if the first clause is simply a propositional variable and the final clause is simply the negation of that same variable, this blatant contradiction will not be detected early. Constraints can help remedy this. Logical clauses can be mapped into linear constraints as is illustrated by the following example.

Consider the clause $x_0 \vee \bar{x}_1 \vee x_2$, where the bar represents negation. Transforming this into a constraint, we have

```
1 <= X[0] + (1-X[1]) + X[2] ;
```

This is an instance of a perfectly general way of mapping clauses to constraints: the positive literals x_i are sent to continuous variables X[i] and the negatives literals $\bar{x}_i$ are sent to (1-X[i]). Suppose this transformation is applied to the clauses in a CNF formula. Then to establish the satisfiability of the original formula, a solution to the constraints must be found that makes each continuous variable 0 or 1. The simple step of representing clauses this way is in fact quite powerful and subsumes an important inference method known as *unit resolution*. In other words, the linear relaxation of the disjunctive program automatically provides an important pruning tool. This point will be developed in Section 10.4.

Setting up these constraints is easy enough, and this can be done with a routine such as

```
setup()
{
    and(int j=0;j<N;j++)
        X[j] <= 1;
    and(int i=0;i<C;i++)
        1 <= sigma(int k=begin[i];k<begin[i+1];k++)
            (
              parity[k]*X[vars[k]] +
                  (1 - parity[k])*(1-X[vars[k]])
            );
}
```

Moreover, when the constraint form of the clauses is given for a constrain-and-generate approach, the and/or loop can be simplified to one that simply tries to bind 0 or 1 to each of the continuous variables in turn. The main procedure can now take the form:

```
2lp_main()
{
    data();
    setup();

    and(int j=0;j<N;j++)       // Loop through variables
        or(int b=0;b<2;b++)    // b for bit
            X[j] == b;         // Truth value of X[j] is b
}
```

Though this code is a marked improvement over the first version, its performance still suffers from the fact that it is rigid in its treatment of the propositional variables. That is, there are two degrees of freedom that are not used: the choice of the next variable to branch on and the direction 0 or 1 in which to branch first. In fact, these two degrees of freedom are exploitable in disjunctive programs because of the commutativity of both union and intersection. We will see several ways of exploiting this flexibility, especially in Chapter 12. We will return to the satisfiability problem for CNF formulas on several occasions. The problem is pervasive and new methods for attacking it are continually being developed; see Barth (1996).

Let us note how the duality between propositional variables and clauses appears in this model. In the first piece of code, the branching strategy is based on clauses: The and loop cycles through the clauses trying to guarantee each clause is satisfied. In the second algorithm, the clauses are encapsulated as constraints in the setup routine. From there the branching is based on the propositional variables: The and loop cycles over the variables giving each an interpretation as false or true. Hence, in this model, duality manifests itself at the level of the search strategy.

9.2 Disjunctive programs

One more point: In mapping literals into constraints, note that the constraint X == 1 corresponds to making the propositional variable true and that the constraint X == 0 corresponds to making its negation true. In this correspondence there is no attempt to introduce != or other strict relational operator.

End of Model 9.2

Data sets for CNF problems can be obtained from DIMACS by anonymous ftp to dimacs.rutgers.edu in the directory pub/challenge/sat/benchmarks/cnf. These sets and othere can also be found via http://dimacs.rutgers.edu/Challenges/index.html.

In a program an intersection of disjunctive sets takes the form of a conjunction of persistent disjunctions of conjunctions of constraints. Because of this, a disjunctive program is said to be in CNF.

Logical formulas can also be put in a form dual to CNF. A formula can be written as a disjunction of conjunctions of literals; this is known as *Disjunctive Normal Form* or DNF. In the case where the disjunctive program consists of a single disjunctive set, the disjunctive program is also said to be in DNF. By applying the distributive laws to intersection and union, any disjunctive program in CNF can be transformed into one in DNF. However, the passage from the original CNF form to the DNF form might prove computationally daunting for the number of terms involved can explode exponentially. The CNF form of a disjunctive program provides a compact description of a formidable search tree; the corresponding DNF form is a description of all the paths leading to the leaf nodes of this tree.

Let us look at a form of disjunctive condition which arises frequently in nonpreemptive scheduling problems. Given two jobs or tasks A and B that are competing for the same resource, it must be decided whether A goes first or B goes first. This is an example of a *temporal requirement* or *temporal constraint*. Suppose that the finishing time of A is represented by the continuous variable A and that of B by B and that the time to perform job A is a and that to perform job B is b. We can model the fact that the two jobs cannot be running at the same time by means of a persistent disjunction

```
either A + b <= B;
or B + a <= A;
```

Let us apply this to the job shop problem of Model 6.3, The Body Shop.

Model 9.3 The Body Shop Recast

In Model 6.3 the randomized hillclimbing code found a solution with a finishing time of 59. The question has come up of finding a schedule that has a finishing time of 58. The straightforward code for this is

```
21p_main()
{
    data();
    setup();
    better_schedule();
}
```

The code for `better_schedule` will stipulate that `Makespan <= 58` and use persistent disjunction to search for a solution that meets this challenge.

```
better_schedule()
{
double target;

    target = 58;
    MakeSpan <= target;

    sequence_jobs();
    min: MakeSpan;

    printf("Makespan %f can be achieved\n",MakeSpan);
}
```

For each pair of tasks that run on the same machine, it must be decided which of the two is run first:

```
sequence_jobs()
{
    and(int m=0;m<M;m++)              // On each machine i
        and(int jb=0;jb<JOBS;jb++) // for each pair of jobs
            and(int k=0;k<jb;k++)  // run only one at a time
                either
                    FinTime[m][jb] + task_time[m][k]
                    <= FinTime[m][k];
                or
                    FinTime[m][k] + task_time[m][jb]
                    <= FinTime[m][jb];
}
```

This code reports back that the problem is solvable. Moreover this code suggests a simple way of determining the optimal finishing time: Keep solving the problem for smaller and smaller values of `best` until an infeasible problem is reached. We can employ unit successive decrements, since the data are all integer and the objective function `MakeSpan` will take an integer value.

9.2 Disjunctive programs

```
best_schedule()
{
double target;

    target = 58;    // Known bound
    and(;;) {       // Indefinite loop
        if not {    // Try to meet challenge
            MakeSpan <= target-1;   // Challenge bound
            sequence_jobs();
            min: MakeSpan;          // Do the best you can
            target = wp(MakeSpan);  // Record new bound
        }
        then break; // Bound too strong, exit loop
    }
    printf("The best possible makespan is %f\n",target);
}
```

This code will in fact report back that the optimal makespan is 55 but will not be in a particular hurry about it! As noted in Chapter 6, job shop scheduling problems are notorious.

End of Model 9.3

In this last model we were able to find a way to determine the optimal solution to a disjunctive programming problem. Because of the fact that the optimal solution must be integer valued in this model, the program could be built around a loop with unit decrements. The same technique would not work for the problem of Model 9.1, Bicycles Inc. Reconsidered, because the objective function will not have an integer value at a solution to the disjunctive programming problem. However, the method could be used to find a solution that is guaranteed to be close to the optimum. In general, finding the optimal solution to a disjunctive programming problem is a challenging enterprise. We will return to this topic in Chapter 11.

Early work on disjunctive programming was done by Greenberg (1968). Later research includes the work of Balas (1985) and Beaumont (1990).

Exercises

9.2.1. Formulate the nonpreemptive version of Model 4.5, Batch Scheduling as a disjunctive linear program. If there is a feasible schedule, find one that offers the shortest makespan. Experiment with new data sets.

9.2.2. Develop your own discrete code for SAT.

9.2.3. Prove every propositional formula is equivalent to one in CNF and to one in DNF.

9.3 P, NP, and co-NP

Spurred by Gödel's work on the Incompleteness Theorems and the Church-Turing analysis of computability, mathematicians built analytical tools for classifying the difficulty of solving problems. These concepts form the basis of the branch of mathematical logic known as Recursion Theory. With the advent of computers, the mathematical models of computation have been sharpened to account for notions such as time and space. These modern developments take place at the intersection of logic, operations research, and theoretical computer science. Here we give only a brief overview of some of the concepts and classifications that have been explored. For further reading, we refer the reader to Garey and Johnson (1979), which is the classic text in the field, and to Papadimitriou (1994).

In this discussion we identify integers with their binary representation as a string of 0s and 1s. In particular, a function from integers to integers can be considered a function from strings to strings. A set of strings is called a formal language or, more simply, a *language*. The length of a string x is denoted $|x|$. Note that the number of strings of length n on an alphabet of m letters is m^n.

A problem can be specified in terms of the language of all strings that represent instances of the problem according to some coding convention. Hence *solving a problem* in a formal sense means providing a function that outputs the solution when given an instance of the problem as an argument.

The definitions that follow depend on an underlying model of the computing machine and the way time and space are measured. The standard model is the multi-tape Turing machine. The good news is that the theory is robust in that the analysis it leads to is the same for all the usual models of computation. The bad news is that for a proper understanding of the subject matter, one has to work through the details of at least one of these models of computation. In some of these models, an algorithm is hard-wired into the circuitry of a computational device, as in the Turing machine case, while in others, it is a program for a fixed universal device. The Turing machine model and its relation to other models are discussed in Aho, Hopcroft, and Ullmann (1974).

An algorithm is said to be a *polynomial time algorithm* if there is a monomial Cn^k such that the run time of the algorithm on input strings of length no greater than n is bounded by Cn^k. If a function can be computed by a polynomial time algorithm, the function itself is said to be a *polynomial time function*.

It is easy to construct examples of functions that cannot be computed in polynomial time. For example, the function to produce the binary representation of 2^n given the binary representation of n as input is not a polynomial time function because the output requires $n+1$ bits; writing these bits requires an exponential number of operations relative to the input size, which is at most $\log_2(n) + 1$.

9.3 P, NP, and co-NP

A *solution* of a mathematical problem is defined to be a function that outputs an appropriate message if the instance of the problem cannot be solved and that produces a string to solve the problem otherwise. A problem that can be solved by a polynomial time function problem is said to be *solvable in polynomial time*.

As an example of a problem solvable in polynomial time, we have that of determining whether a system of linear equations with rational coefficients has a solution in the rational numbers and if so producing a solution. The classical Gaussian elimination procedure provides an engine to do this.

Now let us consider a more specialized kind of problem. A *decision problem* is one that requires a simple YES or NO answer. Solving it is done by means of a function that outputs only two values 0 and 1. A string which is mapped to 1 is said to be *accepted*. An accepted string is an instance of the problem that can be solved successfully. A decision problem can be identified with its set of accepted strings. The class of all decision problems that can be solved in polynomial time is denoted P.

Determining whether a system of linear equations with rational coefficients has a solution in the rational numbers is in P, as we have just noted. The problem of determining whether a set of linear constraints with rational coefficients has a solution in the rational numbers is called the *linear programming problem* or *LP*. The classical techniques of Fourier-Motzkin elimination and of the simplex method do not provide a polynomial time algorithm for this problem. Using the ellipsoid method, Khachian (1979) showed that this problem is in P. In fact, since Khachian's work, there have been many further exciting developments in this field such as the work of Karmarkar (1984) on interior point methods; references for this and other developments are Fang and Puthenpura (1993) and Saigal (1995). The related problem of producing a numerical solution for a solvable instance of LP is also in polynomial time. The passage from the 0-1 function to solve the decision problem to the rational valued function to produce a solution to the constraints can be done by an argument using binary search as in Papadimitriou and Steiglitz (1982). The back-and-forth between the decision problem and related problems is important and will come up often, as we will see.

We use the notation xy for the concatenation of the strings x and y. Let B be a decision problem; then B is said to be in *nondeterministic polynomial time* or NP if there is a decision problem A that is in P and a monomial Cn^k such that

$$x \in B \leftrightarrow \exists y(|y| \leq C|x|^k \wedge xy \in A)$$

Clearly any decision problem that is in P is also in NP. However, a surprisingly large number of frequently encountered combinatorial problems fall into the category of NP problems without, according to our current lights, falling into the class P.

One of the quintessential NP problems is the *satisfiability problem* for CNF formulas of Model 9.2; this problem is also called *SAT*. Here x is the given CNF formula and y is a sequence of 0s and 1s that provide a satisfying assignment to the propositional variables of the formula x.

A celebrated NP problem is the decision problem form of the traveling salesperson problem (TSP): Given an integer k, a set of cities together with the cost of travel between them and a distinguished starting city, is there a way to visit all the cities once and to return to the starting point at total cost $\leq k$. Here x is the list of costs for travel between cities and y is a permutation of the cities that corresponds to a tour whose total cost meets the bound k.

The problem of determining if there is an integer solution to a set of linear constraints is called the *integer linear programming problem* or *ILP problem*; context permitting, it is simply called, the *integer programming problem* or *IP problem*. Model 2.3, Call 911, is an example of an IP problem. When the solution must be a 0-1 solution, the problem is called a *0-1 integer programming problem*. We have seen several examples of this kind of application, most recently Model 9.2, Silicon Logic. When some of the variables are required to have an integer value, the problem is called the *mixed integer linear programming problem*, or more simply, the *mixed integer programming problem* or *MIP*. All of these problems can be shown to be in NP (see Papadimitriou and Steiglitz 1982). For the 0-1 integer programming problem, the result is immediate. The proof for the general MIP or IP case is far from trivial; it must be shown that if a solution exists, then there is one that can be described by a sequence of bits whose length is bounded by a polynomial time function.

Typically NP-Complete problems are ones where verifying that the pair xy is in the underlying language A is virtually trivial. Finding the y that bears witness to the fact that x is in the NP language B is the daunting part. Thus the disjunctive programming problem is easily seen to be in NP: Given the sequence $j_0,\ldots,j_{N-1}$, it is a polynomial time task to check that the following polyhedral set is nonempty

$$F \cap F_{0j_0} \cap \ldots \cap F_{N-1 j_{N-1}}$$

As we saw in Model 9.2, the problem of finding a satisfying assignment for a set of clauses is easily reformulated as one of finding an integer solution to a set of constraints. When there is a polynomial time function that sends instances of one decision problem to instances of another and that preserves acceptance, this function is called a *reduction*; the first problem is said to be *reduced* to the second. In other words, in Model 9.2, we reduced SAT to IP. On the other hand, it can be shown that all NP problems can be reduced to SAT. This is a fundamental result due to Cook (1971); see also Levin (1973). Since reducibility is transitive, this makes SAT, MIP, and IP equivalent in that they are all reducible one to the other and all problems in the class NP can be reduced to each of them. Problems in NP to which all other NP problems

9.3 P, NP, and co-NP

can be reduced are called NP-Complete. As another example of an NP-Complete problem, we have the special case of IP where the constraint set consists of a single constraint; this problem is known as the *integer knapsack problem*. What is more, the refinement of the integer knapsack problem to the *0-1 knapsack problem*, which requires all variables in a solution to be 0 or 1, is again NP-Complete (see Papadimitriou and Steiglitz 1982).

Scheduling problems are also a rich source of NP-Complete languages. As an example, TSP is NP-Complete. The richness of the class of NP-Complete problems and the science of reducing them one to the other were demonstrated in Karp (1972). If a problem can be shown to be NP-Complete, then an algorithm to solve this problem implicitly contains a method to solve all other NP problems as well. This is strong evidence that the problem can be difficult to solve.

Here, the elegant theory runs into a "mathematical wall." The conjecture is that the NP-Complete problems are not in P; in other words, that P is not equal to NP, written $P \neq NP$. However, this conjecture has eluded both proof and refutation for many years. Still, for *practical purposes*, as long as there is no proof that P is equal to NP, we have to proceed as though P is not equal to NP.

The good news is that solvable instances of NP-Complete problems have solutions of reasonable size; the bad news is that the only known way of solving all the NP-Complete problems is by traversing a search tree that is potentially exponential in size. That is, given an instance x of an NP-Complete problem B, the naive solution will involve a search through the space of all possible y such that the pair xy is in the underlying polynomial time language A. But the number of possible strings y to examine is

$$2^{C|x|^k}$$

In the best of all possible worlds, a program would always choose the best branch of a set of alternatives, thus finding a direct path through the search tree to a goal node. This way the witness y needed to verify that x is in B would be found effortlessly. This would be an implementation of true nondeterminism. However, in the workaday world, we must resort to an and/or loop combination such as

```
and(int i=0;i<N;i++)
    either y[i] = 0;
    or y[i] = 1;
```

in order to generate y. The kind of backtracking support for nondeterminism that a logic-based language such as 2LP provides is sometimes called *don't know nondeterminism*.

There are other versions of the satisfiability problem for propositional logic that are NP-Complete. For example, the set of all propositional formulas that are *not* tautologies is NP-Complete. It's in NP because only a nonsatisfying assignment to the variables of the formula is needed to check that it is not a tautology, exactly as for SAT itself. It's complete because SAT reduces to it trivially. One reason that SAT and CNF are singled out among these logical problems is that the simple structure of the CNF formulas makes the NP-Completeness of SAT more surprising. Moreover its simplicity makes SAT easier to reduce to another problem, and, in fact, SAT is an important tool for establishing NP-Completeness of other problems. Indeed, a special case of SAT is NP-Complete, namely *3-SAT* which is the restriction of SAT to formulas with clauses consisting of at most three literals. The fact that SAT is NP-Complete and reduces so naturally to so many applications means that the and/or structure of CNF is intrinsic to the class NP and so the and/or loop combination is a natural part of programming solutions to these problems.

The class P has a natural symmetry in that a language is in P if and only if its complement is also in P. This observation follows at once from the fact that the two problems have the same instances and that changing an output *bit* into *1 - bit* does not take a function out of the class of polynomial time functions.

In contrast, there is an asymmetry in NP problems. If a solution exists, there is a witness that is not exponentially large relative to the data of the problem instance. But to show that no solution exists means that one must show that *no* path through the search tree can lead to a solution. The number of such paths can be exponentially large relative to the size of a problem instance. The complement of an NP problem is said to be in co-NP. An example of a co-NP language is the set of propositional tautologies. In fact, the sublanguage of all propositional tautologies of the form $C \to p$, where C is a CNF formula and p is a propositional variable, is also in co-NP. This is virtually the same as co-SAT, the complement of SAT; to see this, note that showing $C \to p$ is equivalent to showing that $C \cup \{ \sim p \}$ is not satisfiable. It is also an open problem whether or not NP ≠ co-NP.

In Section 1, when we made the connection between persistent disjunction and classical disjunction, we placed ourselves in the co-NP situation because logical tautology is in co-NP. When discussing the relationship between persistent disjunction and the problem solving process, we were working with nondeterminism.

The duality between NP and co-NP can also be seen in the following direct characterization of the class co-NP obtained by negating the defining condition for membership in NP: By definition a decision problem B is in co-NP if there is a decision problem A in P such that

$$x \in B \leftrightarrow \forall y(|y| \le C|x|^k \to xy \notin A)$$

The complement of an NP-Complete problem has the property that every co-NP problem can be reduced to it and so it is called co-NP-Complete.

9.3 P, NP, and co-NP

There is a form of the set covering problem known as the *minimum cover problem* which is NP-Complete: In addition to the subsets one is given an integer k; the instance is accepted if there is a covering with no more than k subsets. Another example of an NP-Complete problem is *k-graph coloring*. An instance of this problem consists of a set of edges and an integer k; the instance is accepted if there is a coloring of the graph with no more than k colors. In Section 5.1 we considered this problem without the restriction k on the number of colors; that form always has a solution. In Section 10.2 we will discuss the NP-Complete version.

Many NP-Complete problems have an optimization form. For example, with k-graph coloring the optimization form is to produce a minimal coloring. An algorithm to produce an optimal coloring can certainly be adapted to solve the decision problem whether there is a k-coloring. That is, the NP-Complete problem can be reduced to the optimization problem. If an NP-Complete problem can be reduced to a given problem, then the given problem is called NP-Hard. The notion of a co-NP-Hard problem is defined similarly.

Therefore the optimization form of an NP-Complete problem can present formidable obstacles, since it will have both NP and co-NP aspects. Finding a solution is an NP task; verifying that the optimal solution has been found is a co-NP task. When code is run to solve a problem like this, at some point in the search process, the optimal solution will have been found and now the program's task has changed. What makes things interesting is that there is no sure way of knowing when the crossover point has occurred.

For problems in NP, local search algorithms such as Lin's algorithm for the TSP stand in contrast to enumeration methods. Local search methods work on the NP side of a problem finding a variety of good solutions. However, such methods cannot be used to prove optimality of an existing solution. The word *prove* is not accidental here. Verifying optimality, like theorem proving, is on the co-NP side of things.

Computational complexity considerations help explain why there is a restriction in 2LP and many other constraint-based programming systems to linear constraints on the continuous variables. For linear constraints, the consistency problem is in P and, as important, there exist efficient practical algorithms for this problem. More complex constraints can easily lead to intractable mathematical problems. For example, the quadratic constraint $x^2 = x$ forces x to be either 0 or 1. So adding quadratic constraints to linear constraints makes the feasibility problem NP-Hard. Since a program will do one thing if the constraints are feasible and another if they are not, we have the situation where the program is employing an NP-Hard problem as an "oracle"; in complexity theoretic terms we have begun to climb the *polynomial time hierarchy* (Garey and Johnson 1979). Of course nonlinear constraints and nonlinear objective functions are encountered all the time in optimization problems, and methods for solving these problems are constantly being developed. By way of example, there is the case where the constraints are linear but the objective function is quadratic; this optimization problem is known as *quadratic programming*.

There are systems, known as MIP solvers, that support linear constraints on both continuous and integer variables. However, supporting even linear constraints on variables that range over the integers means that the constraint solver must be able to handle the general IP problem. On the other hand, a system that treats such constraints enables the programmer to write much more declarative models because much or all of the search process of the constrain-and-generate paradigm can be shifted to the constrain phase; this point will be taken up in Chapter 13. A number of academic and commercial MIP solvers exist that can perform very well on a wide range of applications. These are an important part of the operations research arsenal.

Exercises

9.3.1. The nonpreemptive form of the problem of Model 5.2 is NP-Complete if a solution is sought which meets some given bound. This problem is listed as SS1-3 in Garey and Johnson (1979). It is clearly in NP. Prove NP-Completeness by reducing 3-SAT to it.

9.3.2. Show that the 2-SAT problem is in P.

9.3.3. Determine the optimal solution to Model 5.1.

10

Soundness and Completeness

Two important logical concepts related to mathematical algorithms are *soundness* and *completeness*. In general terms, a problem-solving method is said to be *sound* if it never leads to incorrect solutions of the problems it's given. A method is said to be *complete* if it always leads to a solution if one indeed exists.

The analysis of algorithms and programs leads to a constant back-and-forth between the mathematical picture and the programming picture. Because of this, we will often apply mathematical terms such as soundness and completeness to computer code. The idea is to try to capture the clarity and the correctness of the mathematics in code, and this is not always an easy task.

In the computer science literature, the term *correctness* is used instead of soundness and the term *partial correctness* is used for code that is sound but not guaranteed to terminate, (e.g. see Gries (1981).

10.1 Loop invariants

In this section we introduce some concepts that will help us to develop and analyze search programs. The key notion will be that of the *loop invariant* adapted to the present context. Loop invariants were introduced in Floyd (1967) as a tool for proving programs correct. For a discussion of classical loop invariants, we refer the reader to Gries (1981).

In Model 2.3, Call 911, and Model 3.4, The Committee, the progressive roundoff heuristic is used to find a solution to problems that require integer-valued solutions for linear constraints. These models are situations where all the multivariable constraints are so-called demand constraints; that is, constraints on a set of continuous variables $X[0],...,X[N-1]$ of the form

```
sigma(int i=0;i<N;i++) a[i]*X[i] >= b;
```

with all coefficients a[i] nonnegative. The only other constraints are integer valued upper bounds on the continuous variables. Schematically, both models have an objective function Z and are solved by means of a central loop of the form

```
and(int i=0;i<N;i++) {
    min: Z;
    X[i] == ceil(X[i]);
}
```

In both cases this heuristic leads to a quick and usable solution to a challenging problem. At the risk of being pedantic, let us examine this code and the underlying algorithm in some detail. In particular, let us discuss how this code is both sound and complete.

In the and loop, before the loop variable is incremented from i-1 to i, a sequence of i ceilings $n_0,\ldots,n_{i-1}$ has been produced such that the constraints X[0]==n_0,....,X[i-1]==n_{i-1} are consistent with the initial demand constraints that were set up. In programming terms, what we have here is a *loop invariant* that describes the state of affairs that is preserved by a pass through the body of the loop. From this, it follows that if the and loop is exited successfully, the problem is solved; for when the loop control variable i is incremented to the value N, the loop is exited, and the continuous variables X[0],....,X[N-1] are fixed at integer values that satisfy the demand constraints. We can formalize the loop invariant by commenting the code as follows:

```
// There is a sequence of integers n_0, ..., n_{i-1}
// such that X[0] == n_0, ..., X[i-1] == n_{i-1}
and(int i=0;i<N;i++) {
    min: Z;
    X[i] == ceil(X[i]);
}
```

In this notation, if i is 0, a sequence indexed up to i-1 is interpreted as the empty sequence. This convention does away with the need to make a special case for i==0 and will prove very useful.

We read the loop invariant as saying that when the loop control variable is incremented to i, the conditions in force are those given, namely that X[0],....,X[i-1] have been fixed in a feasible way at integer values. So when the control variable reaches N, the loop is exited, and we conclude that X[0],....,X[N-1] are fixed at integer values that yield a solution to the given problem. In other words, if the loop terminates, a solution has been found. Thus the loop invariant on the and loop asserts the soundness of the code.

Now let us turn to completeness. We claim that the code will find an integer-valued solution to the constraints if one exists. First, if there is to be any solution, the demand constraints and the upper bounds on

10.1 Loop invariants

the continuous variables must be consistent. So let us suppose this to be the case. We are now at the and loop; we can proceed by induction on the loop control variable i to show that each pass through the body of the and loop will be successful. For this we use the following remark: If the constraints

```
X[0]==a₀;   ...   X[i-1]==aᵢ₋₁;  X[i]==aᵢ;
```

are consistent with the demand constraints and bounding constraints, so are

```
X[0]==a₀;   ...   X[i-1]==aᵢ₋₁;  X[i]==ceil(aᵢ);
```

Completeness follows from the fact that in the body of the and loop, a value n_i at which to fix X[i] is chosen such that the constraint X[i]==n_i is consistent with the demand constraints, the bounding constraints, and the previous constraints X[0]==n_0,...,X[i-1]==n_{i-1}.

In programming with constraints and logic, the key loop construct is the and/or loop combination. Very roughly put, the and construct carries the burden of soundness, while the or construct carries the burden of completeness. To illustrate this, let us look at the generic disjunctive program. The scheme of the code for this kind of application is

```
and(int i=0;i<M;i++)
    or(int j=0;j<N;j++)
        cc(i,j);
```

where cc(i,j) is code to generate the conjunction of constraints that defines the feasible region F_{ij}.

In a disjunctive program, as the body of the and loop is entered from the top and the loop variable i is set at 0 or incremented from a previous value, the situation is that for 0,...,i-1 values $j_0,...,j_{i-1}$ have been found such that the calls to cc(0,j_0),...,cc(i-1,j_{i-1}) have all succeeded. Entering this loop invariant as a comment, we have

```
// cc(0,j₀),...,cc(i-1,jᵢ₋₁) successful
and(int i=0;i<M;i++)
    or(int j=0;j<N;j++)
        cc(i,j);
```

When the loop control variable i is incremented to N, the loop is exited, having determined a sequence $j_0,...,j_{N-1}$ such that the constraints generated by the calls to cc(0,j_0),...,cc(i-1,j_{N-1}) are all consistent. These constraints determine a feasible region that provides a solution to the disjunctive program.

Now turning to the or loop, when the control variable j is set to 0 or is incremented from a previous value, the successive calls to `cc(i,0),....,cc(i,j-1)` have all failed to lead to a solution. This is, in fact, an invariant of the or loop, since it holds every time the loop is entered from the top. Writing this loop invariant as a comment, we have

```
// cc(i,0),...,cc(i,j-1) do not lead to a solution
or(int j=0;j<N;j++)
    cc(i,j);
```

There are two different ways of exiting the or loop: success or failure. If a call to `cc(i,j)` succeeds, the or loop is exited successfully. Control returns to the and loop, and its control variable i is incremented to i+1. On the other hand, if the or loop control variable j is incremented to N, then the or loop is exited unsuccessfully. At this point control passes back to the previous invocation of the or loop, the one where the control variable of the and loop was equal to i-1. To work this into the comments, we can write

```
// cc(0,j_0),...,cc(i-1,j_{i-1}) successful
and(int i=0;i<M;i++)
    // cc(i,0),...,cc(i,j-1) do not lead to a solution
    or(int j=0;j<N;j++)      // If j == N, decrement i
        cc(i,j);    // If successful, increment i
```

From an intuitive point of view, the and loop invariant corresponds to an induction hypothesis to the effect that something holds at i-1. The or loop invariant corresponds to the assertion that the hypothesis will hold at i if it possibly can: Otherwise, the search will backtrack. However, this local view at levels i-1 and i does not provide a full proof of completeness of the algorithm. For that, a look down the search tree is needed.

The or loop in the case under consideration cycles through all the alternatives `cc(i,j)` that are given. For our purposes we can suppose that these exhaust all the possible alternatives. Given that, we claim that this code will find a feasible region that satisfies the disjunctive program if a solution exists. In other words, if there is a sequence of integers $j_0,...,j_{M-1}$ such that the sequence of calls `cc(0,j_0),....,cc(M-1,j_{M-1})` will generate a feasible region, then the code will terminate successfully, in effect finding such a sequence. In fact, the sequence that the program will determine is the least such sequence in terms of the lexicographic order on sequences of integers. This least sequence corresponds to the path leading to the leftmost solution in the tree, as is illustrated in Figure 10.1. To verify that this solution will be found, we argue by induction on M, the length of the and loop. Let $j_0,...,j_{M-1}$ be the lexicographically least sequence of integers such that the constraints generated by `cc(0,j_0),...,cc(M-1,j_{M-1})` are feasible. Then for

10.1 Loop invariants

every $j < j_0$, the call cc(0,j) fails to lead to a solution; therefore the call to cc(0,j_0) is made in the program. By the choice of j_0, this call succeeds, and the situation is equivalent to

```
cc(0,j0);
and(int i=1;i<M;i++)
    or(int j=0;j<N;j++)
        cc(i,j);
```

But now the induction hypothesis applies, since we have reduced the length of the and loop by 1.

It is interesting to note that the argument just given is a finite version of a key step in the proof of the Gödel Completeness Theorem for first-order logic, (Gödel 1930). In this proof, a form of the König Infinity Lemma is used to find an infinite path through a finitely branching tree (see Church 1956). The path through the tree yields an interpretation for the formula to be satisfied, much the same as the path through the search tree determines a solution to a disjunctive program. In fact, let us look at the disjunctive programming formulation of the satisfiability problem in Model 9.2, Silicon Logic.

The code consisted of data and setup routines and the following central loop:

```
and(int j=0;j<N;j++)
    or(int b=0;b<2;b++)
        X[j] == b;
```

Introducing the appropriate loop invariants, the commented form of the and/or loop is

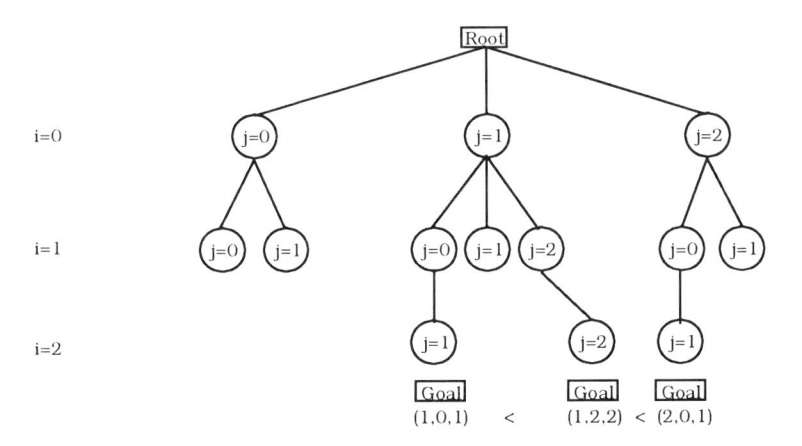

FIGURE 10.1 Lexicographic ordering of the and/or search.

```
       // X[0]==bit_0,...,X[j-1]==bit_{j-1}
       and(int j=0;j<N;j++)
            // X[j]==b-1 does not lead to a solution
            or(int b=0;b<2;b++)//If b==2, fail and decrement k
                X[j] == b;     //If successful, increment j
```

This code will find the least sequence of N bits that satisfies the formula, least in the lexicographic order on sequences of bits of length N. So the path through the tree finds a satisfying assignment for the CNF formula by giving a truth assignment to each propositional variable. One can note the complementary relation between the invariant on the and loop and the invariant on the or loop: The former is a positive assertion, the latter a negative assertion.

We will now make explicit a pair of conditions on the and loop and a pair of conditions on the or loop that ensure soundness and completeness. By employing continuous variables in the satisfiability example of Model 9.2, we are guaranteed that at the top of the and loop, the constraints $X[0]==bit_0,\ldots,X[j-1]==bit_{j-1}$ are consistent with each other and with the initial setup constraints. A search that maintains this kind of consistency is said to be *coherent*. For code built on continuous variables, this is automatic. In working with variables of type int and double, there is no notion of global feasibility or consistency, and so the programmer must take care to provide for it and to monitor it. We discuss such examples in the next section.

A second condition on the and loop also must be taken into account for soundness. In our example, it is critical that the setup constraints and the constraints on X[0],...,X[j-1] are not undone by a pass through the body of the and loop at j. In other words, decisions made at one level on a path in the search are not undone at subsequent levels. A search that maintains this condition is said to be *monotonic*. This kind of monotonicity is provided automatically by the behavior of the type continuous in a disjunctive program but must be attended to by the programmer when variables of other type enter the picture.

Let us turn to the or loop and completeness. The second algorithm of Model 9.2, Silicon Logic, is complete because the persistent disjunction ensures that all available alternatives are kept alive, since they are enumerated in the or loop. The or loop invariant makes explicit the fact that the code is conducting a search through all possible combinations that can lead to a sequence that satisfies the given CNF formula. A search with the property of providing all possibly successful alternatives is said to be *enumerative*. Of course, if the given formula is not satisfiable, then this search will terminate without success, since even a complete and sound algorithm cannot find a solution if none exists. If a solution does exist, then the code will find the one that is least in the lexicographic order.

There is another condition on the or loop that must be taken into account for completeness. For example, in the code for Model 9.2, when

10.1 Loop invariants

the `and` loop control variable is `j`, each time the `or` loop control variable `b` is incremented and the body of the `or` loop is entered, the feasible region is the same as it was for the previous value of `b`. Thus each child of a parent node in the search is presented with the same feasible region as the other children of this node. A search with this property is said to be *fair*. Again, when the situation is not that of a disjunctive program, the program must ensure that each disjunct is presented with the same state of affairs as its siblings. In a vanilla disjunctive program, this state of affairs or more simply *state*, is given by the feasible region and the values of the loop control variables. In other situations the state will consist of the values stored in arrays and individual variables of types `int` and `double` as well.

Let us summarize the discussion thus far on strategies for programming algorithms based on enumeration and tree search, where the object is to find a goal node. In order to be sound, as it progresses through the tree, the search should conform to two vertical conditions: namely, the search should be

1. *monotonic*. Decisions made at depth `j < i` should not be undone at depth `i`.
2. *coherent*. New decisions made at a node must be consistent with decisions made along the path leading to the node.

In order to be complete, a *finite* tree search should conform to two horizontal conditions, namely, the search should be

3. *enumerative*. All possible children of each node should be accounted for.
4. *fair*. Each child of a node should be presented the same state as its siblings.

In addition to these conditions, a depth-first search must of course progress down the tree, and decisions must be made for all necessary depths. Typically this aspect of things will be taken care of by the search program's `and/or` loop structure, and the argument given above that the code for the generic disjunctive program finds the leftmost solution in the tree. There are many other situations where one has to justify that a sufficient depth in the tree will be reached. In particular, in Chapter 12 this will happen with the *injury method*, a depth-first technique that seeks to minimize the depth of the search.

To illustrate some of these notions, let us take another look at the situation encountered in Model 9.1, Motorcycles Inc. Reconsidered. We can comment the central `and` loop as follows:

```
// either Bike[0]==0 or Bike[0]>=lo[0]
//                   ...
// either Bike[i-1]==0 or Bike[i-1]>=lo[i-1]
and(int i=0;i<MODS;i++)
    decide(Bike[i],lo[i]);
```

The disjunctive part of the program is in the `decide` routine. There we formulated the two choices on a model of motorcycle by means of the `either/or` construct rather than by means of a loop. This makes for a bit of a change in the writing of the comments:

```
decide(continuous Bike, double low)    // Version II
{
    // First alternative
    either Bike >= low;

    // Bike >= low does not lead to a solution
    or Bike == 0;

    // If Bike == 0 doesn't lead to a solution, backtrack
}
```

This code satisfies the horizontal and vertical conditions for soundness and completeness. We have used the `decide` routine to branch first on the alternative of meeting the threshold `Bike >= low` and then in the other direction `Bike == 0`. However, this strategy to find a plan is very uninformed and is not likely to lead to an especially good solution. It is true that it will lead to a consistent sequence of decisions of which products to make and which not to make if such a solution exists at all. It is also true that when this determination is completed, the objective function is optimized. However, the determination of which products to include will have been made without taking the objective of maximizing revenue into account. Therefore, we could probably improve on things if we re-optimized the objective function after each decision and guided the search by querying the witness point. In other words, we will evaluate the alternatives of including the model of motorcycle in question in a two valued way: A "good" evaluation is given to a model if, after the objective function is optimized, the witness point at that model of motorcycle is at a value greater than or equal to its threshold; otherwise, that model of motorcycle gets a "bad" evaluation.

```
decide(continuous Bike, double low)    // Version III
{
    max: Z;

    if wp(Bike) >= low;                // Good evaluation
```

10.1 Loop invariants

```
      then
         either Bike >= low;
         or Bike == 0;
      else                          // Bad evaluation
         either Bike == 0;
         or Bike >= low;
}
```

So with this version of the `decide` procedure, we have optimized the objective function and used the current witness point to fix the order in which to consider the alternatives `Bike >= low` and `Bike == 0`. But both options are kept alive, so the central and/or loop of the program satisfies the enumerative condition for tree search. This code is still guaranteed to find a solution if one does exist, and the search has been oriented toward finding a solution that tends to yield a good value for the objective function. In fact, now the projected revenue is $108,400, which is a substantial improvement over the result of Model 9.1.

We obtain a hill-climbing strategy by selecting one alternative while discarding the other:

```
decide(continuous Bike, double low)    // Version IV
{
    max: Z;

    if wp(Bike) >= low;
    then Bike >= low;
    else Bike == 0;
}
```

This hill-climbing search, however, is not guaranteed to find a feasible solution even if one does exist. It can lead to committing to production of items that use up resources that might be needed elsewhere. In fact, with these data, it fails to find a feasible solution. The code no longer satisfies the enumerative condition necessary in this situation for completeness, although it does satisfy the appropriate vertical conditions for soundness.

Local search procedures such as hill-climbing typically are sound but not complete algorithms. As further examples of this kind, the progressive roundoff technique for the set covering problem is sound but not complete if an initial bound `Z <= K` is given and the application is converted into the minimum cover problem, which is NP-Complete. Also, the randomized algorithm of Model 6.3, The Body Shop, is sound but not complete if an initial bound `MakeSpan <= best` is given.

The next model will encounter several difficulties in the search for a solution. Each will have to be met with an appropriate response.

Model 10.1 Corporate Paper

A financial planner is asked to put together a portfolio for an exigent client. The client has $55,000 to invest for the coming fiscal year. Their discussions lead them to plan to invest in AA corporate bonds, AAA corporate bonds, and a money market fund. The investment in the money market fund is expected to yield 2.50% interest. Tables 10.1 and 10.2 give the selling price per hundred and the rate of return for each of the AA and AAA corporate bonds under consideration.

TABLE 10.1 AA Bonds

Bond number	0	1	2	3	4
Price/100	421	686	375	330	400
Rate of Return	.090	.084	.0825	.081	.0825
Bond number	5	6	7	8	9
Price/100	444	522	545	550	275
Rate of return	.081	.091	.09	.083	.0825

TABLE 10.2 AAA Bonds

Bond number	0	1	3	3	4
Price/100	782	982	466	700	700
Rate of return	.076	.075	.07	.0775	.08

The investor wants a mix that will have an expected yield of at least 6.5% overall. The financial planner recommends strongly that the money market be at least 12% of that placed in AA bonds and at least 20% of that invested in AAA bonds. Although the AA bonds listed are more attractive because of their promised rate of return, in order to reduce uncertainty the planner wants the capital placed in AA bonds to be no more than 85% of that placed in AAA bonds. In a similar vein the client wants no more than $7,500 invested in any one security. For reasons of liquidity, bonds can only be purchased in units of 100. Can the planner meet these requirements in an investment portfolio?

This would be a straightforward linear programming problem but for the requirement that the bonds must be purchased in multiples of 100. A first attempt at dealing with this added complication is to formulate and solve the linear program and then to use a roundoff technique. However, here the constraints are not simple resource constraints, so rounding down is not guaranteed to work. In any case let us formulate the model as a linear program and see what happens when we try to roundoff quantities.

Let us set up symbolic constants and data arrays such as

10.1 Loop invariants

```
#define FUND 55000.0
#define AA 10    // Number of AA bonds
#define AAA 5    // Number of AAA bonds
#define PPH 0    // Price per hundred
#define RR 1     // Rate of return
#define LIMIT 7500.0

double dbla[2][AA];     // Data from Table 10.1
double trpla[2][AAA];   // Data from Table 10.2
```

and let us introduce continuous variables

```
continuous DblA[AA],TrplA[AAA],Mm,Return;
```

Here `DblA[i]` will represent the number of hundreds of units purchased of the `i`th AA bond, `TrplA[j]` the number of hundreds of units of the `j`th AAA bond, and `Mm` the amount placed in the money market fund. So in the solution we seek, the witness point at both `DblA[i]` and `TrplA[j]` must be integer valued. We can initialize the arrays with a call to a procedure `data()` as follows:

```
data() {

    dbla = {
        421, 686, 375, 330, 400, 444, 522, 545, 550, 275,
        .09,.084,.0825,.081,.0825,.081,.091,.09,.083,.0825
    };

    trpla = {
        782, 982, 466, 700, 700,
        .076,  .075,  .07,  .0775,  .08
    };
}
```

Thus `dbla[PPH][i]` is the price per one hundred units of AA bond `i` and `dbla[RR][i]` is the rate of return on investment in AA bond `i`. The amount invested in this security is

  `dbla[PPH][i]*DblA[i]`

and the return on this security will be

  `dbla[RR][i]*dbla[PPH][i]*DblA[i]`

The situation for AAA bonds is the same *mutatis mutandis*.
    The routine to set up the constraints on the portfolio will be the following:

```
setup()
{
continuous TotalAA, TotalAAA;      // Auxiliary variables
continuous ReturnAA, ReturnAAA;    // for readability

    TotalAA                        // Investment in AA bonds
    == sigma(int i=0;i<AA;i++) dbla[PPH][i]*DblA[i];
    TotalAAA                       // Investment in AAA bonds
    == sigma(int j=0;j<AAA;j++) trpla[PPH][j]*TrplA[j];

    .85*TotalAAA >= TotalAA;       // Relation of AAA to AA

    TotalAA + TotalAAA + Mm == FUND;  // 3 ways to invest

    Mm >= .12*TotalAA;             // Money market share
    Mm >= .20*TotalAAA;

    ReturnAA == sigma(int i=0;i<AA;i++)     // AA return
                dbla[RR][i]*dbla[PPH][i]*DblA[i];
    ReturnAAA == sigma(int j=0;j<AAA;j++)  // AAA return
                 trpla[RR][j]*trpla[PPH][j]*TrplA[j];

    // Return from all investments
    Return == ReturnAA + ReturnAAA + .025*Mm;

    Return >= .065*FUND;           // Client's requirement
}
```

Our aim is to have the linear programming model assist us in finding a solution to the planner's problem where both the constraints and the integrality requirements on the bond purchases are met. This process will be aided if we tighten the linear relaxation and bound the continuous variables as sharply as possible. The most that can be invested in any one bond is given by the symbolic constant LIMIT, so the number of hundreds of units of AA bond i that can be purchased is bounded by the quotient LIMIT/dbla[PPH][i]. Further, since DblA[i] is to be integer valued in the solution, this can be sharpened to floor(LIMIT/dbla[PPH][i]). The situation is perfectly similar for the AAA bonds. So we can enforce these bounds by calling

```
bounds()
{
    sharp_bounds(AA,dbla[PPH],DblA);    // Tightens model
    sharp_bounds(AAA,trpla[PPH],TrplA);
}
```

where the sharp_bounds routine is

10.1 Loop invariants

```
sharp_bounds(int a, double price[], continuous Bonds[])
{
    and(int i=0;i<a;i++)
        Bonds[i] <= floor(LIMIT/price[i]);
}
```

The most naive roundoff strategy would be to maximize `Return` and then to round down the values of `DblA[i]` and `TrplA[j]` to their floor values. A progressive roundoff strategy would be to loop through the securities, maximizing `Return` each time, and to set the value of `DblA[i]` or `TrplA[i]` to its floor. To carry out this progressive roundoff strategy, we would put together code such as the following:

```
21p_main()
{        // NAIVE AND NOT GUARANTEED TO SATISFY CONSTRAINTS
    data();
    setup();
    bounds();

    progressive_roundoff_loop();
    output();
}

progressive_roundoff_loop()
{
    and(int i=0;i<AA;i++)
        fix_security(DblA[i]);

    and(int i=0;i<AAA;i++)
        fix_security(TrplA[i]);
}

fix_security(continuous Security)
{
    max: Return;
    Security == floor(Security);   // Progressive roundoff
}

output()
{
    printf("The rate of return is %.3f\n",wp(Return)/FUND);
    ...
}
```

Unfortunately, this program will fail to find a solution to the financial planner's problem! The same is true of the naive roundoff strategy. In both cases the strategy hits a wall of infeasibility.

Finding solutions to linear constraints subject to additional logical or integrality requirements is a challenging enterprise. What we need here is a way of providing the progressive roundoff strategy with alternatives to backtrack to if an infeasibility is reached. In other words, the search strategy must be made enumerative.

To enhance the progressive roundoff strategy with backtracking, we will use an `or` loop that will start at the greatest amount that can consistently be invested in a security and, upon backtracking, decrement the amount invested by 100 bonds. To accomplish this, let us design a procedure `try` that takes a continuous variable `Bond` for the bond in question as a parameter:

```
try(continuous Bond)
{
int bottom, top;

    min: Bond;
    bottom = ceil(Bond);    // Lower bound for loop

    max: Bond;
    top = floor(Bond);      // Upper bound for loop

    // Bond==top,...,Bond==k+1 do not lead to a solution
    or(int k=top;k>=bottom;k--)//If k==bottom-1, backtrack
        Bond == k;  // Purchase bonds; return to and loop
}
```

In this routine the loop variable k is initially set so that the security is fixed at its greatest possible purchase amount. Upon backtracking k is decremented by 1 each time, and the security is fixed at the next lower possible purchase amount. Thus the strategy is to be as greedy as conditions will allow. This strategy is based on giving the highest evaluation to the child node that fixes the security at its greatest possible integer value. The `or` loop goes from highest to lowest possible integer value for the continuous variable. Now the `or` loop is enumerative because it cycles through all possibilities. The code is quite straightforward:

```
21p_main()
{
    data();
    setup();
    bounds();
    central_loop();
    output();
}
```

10.1 Loop invariants

```
central_loop()
{
    and(int i=0;i<AA;i++)
        try(DblA[i]);
    and(int j=0;j<AAA;j++)
        try(TrplA[j]);
}
```

If a solution does exist, then the code will find the one that is greatest in the lexicographic order on sequences of nonnegative integers of length AA+AAA, since the or loop goes from top to bottom. In other words, the labeling will be the opposite of that of Figure 10.1. However, this solution will be the first one that is uncovered by the search and so is still the leftmost solution in the search tree.

Indeed, the program finds that the portfolio is feasible and advises placing $22,782 in AA bonds, $26,822 in AAA bonds and $5396 in the money market fund. This yields a return of 7.4%.

The strategy used in this model is *guaranteed* to find a solution if one exists, since the try routine enumerates all possible values that the security can have. Let it be said, however, that this *guarantee* does not always mean that the search will terminate in any reasonable amount of time.

This search strategy has a natural structure. It is a depth-first search but the order in which the children of a node are generated is from most promising down to least promising. This is called *best-child-first search*. In this example, the ordering of the children is simple and is based on a greedy principle; in other situations more complex evaluations of the child nodes are called for.

Now that we found a way to reach a solution to the planner's problem, let us consider ways of finding a better solution faster.

The loop that we have used to fix the continuous variables DblA[i] and TrplA[j] considers the AA bonds before the AAA bonds but within each category simply follows the order given by the data. A natural greedy strategy suggests itself here. The bonds should be considered going from the most profitable to the least profitable. Since all the AA bonds have a higher rate of return than the AAA bonds, this means reordering each category and again looping first over the AA bonds and then the AAA bonds.

To keep track of the sorted securities, let us introduce two global integer arrays:

```
int sortAA[AA],sortAAA[AAA];
```

We can store the indices of securities in these arrays from the most profitable to the least profitable. This can be done by calling a sorting routine for each type of bond:

```
    arrange(AA,sortAA,dbla[RR]);
    arrange(AAA,sortAAA,trpla[RR]);
```

The sorting routine `arrange` can be formulated using, say, a bubble sort as follows:

```
arrange(int a,sort[], double rr[])
{
int temp;

    and(int k=0;k<a;k++)
        sort[k] = k;
    and(int i=a-1;i>=0;i--)
        and(int j=0;j<i;j++)
            if rr[sort[j]] < rr[sort[j+1]];
            then {
                    temp = sort[j];
                    sort[j] = sort[j+1];
                    sort[j+1] = temp;
            }
}
```

The main routine becomes

```
2lp_main()
{

    data();
    setup();
    bounds();

    arrange(AA,sortAA,dbla[RR]);
    arrange(AAA,sortAAA,trpla[RR]);

    new_central_loop();
    output();
}
```

where the `new_central_loop` goes through the bonds in sorted order:

```
new_central_loop()
{
    and(int i=0;i<AA;i++)
        try(DblA[sortAA[i]]);
    and(int j=0;j<AAA;j++)
        try(TrplA[sortAAA[j]]);
}
```

10.1 Loop invariants

This code is likely to find a solution faster, and the one found is more likely to give a higher return on investment. In fact, the program now advises placing $22,786 in AA bonds, $26,830 in AAA bonds, and $5,384 in cash. This yields a rate of return of 7.722%. The optimal solution to the linear relaxation yields a rate of return of 7.742%; thus we know that the solution found is within 1% of the best possible solution. In the next chapter we begin the study of methods to try to find the best solution possible for a disjunctive program. That is, we will discuss methods that yield provably optimal solutions.

The question can be raised whether this trick of sorting the bonds together with progressive roundoff is enough to find a solution; with this data set the answer is still no.

End of Model 10.1

In the above model the basic structure of the code was again an and/or loop where the and loop ran through the bonds and the or loop branched on each of these securities in turn. When a search is organized as an and/or loop in this way, there are two important degrees of freedom: the order in which the and loop runs through the variables to be branched on and the way in which the or loop runs through candidate values for each variable. In the model the order in which the securities were encountered in the and loop was based on the fact that the securities were sorted with the most profitable securities first. Then in the or loop the candidate values for each bond went from its greatest possible value to its least possible value. So in this model we used greedy strategies to organize both of these aspects of the search for a solution. In the chapters to follow, we will consider many different ways of organizing the handling of an and/or loop combination. As we saw in the previous Chapter, this combination plays a privileged role in combinatorial problems.

Thus far we have looked at pure disjunctive linear programs. Most often applications will require an adroit mix of methods. However, more general constrain-and-generate models can be analyzed in terms of horizontal and vertical conditions. As an example, consider the puzzle of Model 8.5, Who Has the Zebra. The burden of the analysis falls on the and/or loop in the permutation routine. Since continuous variables are used, the search is easily seen to be monotonic and fair. The verification that the search is coherent has two parts to it in this example: The continuous variables must be fixed so as to maintain feasibility and for each category, the continuous variables must be fixed at values that will determine a permutation of $0,...,N-1$. The first part of the coherence condition follows from the properties of continuous variables; the second is a consequence of the following invariant on the and loop in the body of the permutation routine:

10 Soundness and Completeness

```
// X[0] == n_0, ..., X[i-1] == n_{i-1}
// n_0, ..., n_{i-1} all distinct, all < N
and(int i=0;i<N;i++)
    or(int h=0;h<N;h++){
        and(int k=0;k<i;k++)
            nint(X[k]) != h;
        X[i] == h;
    }
```

When the and loop is exited, X[0],...,X[N-1] are fixed at N distinct values from among 0,...,N-1; hence all values must be taken, and a permutation is produced.

Finally, the or loop is enumerative in that it attempts to fix X[i] at all possible values h that preserve the invariant on the and loop:

```
// For each j < h either X[n] == j for some n < i
// or X[n] == j does not lead to a solution
or(int h=0;h<N;h++){          // If h==N, decrement i
    and(int k=0;k<i;k++)      // Check h available
        nint(X[k]) != h;
    X[i] == h;                // If successful increment i
}
```

In short, soundness follows from the fact the permutation routine can only generate permutations, and completeness follows from the fact that the permutation routine enumerates all possible successful permutations.

There is a duality relation between soundness and completeness, which is analogous to the duality between *for all* and *there exists*. By reversing the output for a decision problem, an algorithm for the problem yields an algorithm for the complementary problem. If an algorithm for the original decision problem is sound, then it yields a complete algorithm for the complementary problem; if an algorithm is complete, it yields a sound algorithm for the complementary problem. Note that in the definition of *complete* algorithm for a decision problem, it is possible for the problem-solving method to accept a problem instance incorrectly; completeness only requires that it never fail to accept an instance that should be accepted.

Thus a sound algorithm for SAT yields a complete algorithm for co-SAT. Very effective sound but not complete algorithms for SAT have been developed (e.g. see Kamath, Karmarkar, Ramakrishnan, and Resende 1991; Selman, Levesque, and Mitchell 1992). In the latter paper the algorithms employ local search and randomization; local search methods typically provide sound algorithms that are not complete.

On the other hand, a complete algorithm for SAT yields a sound algorithm for co-SAT. There is another way of looking at this: If the object is to show that no path in a tree leads to a goal node, then the vertical conditions for tree search ensure completeness, and the horizontal conditions ensure soundness.

Exercises

10.1.1. Apply loop invariants to the Drop/Add solution of the set covering problem.

10.1.2. Apply comments that describe the loop invariants for the routine `sequence_jobs` of Model 9.3.

10.1.3. The Fibonacci sequence f_n is defined by the recurrences $f_0 = 0$, $f_1 = 1$ and $f_n = f_{n-1} + f_{n-2}$ for $n > 1$. Show using a loop invariant that the following code segment stores f_n in a for $n \geq 1$.

```
i=1; a=1; b=0; c=0;
and(;;)
    if n <= i; then break;
    else { i=i+1; c=b; b=a; a=a+c;}
```

10.1.4. Suppose that `a[N]` is an array of integers that we want to sort by means of a bubble sort: While there are two consecutive elements out of order, switch any such pair. Use a loop variant to prove that the bubble sort terminates and sorts the array.

10.2 Discrete models

In this section we consider some examples of applications which are formulated without the aid of continuous variables.

As noted in Section 9.3, an example of an NP-Complete problem is k-graph coloring. In Model 5.1, Parallel Sessions, we considered the problem of graph coloring without the requirement that the number of colors be bounded by some k. Without this restriction the problem always has a solution, and the code we developed used a greedy algorithm to find a quick-and-dirty solution. There the central piece of code consisted of the routines

```
color_the_graph()
{
    color_first_vertex();
    and(int v=0;v<V;v++)
        c_or(int c=0;c<=nc;c++)
            color_the_vertex(v,c);
}
```

```
color_the_vertex(int v,c)
{
    conflict_free(v,c);

    color[v] = c;
    if nc < c;
    then nc = c;
}
```

Let us suppose that K is a symbolic constant set equal to the given integer k, and let us proceed to develop code, admittedly very naive code, for the NP-Complete version of the problem. In fact, this particular NP-Complete problem has been the object of intense study. For example, a method known as *semidefinite programming* has a been applied to it; this is a technique that uses continuous variables in an extension of linear programming (see Karger, Motwani, and Sudan 1994).

Returning to our naive discrete formulation of the problem, to account for the limit K on the number of colors that can be used, the basic change is to switch from the conditional `c_or` loop to the persistent `or` loop. This will make the search enumerative. To ensure that the search is fair, the number of colors `nc` that have been already used must be the same for each value of the control variable `c` in the `or` loop. For this, we can replace the single integer `nc` by an array

```
int nc[V];
```

and we use this array to stack the numbers of colors used at each level in the search. This stacking technique will protect the value of `nc[v]` and serve to make the search fair.

The array element `colors[v]` is only assigned at level `v` in the search, so this aspect of the code ensures the search is monotonic and fair and can stay unchanged. This time around we cannot be sure that vertices can be colored consistently, since we have to restrict the number of available colors to K. So we transform the `color_the_vertex` routine to

```
try_to_color_the_vertex(int v,c)
{
    conflict_free(v,c);

    color[v] = c;
    if nc[v-1] == c;
    then nc[v] = c+1;
    else nc[v] = nc[v-1];

    nc[v] <= K;      // Respect limit on colors
}
```

10.2 Discrete models

The principal change in the central loop is to transform the c_or loop into an or loop:

```
try_to_color_the_graph()
{
    color[0] = 0; // Color first vertex
    nc[0] = 1;
    // Vertices 0,...,v-1 colored with 0,...,nc[v-1]-1
    and(int v=1;v<V;v++)
        // 0,...,c-1 for vertex v do not lead to a solution
        or(int c=0;c<=nc[v-1];c++)
            try_to_color_the_vertex(v,c);
}
```

With this program the decision problem is solved in the acceptance case by producing a coloring: In the other case the code simply fails to find any solution. When a solution is found, the numbers of colors used is stored in nc[V-1].

In this code, we cycle through the vertices and find colors for them. In Model 5.1 we also looked at the dual search strategy of looping over colors and finding vertices for the colors. This strategy can also be adapted for the k-graph coloring problem.

Let us look at another discrete model to illustrate these points further. For that we now turn to a venerable chestnut of the literature, the N-queens, Lucas (1882). In the process we will describe some of the kinds of things that can be done to make discrete models closer in spirit to continuous generate-and-test models and more efficient in the process.

The model itself is venerable to the point of overexposure. However, it will provide an easily understandable framework for making several important points. To start, it is a situation where there is competition for a discrete set of resources. Assigning a resource to one competitor will create "holes" in the set of resources that are still available for the others. An important feature of continuous models is that each continuous variable ranges over a closed interval [a,b]. In Model 10.1 Corporate Paper, we were able to determine an interval [bottom,top] for each bond so that bottom and top were integral and every integer value in the interval was feasible for fixing the bond. In the next model the feasible integral values for a variable will take the form of a much punctured discrete interval, something like [0,2,4,6]. For applications with this structure, discrete methods can often be more effective than continuous ones. In fact, in Model 8.5, Who Has the Zebra, we already encountered this kind of situation in a hybrid model that used continuous and discrete methods. When a house is assigned to one gentleman, this pokes holes in the sets of houses that can be assigned to the others. The same is true of course for profession and the other categories. We dealt with this, not through the method of disjunctive linear programming, but

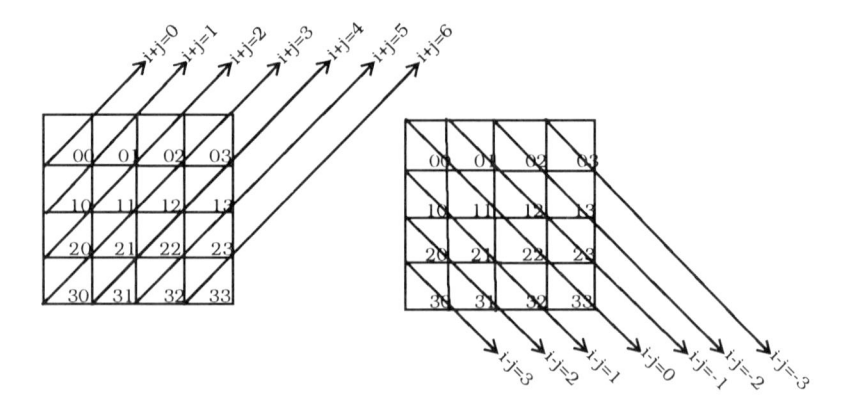

FIGURE 10.2 Left-to-right, upward and downward diagonals.

through a procedural apparatus that generated all possible permutations, always checking whether the next resource was available.

Model 10.2 The N-Queens

Can one place N queens on a N-by-N chessboard so that no two queens are in a position to capture one another? For N = 8 this question and related ones such as determining the total number of solutions have a long history going back to Gauss. In Falkowski and Schmitz (1986), it shown that there is a solution for all N > 3. The proof is an example of how experimentation with computer code can lead to the insight needed for the mathematical result.

To attempt to find a solution by means of a program, we can employ the generate-and-test method. A solution to the puzzle is a placement of the queens so that no two queens share the same row or the same column or the same diagonals. In Figure 10.2, i is used to denote a row, and the relationships between row i, column j and the diagonals are illustrated.

We will formulate two approaches to coding this problem. The first approach will be the more naive and will simply code up the process of placing the j th queen in a square in the j th column and checking that this placement is safe.

To keep track of where each queen is placed, let us represent the situation by means of an array qu[N] of type int where qu[j] will be the row on the chessboard of the queen placed in the j th column. Thus for jj < j < N, the queens in columns jj and j do not share the same row if

```
qu[jj] != qu[j]
```

10.2 Discrete models

They do not share the same left-to-right upward diagonal if

$$j - qu[j] \ne jj - qu[jj]$$

and they do not share the same left-to-right downward diagonal if

$$j + qu[j] \ne jj + qu[jj]$$

The most naive way to find assignments for all the `qu[j]` that would satisfy these conditions would be to cycle through all possible ways of placing the queens on the board until a satisfactory position is reached. We can do a little better if we note that for `jj < j < N`, the tests comparing `qu[jj]` and `qu[j]` can be essayed once `qu[jj]` and `qu[j]` are assigned. Let us encapsulate this in a procedure.

```
test(int jj,j)
{
    qu[jj] != qu[j];
    j - qu[j] != jj - qu[jj];
    j + qu[j] != jj + qu[jj];
}
```

The strategy is therefore to generate the positions for the queens successively and to check if the newest queen placed is in a position to attack one of the previously placed queens.

To generate a position for the queen in column j, we can simply loop through all possible rows:

```
generate(int j)   // j is the and loop control variable
{
    // qu[j]=0,...,qu[j]=k-1 do not lead to a solution
    or(int k=0;k<N;k++)  // If k==N, backtrack to j-1
        qu[j] = k;       // else place j th queen in row k
}
```

The main routine will contain the and loop for proceeding from column to column:

```
#define N 8   // The traditional chessboard
int qu[N];

21p_main()
{
    // Queens 0,...,j-1 placed in rows qu[0],...,qu[j-1]
    // so that no two queens can attack each other
```

```
    and(int j=0;j<N;j++) {
        generate(j);        // Place the j th queen
        and(int jj=0;jj<j;jj++)
            test(jj,j);     // Test safety of j th queen
    }
    output();               // The queen in column j is in row qu[j]
}
```

In this program backtracking will bring control back into the or loop in the `generate` routine to try further alternatives. The restrictions on assignment to a loop variable guarantee that the or loop variable k will not have been changed and will always be incremented correctly. Control thus returns to a subroutine called in the body of the and loop to which the loop control variable j has been passed. The parameter j is not altered by an assignment in the `generate` routine. Therefore we can identify this parameter with the and loop control variable. Another point to make is that the decision on where to place the queens 0,...,j-1 is not undone at level j or subsequent levels in the search; that is, the assignments to qu[0],...,qu[j-1] are not changed by execution of the body of the and loop with control variable equal to j. In other words, the state is recorded in qu[0],...,qu[j-1], and each sibling will be passed the same state.

On the other hand, when backtracking takes place and a new assignment qu[j] = k is about to be made in the or loop in the `generate` routine, the values of qu[j],...,qu[N-1] might not be the same as they were when the assignment to be made was the previous one, qu[j] = k-1. However, the assignment qu[j] = k will overwrite the contents of qu[j], and all the assignments to qu[0],...,qu[j-1] will have been preserved. Therefore the `test` routine performs properly. If it succeeds, the and loop invariant is preserved, since the contents of qu[j+1], ...,qu[N-1] are irrelevant here.

So what we are doing here is using the array qu as a stack; the element qu[j] can only be changed when j is the top level of the stack. In other words, the assignment qu[j] = i pushes an element on the stack, and in backtracking from level jj down to level j, the elements qu[jjj] for j <= jjj <= jj are in effect popped off the stack.

Note that here we have to declare explicitly in the and loop invariant that the sequence qu[0],...,qu[j-1] is consistent in that no queen menaces any earlier queen. Since this kind of global consistency is not automatic in working with variables of type int and double, it must be accounted for explicitly in the program.

The code describes an algorithm that is complete because all alternatives for assigning qu[j] are spanned in the or loop. So this code meets the horizontal and vertical conditions for sound and complete tree search. In the two previous generate-and-test situations, Model 8.2 and Model 8.3, the enumeration descended to the maximum depth of the tree until a solution was found, and no lookahead was used to prune

10.2 Discrete models

nodes; in the present model the test whether it is safe to place a queen in a particular row prunes nodes that can not possibly lead to a solution.

We now develop an approach to this model that speeds up the test phase. In this version we will have to take care that each child is presented with the same state as its siblings. We will use techniques known as *marking* and *trailing*.

When a queen is placed on the board, four "resources" are needed, namely a column, a row, a left-to-right upward diagonal, and a left-to-right downward diagonal. The test for safety when placing a queen is that no previous queen has used any of these resources. By proceeding column by column, the code will ensure that no column is shared by two queens. For the other resources, let us use a marking technique: A resource that is still unclaimed will be labeled FREE, while a resource that has been claimed by the queen in column j will be labeled with j. The rows can be represented by an integer array row, and the fact that the queen in column j occupies row i can be recorded by the assignment row[i] = j. In other words, we use j itself as a label to mark the row as claimed. The upward diagonals can be represented by an integer array up and a queen placed in row i and column j occupies the resource up[i+j]. This can be recorded by the assignment up[i+j] = j. Similarly, for the downward diagonals, we can use an integer array down. To avoid negative indices, we can shift everything by N and have the queen in row i and column j occupy the resource down[i-j+N], which will be signaled by the assignment down[i-j+N] = j. The additional declarations are

```
#define N 8
#define FREE N
int row[N];
int down[2*N];
int up[2*N];
int top;    // Top of the trail stack
```

Any unoccupied diagonal will be labeled FREE. In fact, to start the program we can label all resources as FREE:

```
initial_markings()
{
    and(int k=0;k<N;k++)
        row[k] = FREE;
    and(int k=0;k<2*N;k++)
        down[k] = FREE;
    and(int k=0;k<2*N;k++)
        up[k] = FREE;
    top = -1;       // Initial stack is empty
}
```

The simplification in the search will come in testing whether it is safe to place the queen in the `j`th column in the `i`th row. Indeed, the test code now becomes

```
check_free(int i,j)
{
    row[i] == FREE;         // Check row free
    down[i-j+N] == FREE;    // Check diagonals free
    up[i+j] == FREE;
}
```

This way we have replaced the loop of tests

```
        and(int jj=0;jj<j;jj++)
            test(jj,j);    //Test safety of j th queen
```

with a single test.

When the `j` th queen is placed in row `i`, the three resources must be marked with the label `j`; for this we call `label(i,j,j)` where

```
label(int i,j,tag) // row i, column j, label tag
{
    row[i] = tag;
    down[i-j+N] = tag;
    up[i+j] = tag;
}
```

However, when backtracking takes place and the placement of a queen in row `i` and column `j` is undone, these resources must be freed in order for the `check_free` tests to be correct. To reopen the resources labeled by the call `label(i,j,j)`, the call `label(i,j,FREE)` must be made.

To summarize, this strategy for solving the *N*-queens can be coded as follows:

```
21p_main()
{
    initial_markings();
    solve_queens();
    output();   // Left to reader
}

solve_queens()
{
    // For jj < j, the queen in column jj is
    // safely placed in the row and diagonals labeled jj
    // All unoccupied rows and diagonals are labeled FREE
    // top is equal to j-1
```

10.2 Discrete models

```
        and(int j=0;j<N;j++)    // j th column
            // Placing queen j in rows 0,...,i-1
            // does not lead to a solution
            or(int i=0;i<N;i++){    // If i==N, decrement j
                and(int k=top;k>=j;k--)    // Pop trail stacks
                    label(qu[k],k,FREE);
                check_free(i,j);    // Check row i and diagonals
                label(i,j,j);       // Push trail stacks
                qu[j] = i;          // Record placement
                top = j;            // Advance top of trail stack
            }
}
```

In the `solve_queens` code, we had to take care that each child of a parent node is presented with the same state as its siblings. This is necessary to preserve the clause of the and loop invariant that asserts that only occupied rows and diagonals are marked:

```
        // All unoccupied rows and diagonals are labeled FREE
```

In other words, fairness had to be addressed specifically.

The search is monotonic because, if a row or diagonal is marked at level j, it cannot be changed at later levels down the same path. The search is enumerative because the or loop cycles through all possible rows in order to place the jth queen. The search is coherent because the data structures and markings are set up so as to ensure that only consistent choices are made and that no queen can attack another.

Let us make some remarks of general import about this model. In the first coding of the application, we used the array qu as a stack to store the partial solutions to the puzzle that the code was generating. Popping the stack did not have to be done explicitly, since it was enough to reset the level j by overwriting the previous contents. In the second version of the code, we have three data structures, row, down, and up, that are used to maintain the state of the partial solutions. However, this time if a change is made to one of these arrays and this change does not put the search on a path leading to a solution, the change is undone in order to restore the array to its proper state. But rather than keeping a stack of row arrays, a stack of down arrays, and a stack of up arrays, what we do is keep track of the changes made to row, down, and up. These changes are undone by the call

```
            and(int k=top-1;k>=j;k--)
                label(qu[k],k,FREE);
```

This technique is called *trailing*. This technique is stack based; indeed, what we are doing is stacking the records of the changes to the arrays row, down, and up and undoing them when popping these records. The stack of changes to be undone is kept explicitly in the structure qu which is called a *trail stack*.

In this puzzle, in both versions of the code, the enumeration of the search tree was simplified by running tests on the placement of the jth queen before continuing on to the next queen. When such a test fails, the entire tree beneath that node is pruned. However, the enumeration is not compromised since no goal node has been pruned. This is similar in effect to the lookahead that the linear relaxation brings to constrain-and-generate models. For discrete applications there are notions of global consistency such as *arc consistency* and *k-consistency*; for more on these developments, a place to start is Mackworth and Freuder (1993). Maintaining this kind of consistency is an analog for the discrete case of maintaining linear consistency in the continuous constrain-and-generate paradigm.

There are several variations on this puzzle. For example, the size N of the chessboard can be changed. The simple backtracking code we have developed will have difficulty for N equal to 30, even if written in a language such as C and compiled for optimal performance.

A method for speeding up this kind of code is not to progress in the order $j=0,...,N-1$ but rather to choose the next queen to place differently. Randomization is one idea. Here, the simplest thing to do is take the backtracking code and to go across the columns in random order and to go across the rows in random order. This randomizes both the horizontal and vertical directions of the search, but soundness and completeness are preserved. Another way is to choose j for which the number of available rows is minimal; this is an example of a *first-fail* strategy (Haralick and Elliot 1980). In general, it is a good strategy to branch on the most "embattled" variable, by which we mean the most constrained variable.

Leaving the programming to the exercises, let us illustrate by example how the first-fail strategy can serve in this kind of application. The idea is to work on the column whose queen has the fewest possible squares available. Breaking ties by choosing the first column and the first row available, we immediately reach the state of Figure 10.3(a). Clearly the most embattled column is 5, counting as usual from 0. Placing the queen in the only available row leads to this state of Figure 10.3(b). Continuing, we find that the position of Figure 10.3(a) is untenable, and this is done with no backtracking *en route* to detecting the inconsistency. This contrasts mightily with the behavior of the above code in this situation. To utilize this strategy, one makes heavy use of the marking and trailing technique.Yet another interesting tactic is to use a local search algorithm (see Minton, Johnston, Philips, and Laird 1990). Finally, let us note that there is an obvious duality in this puzzle between the rows and columns of the chessboard. Interestingly, even this simple a duality can be put to good use (see Jourdan 1995).

10.2 Discrete models

In Chapter 12 we treat an analog of the first-fail strategy for continuous models, the *injury method*. In Chapter 11, we consider the variant of the N-queens puzzle which requires that all solutions be found.

End of Model 10.2

In both programs for the above model, the state maintained is a read-only list of the decisions that have been made up to the current level; these decisions cannot be changed at higher levels but can be read there so as to maintain consistency. In the artificial intelligence literature, this kind of search is known as *state space search*. The term *goal state* is used to denote a state that yields a solution to the problem at hand. Optimization and state space search are the subject of Section 11.4 and much of Chapter 14.

Exercises

10.2.1. Rewrite the code for Model 8.2 in and/or loop form. Apply comments that describe the appropriate loop invariants.

10.2.2. Add the machinery of Exercise 8.2.3 to the N-Queens code.

10.2.3. Write code to solve the problem of the Knight's Tour: Can the knight visit every square of an N-by-N chess board exactly once, starting in a corner and only making legitimate moves. Hint: At the outset, build a table to implement a heuristic that favors moves towards the boundary of the board. Compare the solution found by your code with the solution you yourself would find working with paper and pencil on the 8-by-8 problem.

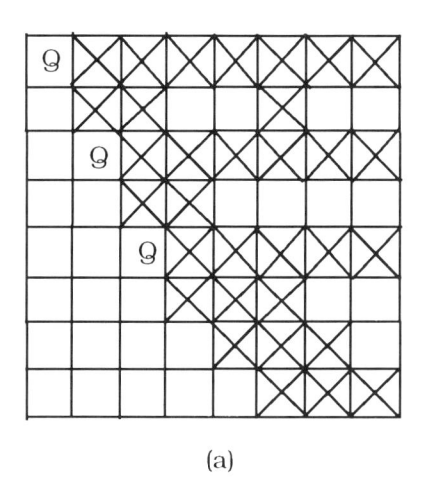

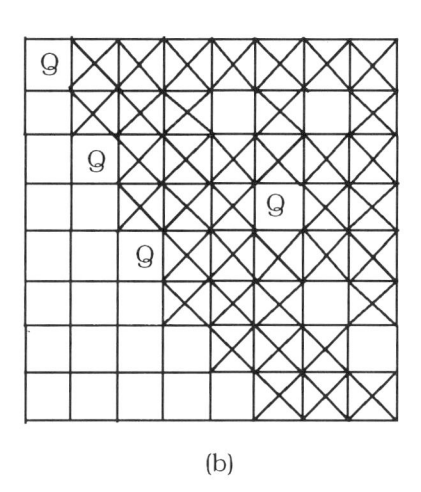

(a)   (b)

FIGURE 10.3 First-fail applied to the eight queens.

10.3 Recursion and nesting

In this section we address the question of how a tree search of the kind we have been discussing is coded in traditional programming languages. For discrete applications the functionality of persistent disjunction can be captured by conditional disjunction and recursion. To illustrate this here, let us write recursive code for the *N*-queens application that does not use persistent disjunctive constructs. This recursive code will follow the second algorithm given in Model 10.2. Here too care must be taken that the search conform to the vertical and horizontal conditions that we have been applying to the and/or loop combination.

The declarations are the same as before: The main procedure is changed to

```
2lp_main()
{
    initial_markings();
    if rec_solve_queens(0);
    then output();
    else printf("No solution found\n");
}
```

The following recursive procedure to solve the puzzle is derived in a straightforward manner from the solve_queens procedure in Model 10.2. In lieu of the loop invariant on the and loop, we have preconditions and postconditions on the recursive routine rec_solve_queens.

```
// For jj < j, the queen in column jj is safely
// placed in the row and diagonals labeled jj
// These labels will not be changed by the call to
// rec_solve_queens(j)
// All unoccupied rows and diagonals are labeled FREE
// The procedure call succeeds if N-j more queens can be
// safely added; otherwise the call fails
// The global variable top is equal to j-1
rec_solve_queens(int j)
{
    if j == N;      // Solution found
    then return;
    inner_loop(j);
}

inner_loop(int j)
{
    // Placing queen j in rows 0,...,i-1
    // does not lead to a solution
```

10.3 Recursion and nesting

```
    c_or(int i=0;i<N;i++){        // If i==N, decrement j
       and(int k=top;k>=j;k--)    // Pop trail stack
          label(qu[k],k,FREE);
       check_free(i,j);           // Check row i and diagonals
       label(i,j,j);              // Push trail stacks
       qu[j] = i;                 // Record placement
       top = j;                   // Reset top of trail
       rec_solve_queens(j+1);     // Succeed if this
                                  // recursive call succeeds
                  // Else go to top of loop and increment i
    }
}
```

Here we have formulated vertical conditions on the entry into the recursive procedure call `rec_solve_queens(j)` that correspond to the invariant on the and loop in the earlier code. The invariant conditions on the c_or loop are virtually the same as the horizontal conditions on the or loop in the earlier code.

However, the simple header `c_or(int i=0;i<N;i++)` of the c_or loop is also hiding a recursion. In cases where the horizontal movement across the tree is not so tidy, this part of the search will be coded by means of a recursion rather than a loop. In the example at hand, this means rewriting this loop as follows:

```
recursive_inner_loop(int i,j)
{
    i != N;    // Fail if N is reached
    and(int k=top;k>=j;k--)       // Pop trail stack
       label(qu[k],k,FREE);
    c_either {
       check_free(i,j);           // Check row i and diagonals
       label(i,j,j);              // Push trail stacks
       qu[j] = i;                 // Record placement
       top = j;                   // Reset top of trail
       rec_solve_queens(j+1);     // Succeed if this does
    }
    or recursive_inner_loop(i+1,j); // Else increment i
}
```

The call to `inner_loop(j)` must be replaced by a call to `recursive_inner_loop(0,j)`.

The and/or structure intrinsic to so many applications has been transformed into a double recursion. In fact, this double recursive structure underlies the WAM architecture used by the 2LP interpreter and other systems for programming with logic. The WAM unpacks the double recursion and optimizes the management of the resulting stacks (see Maier and Warren 1988).

The recursive approach can also be used for disjunctive linear programs and other situations involving continuous variables. For the record, let us rewrite the code for Model 9.2, Silicon Logic, this way:

```
2lp_main()
{
    data();
    setup();
    rec_satisfiability(0);
}

// X[0]==bit_0,...,X[j-1]==bit_{j-1} consistent with setup
// The call will succeed if an extension
// X[j]==bit_j,...,X[N-1]==bit_{N-1} can be found
rec_satisfiability(int j)
{
    if j == N;         // Solution found
    then return;

    // X[j]==b-1 does not lead to a solution
    c_or(int b=0;b<2;b++) { //If b==2, fail and decrement j
        X[j] == b;              //If successful, increment j
        rec_satisfiability(j+1);   // and succeed if this
                                   // recursive call does
    }
}
```

With this kind of code, sidetracking and recursion perform the function of deep backtracking. However, the code still relies on the behavior of the type `continuous` to restore the feasible region as the sidetracking takes place. We will see other examples of recursion used with continuous models in Sections 12.2 and 12.3.

In the two examples just considered, the recursive code highlights the strategy involved: Given a partial solution to the problem, take one more step, and then ask the recursion to continue the job.

Recursion is a powerful tool. Let us note, however, that when we use recursion to simulate persistent disjunction the simulation is restricted to the life of the top level recursive procedure. When the original call to the recursion returns, it is not possible to backtrack again into that code. In other words, this simulation of nondeterminism is not robust the way choice points are.

In our next example, we will see how recursion is most natural when there are several nestings of and/or constructs. Employing recursion in this setting lets the language interpreter's stack do much of the bookkeeping work. This is especially useful when the underlying support structures are not stacks represented as simple linear arrays, but are more complex tree structures.

10.3 Recursion and nesting

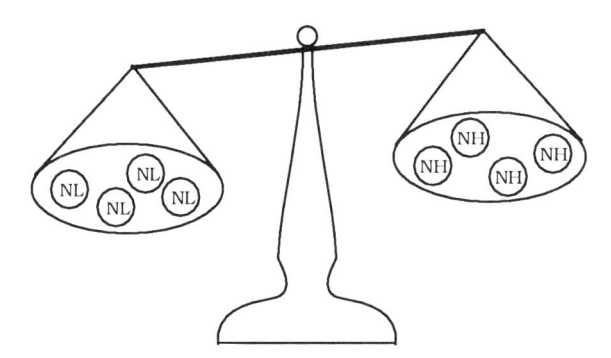

FIGURE 10.4 A scale with four coins on each side.

The next model dates back to medieval times and is narrated in the historical present. It is reported to be a lost tale from the *Arabian Nights*.

Model 10.3 The Twelve Coins

The magician at the court of Samarkand is presented with 12 gold coins and told that one of them is of a slightly different weight from the others, which are all identical, perfect coins. A balance scale is present, and the magician is challenged to determine which coin is imperfect and whether it is heavy or light. To do this, all he can use is the scale, and he can use the scale at most three times. When the magician balks and claims it is too hard, Queen Scheherazade jumps up and says that the hero of her tales solves this puzzle in one of his adventures and, what's more, does it by weighing exactly four coins in each tray each time. Is the queen right or is she spinning yarn 1002?

Let us put on our thinking caps and analyze the key elements involved here. At the beginning all the coins can be labeled as Unknown. What information is gathered when a weighing is made? The weighing can have one of three outcomes: The scale can tip down on the left side, stay balanced, or tip down on the right side. If the scale stays balanced, what we now know is that every coin that was weighed is a perfect coin, but what we knew about the unweighed coins has not changed. If the scale tips down to the left, then we know that all coins that were on the left side can not be light coins, that the coins that were on the right are not heavy coins, and that the unweighed coins are all perfect coins. The corresponding conclusions can be made if the balance tilts the other way. So in addition to Unknown, there are three other labels that can be affixed to coins: Perfect, NotHeavy, and NotLight. As the result of a weighing, these labels will change, and the number of coins in each category will be updated. Let us also note that if a coin labeled NotHeavy is on the side of the scale that tips down, then this coin must be perfect, and similarly for a NotLight coin that is on the side that tips up. So abstractly, before each weighing, what we are presented with is a certain number of coins that are not heavy, a certain number of coins that are not light, a certain number of perfect coins, and a certain number of coins that are unknown. Let us denote these numbers by nh,nl,p,u. The sum nh + nl + p + p is 12, of course. This four-tuple of numbers completely describes the current state of affairs. The result of a weighing will be to update these four numbers, leading to a new state.

Let us make this "scale arithmetic" a bit more precise. Suppose that we are given numbers nh,nl,p, and u that sum to 12 and that are the numbers of coins in each of the four categories. Then suppose that on the left tray we place nh1 coins that are not heavy, nl1 coins that are not light, p1 perfect coins, and u1 unknown coins. Let us also suppose that on the right tray we place nh2 coins that are not heavy, nl2 coins that are not light, p2 perfect coins, and u2 unknown coins. For this to be possible, we must have the inequalities nh1 + nh2 <= nh, nl1 + nl2 <= nl, p1 + p2 <= p and u1 + u2 <= u, and we must also have the equalities nh1 + nl1 + p1 + u1 == 4 and nh2 + nl2 + p2 + u2 == 4. If the scale tips down to the left, the new state is given by the four numbers

```
nh2+u2,            // New number of coins known to be NotHeavy
nl1+u1,            // New number of coins known to be NotLight
(nh-nh2)+(nl-nl1)+(u-u1-u2),    // New number of Perfects
0                  // New number of Unknowns
```

For the scale to be able to tilt down on the left side, it must be possible either for a heavy coin to be on the left side or for a light coin to be on the right side. This gives rise to a simple test that can be made before proceeding to analyze this case:

```
can_tip_left(int nh1,u1,nl2,u2)
{
    not {nh1 == 0; u1 == 0; nl2 == 0; u2 == 0;}
}
```

A perfectly similar analysis applies to the other two possible outcomes of a weighing leading to routines, `can_tip_right` and `can_balance`.

Our strategy will be to use a recursive method. The recursion will be on the number of weighings. Each weighing will push us up one level in the recursion. Given a weighing level w and state values nh,nl,p,u, the recursive function `solve_coins` will loop through all possible weighings; for each weighing and each possible outcome of the weighing, a recursive call at the next level will be made. For example, in the case of the scale's staying balanced, this call will be

10.3 Recursion and nesting

```
solve_coins(
    w+1,                            // New level
    nh2+u2,                         // New NotHeavy
    nl1+u1,                         // New NotLight
    p+nl-nl1+nh-nh2+u-u1-u2,        // New Perfects
    0 );                            // New Unknowns
```

What makes this model different from the others we have been considering is that when a weighing is made, we do not simply go on to a single next stage. This time the next stage splits into three parts, one for each possible outcome of the weighing. So after the weighing is made, the recursive call to `solve_coins` will be made three times, one for each outcome. In logical terms this is a conjunction of recursive calls.

The recursion starts at level 0 and the initial call will be

```
solve_coins(0,0,0,0,COINS);
```

The recursion will terminate successfully, if at level 3 or before the number of perfect coins is 11 and the number of unknown coins is 0. If the level reaches 4, then no solution has been found and the recursive call will fail.

The program will begin with symbolic constants for the number of coins and the number of weighings:

```
#define COINS 12
#define WGHS 3
```

The routine to check that a solution has been reached can be

```
solution_found(int p,u)
{
    p >= COINS - 1;    // At least all coins but one perfect
    u == 0;            // No coins unknown
}
```

The main routine and the recursion can take the form

```
21p_main()
{
    if solve_coins(0,0,0,0,COINS);
    then printf("A solution strategy weighing 4 vs 4
                      each time indeed exists\n");
}
```

```
solve_coins(int w,nh,nl,p,u)
{
    if solution_found(p,u);
    then return;

    w < WGHS;   // Weighing still possible ?

    // Select coins for weighing
    c_or(int nh1=0;nh1<=nh;nh1++)       // NotHeavies on left
    c_or(int nh2=0;nh2<=nh-nh1;nh2++)// NotHeavies on right
        c_or(int nl1=0;nl1<=nl;nl1++)       // and so on
        c_or(int nl2=0;nl2<=nl-nl1;nl2++)
            c_or(int p1=0;p1<=p;p1++)
            c_or(int p2=0;p2<=p-p1;p2++)
                c_or(int u1=u;u1>=0;u1--)
                c_or(int u2=u-u1;u2>=0;u2--)
    {
        four(nh1,nl1,p1,u1,nh2,nl2,p2,u2);// 4 to a tray ?
        if can_tip_left(nl1,u1,nh2,u2);
        then
        solve_coins(w+1,
                nh2+u2,nl1+u1,p+nl-nl1+nh-nh2+u-u1-u2,0);
        if can_balance(p1,u1,u2,nl1,nl2,nh1,nh2,p);
        then
        solve_coins(w+1,nh-nh1-nh2,nl-nl1-nl2,
                    p+nh1+nh2+nl1+nl2+u1+u2,u-u1-u2);
        if {
            can_tip_right(nl2,u2,nh1,u1);
            not same_as_tip_left(nl2,u2,nl1,
                                        u1,nh1,nh2,p1,p2);
        }
        then
        solve_coins(w+1,
            nh1+u1,nl2+u2,p+nl-nl2+nh-nh1+u-u2-u1,0);
    }
}
```

The remaining two procedures in the body of the nested loops are straightforward. First, there is the check that four coins are placed in each tray:

```
#define FOUR 4
four(int nh1,nl1,p1,u1,nh2,nl2,p2,u2)
{
    FOUR == nh1 + nl1 + p1 + u1;   // 4 coins in each tray
    FOUR == nh2 + nl2 + p2 + u2;
}
```

10.3 Recursion and nesting

Second, we have included a test to check if the results of the scale's tipping down to the right were perfectly symmetric with those obtained in the case the scale tipped the other way. This test is of course not strictly needed but does eliminate some amount of unnecessary work. The code for this is similar to the `can_tip_left` routine above:

```
same_as_tip_left(int nl2,u2,nl1,u1,nh1,nh2,p1,p2)
{
    nh1+u1==nh2+u2;  nl2+u2==nl1+u1;
}
```

As pointed out above, in this model a conjunction of recursive calls is made each time the juxtaposed `c_or` loops find a possible weighing. The juxtaposition of the eight `c_or` loops is itself a conjunction of disjunctions. This combination of conjunction, conditional disjunction, and recursion captures the nested and/or structure associated with this model. Without recursion it would be a nesting of conjunctions of persistent disjunctions. However, even with only three weighings, it is awkward to bypass the recursion and write things out explicitly.

Not surprisingly, the program confirms that the queen is right. If one wants to use this code to demonstrate one's skill in an interactive situation, one can begin by weighing any eight of the coins provided by one's adversary and asking the adversary for the outcome. If the scale balances, say, then one can run the code with the number of weighings set to 2 and with initial call `solve_coins(0,0,0,8,4)`. Continuing this way will, in the end, isolate the bad coin. Alternatively, one can build a tree in which the complete strategy is encoded.

Writing out the eightfold `c_or` loop can strike one as awkward. One reaction is to find a way to use recursion to write this code, and the reader is invited to pursue this in the exercises. A better way perhaps would be through a macro facility that enables one to define short-hand for each pair of `c_or` loops.

The code satisfies the horizontal and vertical conditions for sound and complete tree search. The vertical conditions can be formulated as pre- and postconditions on the recursive routine `solve_coins`. The horizontal conditions can be formulated as loop invariants on the nested `c_or` loops.

End of Model 10.3

Exercises

10.3.1. Try to figure out how Scheherazde's hero, Sinbad, knew it was enough to weigh four coins in each tray in an era without computers (see Greenblatt 1965).

10.3.2. If the routine to check for a solution to the 12 coins puzzle is changed to

```
solution_found(int p,u)
{
   p == COINS - 1;    // At least all coins but one perfect
   u == 0;            // No coins unknown
}
```

the code runs the same way. Explain.

10.3.3. Explain what changes if only the first disjunctive loop in solve_coins is a c_or loop and all the others are or loops. Also explain why the code will correct but terribly slow if all the disjunctive loops are or loops.

10.3.4. Write recursive code in a traditional programming language for the Knight's Tour of Exercise 10.1.2.

10.3.5. In a traditional programming language, write recursive code for the N-Queens problem. Test this code on values of $N \geq 30$. Then write recursive code that uses randomization, local search, and a bit of backtracking to try to find a solution. Compare with the previous code.

10.4 Unit resolution

In Model 9.2, Silicon Logic, we took the approach to determining whether a CNF formula was satisfiable by explicitly building a truth assignment that made the formula true. The method is called *semantic* because it constructs an interpretation that gives meaning (true or false) to the propositional variables and thereby to the formula itself. If the search for an assignment to satisfy a formula fails, the formula is unsatisfiable and its negation is a propositional tautology. As a technique for proving formulas to be propositional tautologies, the semantic method reflects the fact that the set of propositional tautologies is in co-NP.

There is another approach to finding out if a formula is satisfiable or if its negation is a tautology that does not refer to satisfying assignments but that works directly on the syntax of formulas. This is the method of *formal proof*. A formula that has a proof is called a *theorem*. If all tautologies are theorems of a formal proof system, the system is complete. If all theorems of a formal proof system are tautologies, the system is sound.

The traditional systems of formal proof used in axiomatic mathematics consist of axioms and rules of inference such as *modus ponens*. These systems tend not to lend themselves to computer implementation. Rather, a more readily coded technique known as *resolution* is

10.4 Unit resolution

employed. Let us briefly describe the method of resolution for proving theorems in propositional logic. We begin with some definitions.

Recall that a literal is a simple kind of clause, one composed of a propositional variable p or its negation $\bar{p}$. If λ is a literal of the form $\bar{p}$, then $\bar{\lambda}$ is defined to be the literal p.

The basic step in a resolution proof is simple. Suppose that $\lambda_1,\ldots,\lambda_n,\mu_1,\ldots,\mu_k$ are literals and that λ is a literal. If we have produced the formulas

$$(\lambda_1 \wedge \ldots \wedge \lambda_n) \rightarrow \lambda$$
$$\lambda \rightarrow (\mu_1 \vee \ldots \vee \mu_k)$$

we can introduce the formula

$$(\lambda_1 \wedge \ldots \wedge \lambda_n) \rightarrow (\mu_1 \vee \ldots \vee \mu_k)$$

In clause form, the first formula is

$$\bar{\lambda}_1 \vee \ldots \vee \bar{\lambda}_n \vee \lambda$$

and the second formula is

$$\bar{\lambda} \vee \mu_1 \vee \ldots \vee \mu_k$$

The derived formula, known as the *resolvent*, takes the clause form

$$\bar{\lambda}_1 \vee \ldots \vee \bar{\lambda}_n \vee \mu_1 \vee \ldots \vee \mu_k$$

Note that the resolvent is composed of the literals from the first two clauses less the opposing pair $\lambda, \bar{\lambda}$.

It is convenient to write clauses $\lambda_1 \vee \ldots \vee \lambda_n$ as sets $\{\lambda_1, \ldots, \lambda_n\}$ to emphasize the commutativity and idempotence of disjunction. Given two clauses $\{\lambda_1,\ldots,\lambda_n\}$ and $\{\mu_1,\ldots,\mu_k\}$ such that λ_i is equal to $\bar{\mu}_j$, the new clause $\{\lambda_1,\ldots,\lambda_{i-1},\lambda_{i+1},\ldots,\lambda_n,\mu_1,\ldots,\mu_{j-1},\mu_{j+1},\mu_k\}$ is a resolvent of the first two. The resolvent can then be added to the current set of clauses and new resolvents can be sought. Clearly, if a set of clauses is satisfiable, then adding a resolvent yields another satisfiable set of clauses. Adding a resolvent to a set of clauses is called a *resolution step*.

A clause with one literal is a *unit clause*. A contradiction can be detected whenever we have two unit clauses of the form $\{p\}$ and $\{\bar{p}\}$. Resolving these two yields the empty clause $\{\}$, which is the sign that a contradiction has been reached.

Given a set C of clauses, a *resolution proof* is a sequence $C = C_1,\ldots,C_n$ such that each C_{i+1} results from C_i by adding a resolvent of two clauses in C_i and such that C_n contains the empty clause. Clearly resolution is sound in that if the empty clause is reached, the original clauses could not have been satisfiable.

The following completeness theorem is due to Robinson (1965). We give a compact version of the proof; for a fuller exposition, we refer the reader to Lewis and Papadimitriou (1981).

Theorem *Resolution is sound and complete.*

Proof We have already noted that resolution is sound. To prove completeness, one proceeds by induction on the number n of propositional variables. If the number is 1, the result is immediate. Assume that the result to hold for $n - 1$ variables, and let C be an unsatisfiable set of clauses in n variables. There must be a variable p such that p and $\bar{p}$ occur in different clauses in C: Otherwise, the set of clauses would be satisfiable. We divide the problem into two subproblems, one where p is made to be true and one where $\bar{p}$ is made to be true. In the first, we form a new set C(p) by deleting all clauses containing p and deleting $\bar{p}$ from all the remaining clauses. By assumption, the set C(p) is not satisfiable; by the induction hypothesis, the empty clause can be derived from C(p). But then the clause $\{\bar{p}\}$ can be derived from C using exactly the same steps. In a similar manner we can form the set C($\bar{p}$) and conclude that the unit clause $\{p\}$ can be derived from C. But then these two unit clauses yield the empty clause. ∎

In going from the semantic method to the syntactic method, what we have done is to replace one search problem by another: The co-NP search to show there is no satisfying assignment for the negation of a formula has been transformed into a search for a sequence of resolvents that yield the empty clause. The resolution proof picture has an intuitive appeal in that the search is leading to a proof, while the co-NP view has us proving something does not exist. But the resolution method does not mean the problem of checking tautologies is in NP; the length of the shortest resolution proof can be exponentially long compared to the size of the formula (see Haken 1985). As an example, we have the pigeonhole principle which is described in Section 13.2. Of course there are more sophisticated ways of implementing resolution-based theorem provers than the simple vanilla variety that we are discussing here. However, for each such method, a challenge problem has been found on which the method exhibits exponential behavior.

Computer scientists worry about space. The basic resolution step can add a large clause to a growing set of clauses. However, when one of the resolved formulas is a unit clause, the situation is different, for rather than adding a clause, we can view this as shortening an existing clause. A resolution step where one of the clauses is a unit clause is called a *unit resolution step*. A resolution proof in which all resolutions are unit resolutions is called a *unit resolution proof*. Unit resolution is sound of course, but incomplete, as the following inconsistent quartet of clauses illustrates:

10.4 Unit resolution

$$\{p,q\} \; \{\bar{p},q\} \; \{\bar{p},\bar{q}\} \; \{p,\bar{q}\}$$

To extend unit resolution to a complete method, one can add a branching mechanism to obtain a blend of syntactic and semantic methods. There is an important family of proof methods based on unit resolution and branching; these are known as Davis-Putnam methods, see Davis and Putnam (1960) and Loveland (1978). In fact, the second algorithm of Model 9.2 is a very naive example of this kind of method. (Improvements will follow from the discussions in Chapters 12 and 13.) As noted in Section 10.1, this kind of method is sound and complete.

By the *unit resolution problem* or *UR*, we mean the language of CNF formulas, which can be shown to be inconsistent by means of a unit resolution proof. We also have

Theorem *If a CNF formula is in UR, then its translation into linear constraints yields an infeasible linear program.*

Proof The proof of this is by a bottom-up argument. The unit clauses translate to constraints that fix the corresponding continuous variables at 0 (for a negative unit clause) or 1 (for a positive unit clause). Then if a unit clause can be derived from one of the initial unit clauses and some clause with two literals, the corresponding variable must also be frozen at 0 or 1. Proceeding inductively, one sees that the continuous variable corresponding to every derived literal must be frozen at 0 or 1. If unit resolution leads to the empty clause, this means that some continuous variable will have to take both values 0 and 1, which is absurd. ∎

For all its virtues, linear programming itself does not automatically add anything to the theorem proving arsenal beyond unit resolution. We have the following remark of Jeroslow (1989):

Remark Let C be a set of nonunit clauses on the propositional variables $p_1,...,p_n$. Then, if these clauses are translated into constraints on the continuous variables $x_1,...,x_n$, the constraints will be consistent, and $x_1 = ... = x_n = .5$ will provide a feasible solution.

However, as we will see in Section 12.1, the linear relaxation and the witness point can help guide the search process in an attempt to find a solution.

A clause with at most one positive literal is called a *Horn clause*. Unit resolution is complete for detecting the unsatisfiability of a set of these clauses. This can be proved in a way very similar to the previous theorem; it is left as an exercise.

A rule of the form

$$(p_1 \wedge ... \wedge p_n) \to q$$

takes the clause form

$$\overline{p}_1 \vee \ldots \vee \overline{p}_n \vee q$$

which is a Horn clause. Rules of this form arise in expert systems and designers strive to keep expert system rules in this form. In the following model we have a (very small) expert system that is composed of rules that are Horn clauses.

Model 10.4 Expert Automotive Repair

A highly automated auto repair shop has an expert diagnostic system to determine what is causing a malfunction in a driver's car. The system is based on a set of rules; the fragment composed of 14 rules dealing with problems starting the car is given in the routine knowledge_base below. These are all propositional Horn clause rules. The battery is checked routinely, and if the battery is not dead, the possible diagnoses the subsystem is capable of reaching are 3 in number: It can be the electrical system, the choke, or the solenoid switch.

The rules of the knowledge base can be translated into linear constraints. In addition to the 3 diagnoses, there are 23 different checks that are involved. The declarations of continuous variables are

```
#define DIAGNOSES 3
#define CHECKS 23
continuous X[CHECKS];
continuous Diag[DIAGNOSES];
```

To enter the rules, we can introduce a collection of macro definitions that will translate the rules directly into constraints:

```
#define Rule 1 <= 1 -
#define yields +
#define plus + 1 -

#define electrical_problem Diag[0]
#define choke Diag[1]
#define solenoid_switch Diag[2]

#define good_gas_feed X[0]
#define start_and_die X[1]
#define batteryOK X[2]
#define fusesOK X[3]
#define misstart X[4]
#define radioOUT X[5]
#define fusesBAD X[6]
#define fusesNEW X[7]
```

10.4 Unit resolution

```
#define bad_weather X[8]
#define bucking X[9]
#define signals_flicker X[10]
#define cough X[11]
#define backfire X[12]
#define gag X[13]
#define NObackfire X[14]
#define gas_line_clean X[15]
#define fuel_filter_clean X[16]
#define carburetorOK X[17]
#define fuel_injectionOK X[18]
#define kicks_over X[19]
#define dies X[20]
#define extreme_cold X[21]
#define heavy_rain X[22]
```

The rules are summarized in the following procedure:

```
knowledge_base()
{
    Rule    good_gas_feed plus start_and_die plus batteryOK
            yields electrical_problem;
    Rule    fusesOK plus misstart plus radioOUT
            yields electrical_problem;
    Rule    fusesBAD plus fusesNEW
            yields electrical_problem;
    Rule    bad_weather plus good_gas_feed
                plus misstart plus bucking
            yields choke;
    Rule    good_gas_feed plus start_and_die
            yields choke;
    Rule    start_and_die plus signals_flicker
                plus bad_weather
            yields solenoid_switch;
    Rule    misstart plus good_gas_feed plus batteryOK
            yields solenoid_switch;
    Rule    cough plus backfire
            yields misstart;
    Rule    gag plus NObackfire
            yields misstart;
    Rule    gas_line_clean plus fuel_filter_clean
                plus carburetorOK
            yields good_gas_feed;
    Rule    gas_line_clean plus fuel_filter_clean
                plus fuel_injectionOK
            yields good_gas_feed;
    Rule    kicks_over plus misstart plus dies
```

```
                yields start_and_die;
    Rule        extreme_cold
                yields bad_weather;
    Rule        heavy_rain
                yields bad_weather;
}
```

In using the expert system, first the driver is posed a set of questions through a user interface; this information is translated into a set of positive unit clauses that are added to the clauses of the knowledge base. Then a checklist is run through by an apprentice mechanic, and these results are also added to the knowledge base as unit clauses. To reach a recommendation, each of the possible diagnoses is considered in turn, and the expert system tries to deduce the diagnosis from the augmented knowledge base:

```
21p_main()
{
    knowledge_base();
    driver_information();
    apprentice_input();
    and(int i=0;i<DIAGNOSES;i++)
        if not Diag[i] == 0;    // Is negation inconsistent ?
        then diagnosis(i);      // If so, conclude diagnosis
}
```

The diagnoses that can be given are DIAGNOSES in number. If diagnosis i is made, the following procedure will print it out:

```
diagnosis(int i)
{
string diagnoses[DIAGNOSES];

    diagnoses = {
        "Electrical system; bad news",
        "The automatic choke; it could be worse",
        "The solenoid switch; this part must be ordered"
    };
    printf("\nThe diagnosis is\n %s\n",diagnoses[i]);
}
```

In this particular instance, the driver's responses to the questions posed lead to the following propositions:

10.4 Unit resolution

```
#define reported ==1

driver_information()
{
    cough reported;
    backfire reported;
}
```

The apprentice's check determines further important information:

```
#define determined ==1

apprentice_input()
{
    gas_line_clean determined;
    fuel_filter_clean determined;
    fuel_injectionOK determined;
    batteryOK determined;
}
```

For our hapless driver the system makes a recommendation that might keep the car in the shop for several days.

The diagnosis is:
The solenoid switch; this part must be ordered

End of Model 10.4

The beauty of unit resolution is that it makes it relatively easy to explain the sequence of reasoning steps that were taken to reach the conclusion. This is extremely important for expert systems. When branching or more elaborate resolution steps are required, extracting explanations demands somewhat more attention.

If a rule has the form $(p_1 \wedge ... \wedge p_n) \rightarrow (q \vee r)$, then it is not a Horn clause. With rules like this in the system, unit resolution will not be complete, and so some branching process will be needed. The way a disjunctive conclusion such as $q \vee r$ typically arises in an expert system, however, is that each branch of the conclusion will give rise to a different Horn clause rule modified by a certainty factor. For example, we might have $(p_1 \wedge ... \wedge p_n) \rightarrow q$ (50%) and $(p_1 \wedge ... \wedge p_n) \rightarrow r$ (40%). We are now in the realm of probabilistic logic. In practice, these situations are often dealt with by means of a *certainty calculus* such as that of MYCIN (Shortliffe 1976).

In this last model the knowledge base consisted of a set of propositional formulas. In many situations it is not possible to represent the knowledge base so simply and a form of first order logic must be used. This requires a much more elaborate deductive mechanism.

We conclude this section with some complexity theoretic remarks. First, the language UR is easily seen to be in P. As we noted in Section 9.3, LP is also in P. There is a notion of reduction that serves to analyze the class P of polynomial time problems. We require a pair of definitions.

A function f is said to be *computable in log space* if it there is an integer k and a Turing machine with a read-only input tape and a write-only output tape such that $y = f(x)$ can be computed using no more than $k^*\log(|x|)$ squares on the work tapes. A problem in P is said to be *P-Complete* if every problem in P can be reduced to it by means of a log space computable function.

We have the following result of Jones and Laaser (1976):

Theorem *UR is P-Complete.*

The proof of the theorem directly models polynomial time computations by means of Horn clauses and invokes the fact that unit resolution is complete for detecting the unsatisfiability of a set of these clauses.

Since the mapping of clauses into constraints of Model 9.2 is computable in log space, we have a result of Dobkin, Lipton, and Reiss (1979):

Theorem *LP is P-Complete.*

This theorem helps explain the universality of linear programming. It also helps explain why one constantly searches for better performing special algorithms for particular families of linear programs; one wants to avoid the overhead of solving everyone else's problems! In fact, we have already seen examples where linear programming subsumes other algorithms such as the greedy method for the linear knapsack of Model 2.4, Jackson's preemptive strategy of Model 4.3, and the Perceptron Learning Theorem of Section 5.3. However, it is this very universal character of linear programming that makes it such a widely applicable tool.

Exercises

10.4.1. We have another example from Lewis Carroll. We are given the following information:

> When I detest an animal, I avoid it.
> No animals are carnivorous unless they prowl at night.
> No cat fails to kill mice.
> Every animal that likes me is in this house.
> Kangaroos are not suitable for pets.
> Only carnivorous animals kill mice.
> I detest animals that do not like me.
> Animals that prowl at night always love to gaze at the moon.
> Every animal that loves to gaze at the moon is suitable for a pet.

10.4 Unit resolution

We are asked to prove on the basis of these statements that "I always avoid a kangaroo." A translation of this statement into propositional logic is $kanga \to avoid$ or $\overline{kanga} \lor avoid$. Negating this formula we have $kanga \land \overline{avoid}$. This is in CNF and consists of two clauses, $kanga$ and $\overline{avoid}$. Show that there is a unit resolution proof of $kanga \to avoid$ by translating the clauses into constraints so as to find an infeasible linear program.

10.4.2. Let F be a set of formulas, and let f be a formula. Then we say that f is a *logical consequence* of F if every assignment that satisfies F also satisfies f. Let H be a set of Horn clauses. Suppose that the disjunction $p \lor q$ is a logical consequence of H. Show that either p is a logical consequence of H or q is a logical consequence of H. Generalize. Comment on what this says about Horn clauses and intuitionistic and classical disjunction.

10.4.3. Show that unit resolution is complete for Horn clauses. Hint: Think bottom-up.

11

Depth-First Branch-and-Bound Search

In the constrain-and-generate and generate-and-test applications we have been treating so far, the emphasis has been on finding a reasonably good solution. In this chapter we will be looking for the best possible solution. Such an application calls for a two-sided look at the situation. On the one hand, we have to generate solutions to the constraints and logical requirements of the application, and on the other hand, we must determine which of these solutions yields the best possible value for the objective function. In this chapter we approach this new problem using depth-first search. In the next two chapters, we will go further into fine tuning depth-first search strategies and into ways of trying to keep search to a minimum. In Chapter 14 we look at alternative methods of search.

11.1 Finding all solutions

Let us first address the issue of generating not just one but all solutions to a generate-and-test or constrain-and-generate application. With the application, we can associate a tree where branching in the tree corresponds to choice points that are created during code execution. The goal nodes of this tree correspond to solutions of the constraints and logical requirements.

In order to reach all the goal nodes, the code must backtrack after each solution is found and proceed to find another one. This can be carried out in 2LP by means of the find_all construct whose syntax is

    find_all <statement>

The semantics of this construct are that after <statement> succeeds, backtracking takes place and the next solution to <statement> is sought. This process continues until all solutions have been found.

When no more solutions are possible, the `find_all` exits successfully without having changed the original feasible region. However, if <statement> itself never succeeds, then the `find_all` <statement> also fails.

The reader has doubtless asked what was cut from Lewis Carroll's prose and replaced with three dots in the statement of the puzzle of Model 8.3, Salt and Mustard:

> The problem is to discover whether these rules are compatible ....

In fact, it was a request to determine all solutions to the puzzle:

> and if so what arrangements are possible.

Let us meet the author halfway and determine the total number of different solutions. This can be done by wrapping the call to `solve_salt_and_mustard` in the `find_all` construct as follows:

```
21p_main()
{
int count;

    count = 0;
    find_all {
        solve_salt_and_mustard();
        count = count + 1;
    }
    printf("The total number of solutions is %d\n",count);
}
```

It turns out, however, that the solution we found with the previous code is the only solution, and in fact, we have met the author all the way!

Model 11.1 All the N-Queens Solutions

Let us consider the problem of finding the number of possible solutions to the N-queens puzzle. For this we can simply wrap the code from Model 10.2 with the `find_all` construct:

```
21p_main()
{
int count;

    count = 0;
    find_all {
        initial_markings();
```

11.1 Finding all solutions

```
        solve_queens();
        count = count + 1;
    }
    printf("The number of solutions found is %d\n",count);
}
```

With N equal to 10, when the `find_all` is placed around the code in the main procedure this way, 724 distinct solutions are found.

In this example the loop invariants can be updated to reflect the fact that solutions are being found as the code proceeds. The principal change is that the `and` and `or` loops can be exited successfully many times.

```
solve_queens()
{
    // For jj < j, the queen in column jj is
    // safely placed in the row and diagonals labeled jj
    // All unoccupied rows and diagonals are labeled FREE
    and(int j=0;j<N;j++)      // If j==N, new solution found
        // top is equal to j-1
        // Placing queen j in rows 0,...,i-1
        // does not lead to a new solution
        or(int i=0;i<N;i++){   // If i==N, decrement j
            and(int k=top;k>=j;k--)    // Pop trail stacks
                label(qu[k],k,FREE);
            check_free(i,j);       // Check row i and diagonals
            label(i,j,j);          // Push trail stacks
            qu[j] = i;             // Record placement
            top = j;               // Reset top of trail stack
        }
}
```

In Figure 11.1 the tree for the 4-queens problem is illustrated. In terms of tree search, wrapping the code to search a tree in a `find_all` enumerates the solutions from the leftmost to the rightmost, admittedly a brute force approach to determining the number of solutions. For more informed approaches to the "all solutions" version of the N-queens problem, see Reingold, Nievergeld, and Deo (1977).

In Section 10.3 we noted that simulating persistent disjunction by means of conditional disjunction and recursion does not create choice points that can be returned to after the recursion completes. For finding all solutions, this means that one cannot simply wrap the recursion in the `find_all` construct. Instead, one must never let the recursive procedure succeed. With the N-queens example, this means changing the escape clause in the `rec_solve_queens` procedure to

```
    if j == N;   // Solution found
    then {
        count = count+1;    // Increment count
        0 == 1;     // Fail and go back for next solution
    }
```

yielding a new routine which, for want of a better name, we can call `continual_rec_solve_queens`.

We change the main procedure to

```
21p_main()
{
    initial_markings();
    if continual_rec_solve_queens(0); // This recursion
                // returns fail after all solutions are found
    then printf("Something is rotten in the state of DK");
    else printf("A total of %d solutions found\n",count);
}
```

Note that the `find_all` construct is not needed here; the recursion has the forced backtracking mechanism built in. This recursive code can readily be written in a traditional programming language.

End of Model 11.1

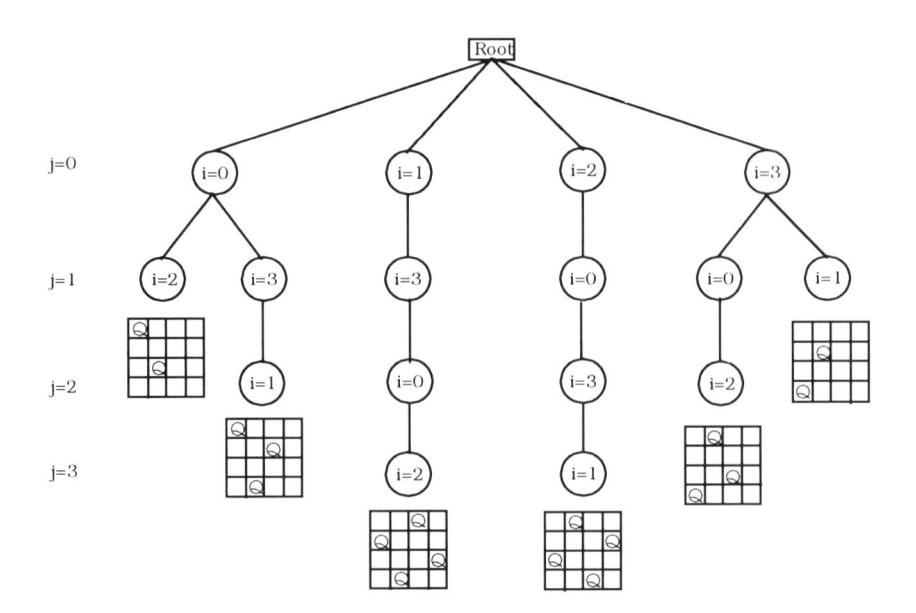

FIGURE 11.1 Nodes visited in the 4-queens problem.

11.1 Finding all solutions

The approach in this last model for finding all solutions is bovine: We simply let the machinery for finding a solution by enumeration continue on even after solutions are found, thereby taking the entire search tree into consideration. On examples with large search space, this method will prove unwieldy. However, one cannot expect the problem of finding all solutions to lend itself to shortcuts or other magic bullets. In fact, for NP-Complete problems, the same is already true of the problem of finding the number of solutions or even of determining whether there are at most k solutions to a problem for given k. For if the problem is NP-Complete, then the "at most k solutions" version of the problem is co-NP hard. That is, the co-NP Compete problem of determining whether there are no solutions is reducible to the k solutions problem with $k = 0$.

Let us return to the topic of optimization. The simplest way to find the *best* solution for an application is to generate *all* solutions, recording the optimal value of the objective function at each of these solutions. Then one can simply select the best recorded solution. Let us apply this to Model 9.1, Motorcycles Inc. Reconsidered. Now we want to produce all consistent solutions in order to find the optimal value that the objective function can have while the logical requirements of the model are satisfied. Let us use the first variant of the decide routine of Model 9.1:

```
decide(continuous Bike, double low)    // Version I
{
    either Bike == 0;
    or Bike >= low;
}
```

With this routine, we can rewrite the main procedure to be

```
2lp_main()
{
double best_so_far;      // New identifier to cache solutions
int cnt;                 // To count solutions

    cnt = 0;                       // Initialize the solution count
    best_so_far = 0.0;    // No solution found yet

    data();                                 // As in Model 9.1
    setup();                                // Ditto
    management_constraint();                // Ditto

    find_all {
        and(int i=0;i<MODS;i++)             // Ditto
            decide(Bike[i],lo[i]);
        max: Z;                             // Ditto
        if wp(Z) > best_so_far;             // Best solution so far ?
        then best_so_far = wp(Z);
```

```
        cnt = cnt + 1;               // Increment count
    }

    printf("The projected revenue is %0.f\n",best_so_far);
    printf("The number of solutions found is %d\n",cnt);
}
```

Note that we have introduced `best_so_far` to keep track of better and better values of Z as they are generated. We find that there are three solutions in all and that the best solution is 108,400. This verifies that Version III of the code in Section 10.1 found the optimal solution. Note that the same number of solutions will be generated if we replace the above version of the `decide` routine by the Version II of Model 9.1 or the sound and complete Version III of Section 10.1.

Let's look at another example, Model 10.1, Corporate Paper. Again, we can wrap the search carried out by the program in a `find_all` construct to obtain

```
find_all {
    new_central_loop();
    max: Return;
    if wp(Return) > best_so_far;
    then best_so_far = wp(Return);
}
```

However, this time this simple method will simply not do, since the program will run on and on. The reader is invited to check this out. The reason this does not work is that the search tree is too large because of the number of possibilities generated by the loop over the `try` routine that is called from `new_central_loop`. Encountering this kind of difficulty in the optimization form of a disjunctive program should not surprise us, since, as noted in Section 9.3, the optimization form of an NP-Complete problem is both NP-Hard and co-NP-Hard. In the next section we develop an important method for coping with this.

Exercises

11.1.1. Prove that the puzzle of Model 8.5, Who Has the Zebra, has only one solution.

11.1.2. Use the machinery of Exercise 8.2.2 in the code for all solutions to the N-queens problem.

11.1.3. Explain why `max: Return;` is not needed in the code below which wraps the `find_all` construct around the central loop of Model 10.2:

11.2 Discrete branch-and-bound search

```
find_all {
    new_central_loop();
    max: Return;
    if wp(Return) > best_so_far;
    then best_so_far = wp(Return);
}
```

11.2 Discrete branch-and-bound search

In Section 2.3 and elsewhere we have encountered knapsack problems. In Section 4.2 we developed code for the minimization version of the linear knapsack problem that did not use continuous variables. The linear knapsack problem is an example where both discrete and continuous methods apply. We will take this one step further in this and the following section where we consider an integer knapsack problem, namely, a knapsack problem with the additional requirement that the quantity of each element placed in the knapsack be an integer. As noted in Section 9.3, the integer knapsack decision problem is NP-Complete and so its optimization form is both NP-Hard and co-NP-Hard. To attack this minimization problem, we will develop the depth-first version of a technique known as *branch-and-bound* search. With each node in the search is associated a *bound* that is an estimate of the best possible value the objective function can have at any solution below that node. This bound is compared with the value of the objective function at the best solution found thus far and a node whose bound falls short is pruned. Using the integer knapsack as the example will enable us to analyze branch-and-bound search in the discrete context and then segue to the continuous case.

Consider the following very simple instance of the integer knapsack problem. There are 4 articles whose weights and costs are given in the data routine below. The task is to load the knapsack with items whose total weight comes to at least 66 and to keep the cost of doing this as low as possible. The rub is that an integral number of each item must be placed in the knapsack. As usual, there is an upper limit on the number available of each item. In this example, the lower limit on each item is 0. Since this is a minimization problem, for sake of this discussion, we can assume that the knapsack has unlimited capacity.

We first approach this as a discrete optimization problem. The declarations are

```
#define N 4        // Number of items
#define W 66       // Weight required

int x[N];          // Number of item i to be placed in knapsack
double c[N];       // Cost of item i
```

```
double w[N];      // Weight of item i
int up[N];        // Upper bound on number of item i to use
```

For the branch-and-bound strategy, we will need

```
double best_so_far;// To record objective function values
int cnt;           // To count nodes
int solu_cnt;      // To count the number of solutions
```

To keep track of the objective function in this discrete approach to the application, we will use two arrays of type `double`

```
double g[N];// To store bounds for objective function values
double h[N];// To store lookaheads
```

The data are listed in the assignment statements in the next routine:

```
data()  // From lowest to highest cost/weight ratio
{
    c = { 1.1,1.2,1.3,.3 };    // Cost
    w = { 60,58,50,10 };       // Weight
    up = { 2,1,2,2 };          // Upper bound
}
```

The strategy for this application will be a greedy one: Try to use as much as possible of the items of least cost per unit of weight to meet the demand requirement W. So the items will be considered in order from the least expensive per unit weight up to the most expensive. As a matter of fact, this is the order in which the items are given, as can be seen from the `data` routine. To determine the amount of an item to place in the knapsack, we start with the maximum possible and loop down to 0:

```
            or(int k=up[i];k>=0;k--)
                x[i] = k;
```

As we make decisions, we can test for feasibility by supposing that all the rest of the items go into the knapsack. Having decided how much of items 0,...,i to place in the knapsack, the test for feasibility becomes: Would placing the maximum amount available of each of the remaining items in the knapsack make the total weight surpass the goal W?

```
test_for_feasibility(int i)
{
    sigma(int k=0;k<=i;k++) w[k]*x[k]  //Already in knapsack
    + sigma(int k=i+1;k<N;k++) w[k]*up[k] // Uncommitted
    >= W;    // Volume required
}
```

11.2 Discrete branch-and-bound search

We will keep track of the costs associated with the decisions being made in the array g:

```
update_objective(int i)
{
    if i==0;
    then g[i] = c[i]*x[i];
    else g[i] = g[i-1] + c[i]*x[i];
}
```

The value of the objective function at the end of a successful search for a solution will be stored in the variable g[N-1] and then recorded in best_so_far if it is indeed the best value found so far. At each node at level i in the search, the value of g[i] is called the *bound* associated with the node. This is a lower bound on the possible value that the objective function can take at a solution below this node.

The nodes generated during the search will be counted in the variable cnt. The branching step can be encapsulated in a procedure:

```
branch_on_item(int i)
{
    // x[i]=up[i],...,x[i]=k+1 do not lead to a new solution
    // with g[i] <= best_so_far
    or(int k=up[i];k>=0;k--) {
        cnt = cnt + 1;         // Count node
        x[i] = k;
        test_for_feasibility(i);
        update_objective(i);
    }
}
```

The output routine will print the best value found for the objective function, the number of nodes visited, and the number of different solutions found:

```
output()
{
    printf("The best solution is %.2f\n",best_so_far);
    printf("The number of nodes generated is %d\n",cnt);
    printf("The number of solutions is %d\n",solu_cnt);
}
```

The central loop of the program will go through the items i and branch to select an integer value for x[i].

```
central_loop()
{
    // x[0],...,x[i-1] form part of a feasible solution
    // g[i-1] records c[0]*x[0]+...+c[i-1]*x[i-1]
    and(int i=0;i<N;i++)
        branch_on_item(i);
}
```

The code for the `central_loop` and the other routines clearly satisfies the requisite horizontal and vertical conditions that make for an algorithm that is sound and complete.

The main routine will contain the central loop wrapped in a `find_all` as follows:

```
21p_main()
{
    best_so_far = 1e6;
    data();
    find_all {
        central_loop();
        solu_cnt = solu_cnt + 1;
        if g[N-1] < best_so_far;
        then best_so_far = g[N-1];
    }
    output();
}
```

This code finds the optimal solution to be 1.4. In all it generates a search tree with 78 nodes and finds 47 different solutions to the problem.

The next step is to note that the code can be improved by tightening the initial value of the control variable in the `or` loop. In the spirit of the greedy strategy, we do not want to provide more than is needed. If `up[i]` allows for more of item i than is needed to load the knapsack to level W, we begin the `or` loop at the minimum amount of this item that will load the knapsack to this level:

```
compute_sup(int & top, int i)
{
    and(int k=0;k<up[i];k++)
        if w[i]*k >= W - sigma(int j=0;j<i;j++) w[j]*x[j];
        then {
            top = k;
            return;
        }
    top = up[i];
}
```

11.2 Discrete branch-and-bound search

and change the `branch_on_item` procedure to

```
branch_on_item(int i) // New version
{
int top;    // Storeback variable

    compute_sup(top,i);
    // x[i]=top,...,x[i]=k+1 do not lead
    // to an optimal solution
    or(int k=top;k>=0;k--) {
        cnt = cnt + 1;
        x[i] = k;
        test_for_feasibility(i);
        update_objective(i);
    }
}
```

Now this restriction on the enumeration can eliminate some solutions, but for every solution so eliminated, there is a better solution that is not eliminated. Hence the optimal solution is not eliminated this way. To work this out more closely, suppose that `top` is the value computed by the call to `compute_sup(top,i)`. Then `top` is either `up[i]` or `top` is the least integer such that

```
w[i]*top >= W - sigma(int n=0;n<i;n++) w[n]*x[n];
```

In the former case, nothing has been changed. In the latter case, setting `x[i] = top` and all `x[j] = 0` for `j > i` yields a solution to the integer knapsack with objective function value `sigma(int n=0;n<i;n++) c[n]*x[n] + c[i]*top`. This choice is not eliminated by our change in the code. Any eliminated solution with `x[i] > top`, would clearly have to have a higher objective function value than this one.

This change does mean, however, that we have to restate the enumerative condition on the `or` loop: All alternatives are enumerated in the `or` loop except possibly some that cannot lead to an optimal solution. Thus the algorithm remains complete, and when wrapped in `find_all`, it will still find all optimal solutions and, most probably, some others as well.

With this improvement the search finds 8 solutions and visits 29 nodes as illustrated in Figure 11.2. In the figure goal nodes are labeled underneath with the value of the objective function at that node; nodes that are generated but fail to satisfy the test for feasibility are labeled underneath with an X. A significant number of nodes from the search tree have been pruned, although it is only the test for feasibility that prunes unnecessary nodes. The idea in branch-and-bound search is to prune nodes that cannot lead to solutions that are as good as or better

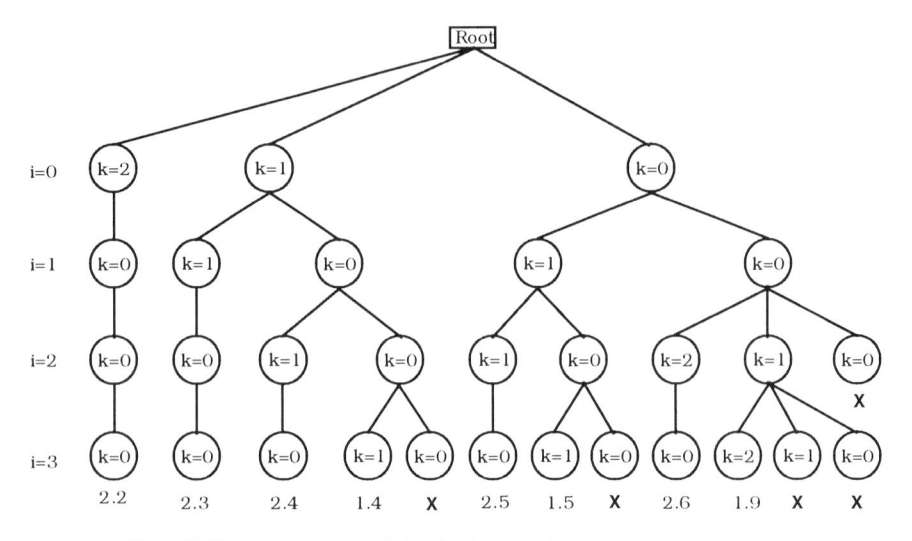

FIGURE 11.2 Nodes visited when using compute_sup.

than the ones already known. To do more pruning this way, we can change the `update_objective` routine to

```
update_objective(int i)  // New version
{
    if i==0;
    then g[i] = c[i]*x[i];
    else g[i] = g[i-1] + c[i]*x[i];
    g[i] <= best_so_far;    // Bounding test
}
```

This improves things further, and the new search tree is pictured in Figure 11.3. Now only 24 nodes are generated and only 2 solutions are found. This time nodes that are labeled X are ones where failure is detected either because the test for infeasibility fails or because the cut-off `g[i] <= best_so_far` is not met.

The or loop in this code now observes a sharper enumerative condition than before:

```
// x[i]=top,...,x[i]=k+1 do not lead to an optimal
// solution or to a new solution with g[i] <= best_so_far
or(int k=top;k>=0;k--) {
    cnt = cnt + 1;
    x[i] = k;
    test_for_feasibility(i);
    update_objective(i);
}
```

11.2 Discrete branch-and-bound search

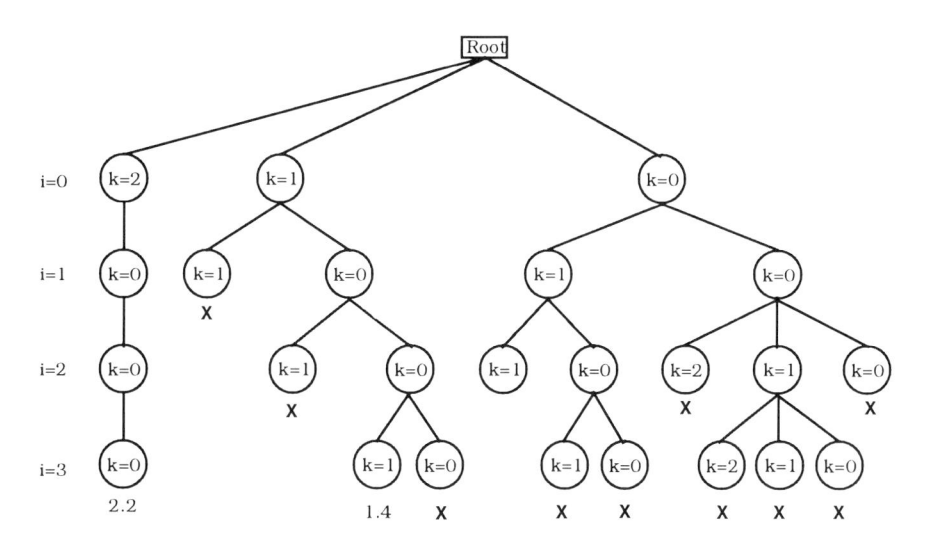

FIGURE 11.3 Nodes visited when using new version of update_objective.

The invariant on the and loop is also sharpened:

```
// x[0],...,x[i-1] form part of a feasible solution
// g[i-1] records c[0]*x[0] +...+ c[i-1]*x[i-1]
// g[i-1] <= best_so_far
and(int i=0;i<N;i++)
    branch_on_item(i);
```

This invariant on the and loop guarantees that when the loop is exited, the objective function g[N-1] is no greater than the most recent value assigned to best_so_far. Hence the new solution found is at least as good as the previous one.

When the search successfully finds a first solution or a new solution, that solution is called the *incumbent*. The incumbent solution is the best one found so far, and it is the one that later candidates must tie or beat.

The initial value of best_so_far has been chosen greater than that at any possible solution. The code is sound and complete, and it will find a solution if a solution exists. Thereafter, as best_so_far changes, the only solutions that will be pruned are those that are cut off by this bound. Therefore the optimal solution to the problem will be among those found by the code.

There is still something else to improve the code that can be done quite easily with the tools at hand. The value stored in g[i] looks back at the costs incurred so far but does not take into account any future costs that have to be incurred. There is a lookahead available here that can provide a lower bound on costs yet to come, namely the value of the objective function of the linear knapsack that results from looking at the tail of the problem. In other words, we will relax the integrality condition on the remaining items to go into the knapsack. To compute this estimate, we can use the greedy strategy of Section 4.2 and apply it to filling what remains to be placed in the knapsack with the items from i+1 to N-1. The following code stores the minimum value of this estimate in the storeback parameter lookahead, which is passed by reference:

```
evaluate(int i, double & lookahead)
{
double scratch,pad;

    scratch = W - sigma(int k=0;k<N;k++) w[k]*x[k];
    lookahead = 0.0;
    and(int k=i+1;k<N;k++) {
        if scratch <= 0.0;      // If nothing left to do
        then return;            // do nothing more
        pad = scratch/w[k];
        if pad > up[k];
        then pad = up[k];
        lookahead = lookahead + c[k]*pad;
        scratch = scratch - pad*w[k];
    }
}
```

That done, we change the update_objective routine to

```
update_objective(int i)    // Revised version
{
    if i==0;
    then g[i] = c[i]*x[i];
    else g[i] = g[i-1] + c[i]*x[i];
    evaluate(i,h[i]);
    // Better bounding test
    g[i] + h[i] <= best_so_far;
}
```

Now we are associating the more realistic bound g[i] + h[i] with each node in the search. What makes this stronger test valid is that the estimate g[i] + h[i] cannot be an overestimate of the value of the objective function at any solution to the problem below this node. This follows from the observation that any solution to the integer knapsack problem is also a solution to the corresponding linear knapsack problem. An evaluation function for a minimization problem that never overestimates the value of the objective function at the best solution

11.2 Discrete branch-and-bound search

beneath a node is said to be an *admissible heuristic*. The requirement that the already incurred costs be no greater than best_so_far has been changed to the requirement that the sum of the already incurred costs plus that of the linear relaxation of the tail of the problem not exceed the value of best_so_far. So the invariant on the or loop is further sharpened to

```
// x[i]=top,...,x[i]=k+1 do not lead to an optimal
// solution or to a new solution with
// g[i] + h[i] <= best_so_far
```

This makes for yet another improvement and leads to a search tree with only 12 nodes, as is shown in Figure 11.4. This time nodes that are labeled X are ones where failure is detected either because the test for infeasibility fails or because the cutoff g[i] + h[i] <= best_so_far is not met

Let us make a final remark on the role of the find_all construct in this application. Even after all the changes to the subroutines and all the improvements in the code, the kernel of the main routine remains the same:

```
find_all {
    central_loop();
    solu_cnt = solu_cnt + 1;
    if g[N-1] < best_so_far;
    then best_so_far = g[N-1];
}
```

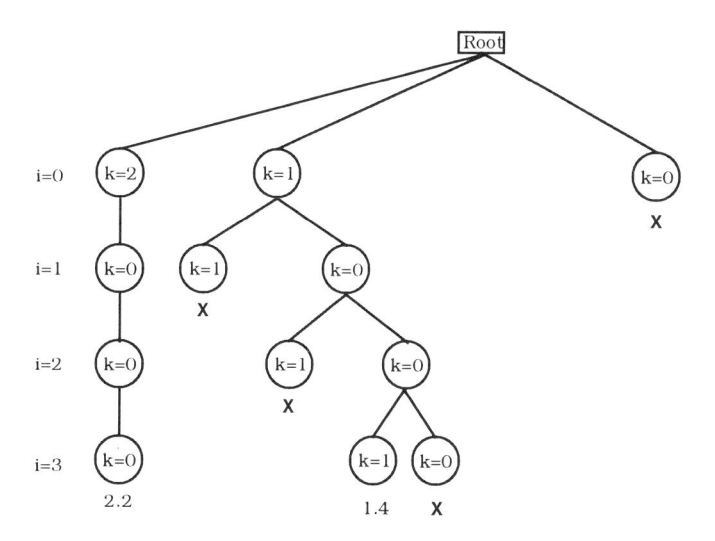

FIGURE 11.4 Nodes visited when using compute_sup and admissible lookahead.

However, we are no longer truly searching for all solutions. Rather, we are using the automatic backtracking of the find_all construct to keep the depth-first search going, but the conditions for a solution are constantly being changed to make the next solution meet the latest bound.

Exercises

11.2.1. Program the optimization version of the graph coloring problem using recursion and conditional disjunction.

11.2.2. Write discrete branch-and-bound code to solve the non-preemptive form of the problem of Model 4.3, Batch Scheduling to optimality. Use Jackson's preemptive schedule in the place of the linear knapsack as a lookahead. Generate new data sets on which to run your code.

11.3 Linear relaxations and branch-and-bound search

In the preceding section we considered the integer knapsack problem as a discrete application. In this section we code it as a disjunctive linear program. The difference is that now the optimum of the linear relaxation of the disjunctive program will furnish the bound at each node and play the role of the estimate g[i] + h[i].

The declarations of the data arrays do not change and we have

```
// The integer knapsack
#define N 4
#define W 66

double c[N],w[N];
double best_so_far;
int up[N];
```

The integer array x[N] is replaced by an array of continuous variables, and we introduce another continuous variable Z for the objective function:

```
continuous X[N],Z;
```

The data routine is unchanged, but now we have a procedure to set up the initial feasible region and to define the objective function:

11.3 Linear relaxations and branch-and-bound search

```
setup()
{
    and(int k=0;k<N;k++)
        X[k] <= up[k];
    sigma(int k=0;k<N;k++) w[k]*X[k] >= W;
    sigma(int k=0;k<N;k++) c[k]*X[k] == Z;
}
```

The branching code is

```
branch_on_item_cont(int i)
{
int top;

    compute_sup_cont(top,i);
    or(int k=top;k>=0;k--){
        cnt = cnt + 1;
        X[i] == k;
    }
}
```

The `compute_sup` code is the same as the discrete version except that `wp(X[j])` must be substituted for `x[j]`.

```
compute_sup_cont(int & top, int i)
{
    and(int k=0;k<up[i];k++)
        if w[i]*k >=
                W - sigma(int j=0;j<i;j++) w[j]*wp(X[j]);
        then {
            top = k;
            return;
        }

    top = up[i];
}
```

The main procedure is now

```
2lp_main()
{
    best_so_far = 1e6;      // A very large number
    data();
    setup();
```

```
    find_all
    {
        and(int i=0;i<N;i++) {
            branch_on_item_cont(i);
            Z <= best_so_far;
        }
        best_so_far = wp(Z);
        solu_cnt = solu_cnt + 1;
    }
    output();    //To be rewritten for continuous variables
}
```

Again, we have 12 nodes and 2 solutions, and the tree is exactly the same as for the previous code, Figure 11.4. In fact, the two programs carry out the same search strategy. The continuous variables provide an automatic test for feasibility, and the constraint `Z <= best_so_far` is equivalent to the test `g[i] + h[i] <= best_so_far`. Thus Z is playing the role of both g and h. In fact, `g[i] + h[i]` is equal to the minimum value of Z, and so it is the minimum value of Z that is the bound associated with a node in the continuous model. The continuous method automatically captures the lookahead feature of the discrete code. On the other hand, we were able to capture the power of the continuous code with discrete methods easily because it is a knapsack problem.

For disjunctive programs and other constrain-and-generate applications, the branch-and-bound technique of forcing the code to constrain the objective function to be as good as the previous solutions found is so basic that it is built into systems that treat this kind of application. The 2LP syntax for this in the minimization case is

```
    find_min: <objective function>;
    subject_to <statement>
```

where <*objective function*> is an affine expression of type `continuous`. In the example of the integer knapsack we have just considered, the main procedure becomes

```
2lp_main()
{
    data();         // As before
    setup();        // ditto

    find_min: Z;
    subject_to
        and(int i=0;i<N;i++)
            branch_on_item_cont(i);
    output();
}
```

11.3 Linear relaxations and branch-and-bound search

This program follows the same itinerary as the previous code.

For the maximization case the keyword is `find_max`. With this construct we can code the branch-and-bound search for the optimal solution to Model 10.1, Corporate Paper, in a very compact way:

```
find_max: Return;
subject_to
    new_central_loop();
```

The program now runs to completion quickly and finds that the optimal return on investment is 7.735%.

Both the `find_max` and `find_min` constructs fail if the `subject_to` statement itself fails: This is the same result as with the `find_all` construct. On the other hand, with the `find_max` and `find_min` constructs, it is not necessary to cache the best value found so far for the objective function; this is done automatically. Another very important point is that upon successful exit, the `find_max` and `find_min` constructs shrink the feasible region to a single point, namely the last incumbent that was found, while `find_all` restores the feasible region to its previous state.

A thing to remember about branch-and-bound solutions of optimization problems is that the analytical tools of reduced costs and ranges are not available for sensitivity analysis. To get some kind of handle on the sensitivity of the solutions to changes in data, "what-if" analysis is always an option. However, even for disjunctive linear programs, the optimum function of Section 7.2 is not necessarily continuous and this should be taken into account when "what-if" techniques are used.

The next model is a more complicated version of the diet problem of Model 1.1. It is based on Bosch (1993) and on Erkut (1994).

Model 11.2 Fast Food

A college student would like to know if he can eat all his meals at the local fast food outlet and still meet the requirements of a balanced diet. In the student newspaper it is reported that a balanced day's food should have at least 100% of the US RDA of vitamins A, C, B_1, B_2, niacin, calcium, and iron. It should also contain at least 55 grams of protein and at most 3000 milligrams of sodium. Furthermore 9 calories come from each gram of fat, and as is well known, at most 30% of one's calories should come from fat. For the student to keep up with his studies and other activities, the diet should provide for at least 2000 calories a day. The student has obtained nutritional and pricing information from the fast food outlet; this information for the student's favorites is given in Table 11.1. The student figures that he can eat at most three servings of any one of these foods in a given day; his other stipulation is that he'll only have milk with cereal and not as a stand-alone drink. He

TABLE 11.1 Prices and Nutrition Facts

Item	Price	Calories	Protein	Fat	Sodium	Vit A	Vit C	Vit B_1	Vit B_2	Niacin	Calcium	Iron
Burger	0.59	255	12	9	490	4	4	20	10	20	10	15
Lean B	1.79	320	22	10	670	10	10	25	20	35	15	20
Big B	1.65	500	25	26	890	6	2	30	25	35	25	20
Fries	0.68	220	3	12	110	*	15	10	*	10	*	2
Nuggets	1.56	270	20	15	580	*	*	8	8	40	*	6
Honey	0.00	45	0	0	0	*	*	*	*	*	*	*
Chef Sal	2.69	170	17	9	400	100	35	20	15	20	15	8
Gar Sal	1.96	50	4	2	70	90	35	6	6	2	4	8
EggSand	1.36	280	18	11	710	10	*	30	20	20	25	15
Cereal	1.09	90	2	1	220	20	20	20	20	20	2	20
Yogurt	.63	105	4	1	80	2	*	2	10	2	10	*
Milk	.56	110	9	2	130	10	4	8	30	*	30	*
Orange J	.88	80	1	0	0	*	120	10	*	*	*	*
Gfruit J	.68	80	1	0	0	*	100	4	2	2	*	*
Apple J	.68	90	0	0	5	*	2	2	*	*	*	4

would like also to find a way to do this as economically as possible. How can this be done?

This is an example of a diet problem with the extra requirement that all purchases be in integral amounts, since the fast food chain doesn't offer fractional hamburgers or other fractional menu items. This means that the continuous variables used to represent the amount of each menu item to be purchased must be integer valued in the solution, which makes it a mixed integer programming problem. To help the student out, we can set up the linear constraints of the underlying linear diet problem, develop a routine for a sound and complete search for integer-valued solutions for the appropriate variables, and then wrap this routine in the `find_min` construct.

To represent the data, we will use an array prices of type `double` to record the Price column of the table and a two-dimensional array food of type `int` to record the information in columns Calories through Iron of the table.

We start with a definition of a symbolic constant and the declaration of a data array:

11.3 Linear relaxations and branch-and-bound search

```
#define ITEMS 15
double prices[ITEMS];
```

Then we introduce symbolic constants for the nutritional categories and declare the array `food`:

```
#define CALOR 0
#define PROT  1
       ...
#define IRON 10

int food[ITEMS][IRON+1];
```

The unknowns of the problem are the number of servings of each menu item to order during the day. For that we can use an array of continuous variables:

```
continuous Serv[ITEMS];
```

For the objective function and for formulating the constraint on calories from fat, we declare the following continuous variables:

```
continuous Cost;
continuous Fat, Calories;
```

Finally, we have symbolic constants for some of the parameters of the application:

```
#define CEREAL 9    // Index of cereal among the food items
#define MILK  11    // Index of milk among the food items
#define LIMIT  3    // Number of servings allowed
```

The data arrays `prices` and `food` are both initialized from Table 11.1:

```
data()
{
    // First Column of Table 11.1
    prices = { .59,1.79,1.65,.68,1.56,0.0,2.69,1.96,
                            1.36,1.09,.63,.56,.88,.68,.68 };
    // Columns 2-12 of Table 11.1
    food = {
        255,12,9,490,4,4,20,10,20,10,15,
            ...
        90,0,0,5,0,2,2,0,0,0,4
    };
}
```

The nutritional requirements can be set up as a collection of linear constraints in a procedure such as the following:

```
setup()
{
    and(int j=0;j<ITEMS;j++)   // Limit on number of servings
        Serv[j] <= LIMIT;

    Serv[MILK] == Serv[CEREAL];    // Milk and cereal link

    Calories >= 2000;       // Minimum calories

    // At least 100% of these nutrients
    and(int i=VITA;i<=IRON;i++)
        sigma(int j=0;j<ITEMS;j++)
            food[j][i]*Serv[j] >= 100;

    // At least 55 grams of protein
    sigma(int j=0;j<ITEMS;j++)food[j][PROT]*Serv[j] >= 55;

    // At most 3000 mg of sodium
    sigma(int j=0;j<ITEMS;j++)
        food[j][SODIUM]*Serv[j] <= 3000;

    // Grams of fat
    Fat ==
    sigma(int j=0;j<ITEMS;j++) food[j][FAT]*Serv[j];
    // Total calories
    Calories ==
    sigma(int j=0;j<ITEMS;j++) food[j][CALOR]*Serv[j];
    // No more than 30% of calories from fat
    // given that each gram of fat has 9 calories
    9.0*Fat <= .30*Calories;

    // Objective function
    Cost ==
        sigma(int j=0;j<ITEMS;j++) prices[j]*Serv[j];
}
```

The simplest code for a sound and complete search is an `and/or` loop:

```
        and(int i=0;i<ITEMS;i++)
            or(int k=0;k<=LIMIT;k++)
                Serv[i] == k;
```

Wrapping this in the `find_min` construct leads to the routine

11.3 Linear relaxations and branch-and-bound search

```
find_least_expensive_meal()
{
    find_min: Cost;
    subject_to
        and(int i=0;i<ITEMS;i++)
            or(int k=0;k<=LIMIT;k++)
                Serv[i] == k;
}
```

The main procedure simply puts things together:

```
21p_main()
{
    data();
    setup();
    find_least_expensive_meal();
    output();    // Left to reader
}
```

The code shows that the student can indeed eat all his meals on a given day at the fast food outlet at a cost of $8.71. In fact, more complete output is

The total cost is 8.71
Fat is 61.00
Calories are 2045.00
Buy 3.00 of item burger
Buy 2.00 of item fries
Buy 3.00 of item honey
Buy 3.00 of item cereal
Buy 1.00 of item yogurt
Buy 3.00 of item milk

End of Model 11.2

A branch-and-bound search can be often considerably helped by having an initial bound on the value of the objective function. An initial bound like this is called a *bluff*. The bluff will prevent the search from coming up with poor solutions that can be reached early in a depth-first search. Sometimes a bluff is known from previous experience or other familiarity with the application. Another way to look for a good bluff is to use a heuristic strategy to get a quick-and-dirty solution and to use this result as an initial bound on the objective function for a branch-and-bound search. A hill-climbing technique or a best-child-first search can often be used to dive for a first solution.

Our next model will involve sparse data, something we have encountered several times. In the discussions preceding Model 3.4, The

Committee, and Model 4.4, The Grid, we considered routines for entering data given in this form. In both cases the constraints were set up altogether at the beginning of the application. The basic setup routine for <= constraints was

```
sparslteq(continuous X[],int nnz[],indices[],
                            double cffs[],rhs[],int m)
{
int begin,end;

    end = 0;
    and(int i=0;i<m;i++) {
        begin = end;
        end = end + nnz[i];
        if begin < end;
        then
        sigma(int j=begin;j<end;j++) cffs[j]*X[indices[j]]
        <= rhs[i];
    }
}
```

In this code the assignments to begin and end for loading the ith constraint depend on the fact that the previous constraints have been loaded in order. In working with logic and backtracking, constraints will often be loaded in a more random fashion and even undone when backtracking occurs. To deal with this, we often have to write the code so as to load a given constraint directly without relying on the constraints' being loaded in a fixed order. One simple trick, one that we have already used in the build_data_structures routine of Model 5.1, Parallel Sessions, is to reset the array nnz[M] after the usual initialization by

```
    and(int i=1;i<N;i++)
        nnz[i] = nnz[i] + nnz[i-1];
```

Now for each constraint i>0, the indices stored in the array indices starting at index nnz[i-1] but strictly before nnz[i] are the indices of the variables in the constraint with nonzero coefficients. So we can now load the ith constraint alone by a call on a routine such as

```
onelteq(int i, continuous X[],int nnz[],indices[],
                                double cffs[],rhs[])
{
int begin,end;

    if i == 0;
    then begin = 0;
    else begin = nnz[i-1];
```

11.3 Linear relaxations and branch-and-bound search

```
        end = nnz[i];
        if begin < end;
        then
        sigma(int j=begin;j<end;j++) cffs[j]*X[indices[j]]
        <= rhs[i];
}
```

We will have the occasion to use both devices for sparse data in the next model.

Model 11.3 Mountaineering

A camping goods store must plan its product line in mountaineering equipment for the coming season. The store can purchase its needs from various suppliers and has a good idea of up to how many units of each item it can sell and at what price. Wholesale prices can be locked in with suppliers provided that the store agrees to take a certain minimum amount of each item purchased. The store can commit to purchasing up to $100,000 of equipment wholesale. However, articles complement each other or clash, and so decisions must be made on which articles and combinations of articles to carry. Certain items are simply too close in price, style or function to one another for more than one of them to be carried. On the other hand, certain groups of articles are so important that at least some minimum number of items from among them must be carried. Accessory items are sometimes sold alone but more frequently in conjunction with the sale of other items. Naturally the store has data as to the number of times an accessory will be purchased to accompany a particular principal item. An item might serve as an accessory to different principal items, and so an accessory item should be ordered and stocked at least in proportion with the principal items it can accompany. What is needed is a plan for making wholesale purchases that can be expected to provide as high a profit as possible for the store.

The data are given in the following tables. Table 11.2 gives the minimum number of units of each item that can be purchased wholesale, the maximum amount of units of the item that can be sold, its unit wholesale cost, and the profit its sale can be expected to bring in. Table 11.3 gives the items that are accessories, the list of principal items they can be sold with, and the ratio of sales of the accessory to the principal items. Table 11.4 lists the sets of incompatible items. Finally, Table 11.5 lists various sets out of which at least some minimal number of items must be carried.

To start, let us introduce the symbolic constants

```
#define N 20              // Number of articles
#define MAXCOST 100000    // Available capital
```

TABLE 11.2 Individual Item Data

Item	Min/Max	Cost	Profit	Item	Min/Max	Cost	Profit
0	100/300	20	45	10	100/300	12	10
1	200/400	21	11	11	200/400	3	11
2	300/500	20	10	12	300/500	2	12
3	300/600	14	14	13	300/600	4	24
4	200/600	13	13	14	200/600	3	13
5	100/500	2	12	15	100/500	2	18
6	200/400	15	15	16	200/400	5	25
7	100/400	6	6	17	100/400	6	6
8	100/700	1	8	18	100/700	1	8
9	120/520	9	9	19	120/520	9	9

TABLE 11.3 Accessories and Principal Items

Accessory	Number of Principal Items	Principal Items	Ratios
8	3	2, 3, 4	.5, .5, .6
9	2	5, 7	.22, .44
17	3	10, 11, 12	.55, .66, .77
18	3	13, 14, 15	.33, .44, .55
19	2	15, 16	.33, .66

TABLE 11.4 Incompatible Items

Item	Number of Incompatible Items	Incompatible Items
0	2	1, 2
1	2	5, 6
2	4	6, 11, 12, 13
3	4	4, 13, 15, 16
10	2	16, 2
11	2	15, 6

We use an array Article[N] of continuous variables to represent the number of units of item i to be purchased for retail sale in the store. Let lo[N] be an array of type double to store the lower bounds on these articles if they are purchased at wholesale, let hi[N] record the upper limits on the number of units of each item that can be sold, let cost[N]

11.3 Linear relaxations and branch-and-bound search

TABLE 11.5 Required Items

Set	Number of Items in Set	Items in Set	Minimum Required
0	3	0, 1, 2	500
1	4	3, 4, 5, 6	400
2	4	1, 4, 9, 11	300
3	4	12, 16, 17, 18	200

contain the wholesale prices of the items, and let `profit[N]` hold the expected profits to be made:

```
continuous Article[N], Z;
double lo[N],hi[N],cost[N],profit[N];
```

As usual, Z is an additional continuous variable introduced for the objective function.

With each accessory item, we have associated a list of principal items that it can serve as an accessory for. These are sparse data and can be handled as in Chapter 3:

```
#define ACCS 5     // Number of accessory articles in 10.3
#define TOTPRIN 13 // Number of principal articles in 10.3
int accs[ACCS],prinsz[ACCS];
int prinindices[TOTPRIN],ratio[TOTPRIN];
```

These arrays will store the indices of the accessories, the sizes of the associated sets of principal items, the indices of the associated principal items, and the ratio of expected sales of the accessories to the principal items, respectively.

The constraints of the application that can be set down at the outset are the upper bounds on the items, the objective function, the lower bounds on the number of items that can be carried from among certain clusters, and the constraints that state that the number of units of each accessory item is at least equal to the weighted sum of the principal items associated with it.

The upper bounding constraints on `Article[i]` are straightforward as are the cost constraint and the objective function:

```
basic_constraints()
{
    and(int i=0;i<N;i++)
        Article[i] <= hi[i];
    sigma(int i=0;i<N;i++) cost[i]*Article[i] <= MAXCOST;
    Z == sigma(int i=0;i<N;i++) profit[i]*Article[i];
}
```

We can set up the constraints on the relation between principal articles and their accessories by means of the following code:

```
acc_constraints()
{
int begin,end;

    end = 0;
    and(int i=0;i<ACCS;i++) {
        begin = end;
        end = end + prinsz[i];
        Article[accs[i]] >=
            sigma(int j=begin;j<end;j++)
                    ratio[j]*Article[prinindices[j]];
    }
}
```

We come to the issue of coding the sets of items from which at least some minimum quantity must be carried. Again, we can use the sparse data technique. Let MUSTSETS be a symbolic constant for the number of such sets, and let TOTMUST be a symbolic constant for the sum of the sizes of these sets. We introduce integer arrays

```
int mustsz[MUSTSETS],mustindices[TOTMUST];
```

to store the sizes of the sets and the elements of the sets. We will store the lower bounds on the items from these sets in an array lwbnds[MUSTSETS] of type double. The requirement that at least lwbnds[k] item from the kth of these sets be included can be coded by a procedure with an and loop:

```
mustcarrysome()
{
int begin,end;

    end = 0;
    and(int k=0;k<MUSTSETS;k++) {
        begin = end;
        end = end + mustsz[k];
        lwbnds[k] <=
            sigma(int j=begin;j<end;j++)
                    Article[mustindices[j]];
    }
}
```

Having considered the constraints of the model that can be set up at the outset, let us turn to the logical requirements in this application:

11.3 Linear relaxations and branch-and-bound search

1. Articles that are carried cannot be purchased in quantities under a certain threshold.
2. Certain articles exclude others.

Note that this application does not call for the solution to provide integer values, since its scale allows for roundoff.

For the first requirement, we can loop through the articles and decide the fate of each in turn. Moreover, when deciding to include an item, we can also force the decision to include none of the items that are incompatible with it, thus taking care of the second requirement. Working bottom-up, let us turn to the mechanics of handling this second requirement.

We again can use the usual device for storing sparse data, and we introduce a symbolic constant TOTEX for the total number of items that are eligible to be excluded. To avoid introducing an array for the items which exclude other items, we can associate the empty list with those that do not. So we will only need the arrays

```
int exsz[N], exindices[TOTEX];
```

Since we are now treating the logical requirements of the model, we will follow the alternative discussed above to deal with random access and backtracking. After the usual initialization, we will reset the `exsz` array so that for each item `i>0`, the indices stored in `exindices` starting at index `exsz[i-1]` but strictly before `exsz[i]` are the items incompatible with item `i`:

```
random_access()
{
    and(int i=1;i<N;i++)
        exsz[i] = exsz[i] + exsz[i-1];
}
```

If item `i` is to be carried, then given `begin` and `end`, we can exclude the incompatible items by means of an and loop of constraints that fix the incompatible items at 0:

```
and(int j=begin;j<end;j++)
    Article[exindices[j]] == 0;
```

If `begin` is equal to `end`, the list of incompatible elements is empty; in this case the and loop is vacuous and so is exited immediately with success.

The four subroutines we have developed up to this point can be grouped together is a setup routine:

```
setup()
{
    basic_constraints();
    acc_constraints();
    mustcarrysome();
    random_access();
}
```

Continuing to work bottom-up, let us write the routine to be called if it is decided to carry an item. This routine will need the item, its lower bound, and the indices into the `exindices` array for this item:

```
carry(continuous Item, double low, int begin,end)
{
    Item >= low;
    and(int j=begin;j<end;j++)
        Article[exindices[j]] == 0;
}
```

Things are simpler if it's decided not to carry an item:

```
donotcarry(continuous Item)
{
    Item == 0;
}
```

At this point we can put together a hill-climbing routine to try to generate a good bluff. First, we will use a one-ply lookahead to evaluate which decision to make:

```
quick(continuous Item, double low, int begin,end)
{
double max_when_in,max_when_out;

    if not {
        donotcarry(Item);
        max: Z;
        max_when_out = wp(Z);
    }
    then {
        carry(Item,low,begin,end);
        return;
    }
    if not {
        carry(Item,low,begin,end);
        max: Z;
        max_when_in = wp(Z);
    }
```

11.3 Linear relaxations and branch-and-bound search

```
        then {
            donotcarry(Item);
            return;
        }
        if max_when_in > max_when_out;
        then carry(Item,low,begin,end);
        else donotcarry(Item);
}
```

Then the hill climb can be carried out by calling this quick decision maker for each item in turn:

```
hill_climb(double & bluff)
{
    and(int i=0;i<N;i++)
        if i == 0;
        then quick(Article[i],lo[i],i,exsz[i]);
        else quick(Article[i],lo[i],exsz[i-1],exsz[i]);
    max: Z;
    bluff = wp(Z);
}
```

With the routines in place that are to be called when deciding whether to carry an item, we can now also develop the principal logical loop of the program. This loop will determine in turn the fate of each of the items under consideration, taking care to exclude incompatible items. Thus the logical requirements of the model will be met. Since the procedure called when an article is to be carried requires four parameters, these parameters can also be passed to the `decide` procedure in a loop as follows:

```
search()
{
    and(int i=0;i<N;i++)
        if i == 0;
        then decide(Article[i],lo[i],i,exsz[i]);
        else
        decide(Article[i],lo[i],exsz[i-1],exsz[i]);
}
```

If one is optimistic, the `decide` procedure can be

```
decide(continuous Item, double low, int begin,end)
{
    either carry(Item,low,begin,end);
    or donotcarry(Item);
}
```

or, if one is in a more pessimistic mood,

```
decide(continuous Item, double low, int begin,end)
{
    either donotcarry(Item);
    or carry(Item,low,begin,end);
}
```

Clearly for both versions of the `decide` procedure, the search strategy satisfies the horizontal and vertical conditions for an enumeration. Hence the routine `search` will find a solution to the logical requirements of the application. Wrapping this routine in the `find_max` construct means that the search will continue and the best possible solution will be located. Note that after the best possible solution is found, the code will continue to look for better solutions. This will have the effect of verifying that the best solution has indeed been found.

The main routine can take the form:

```
2lp_main()
{
double bluff;

    data();      // As in Tables 11.2 - 11.5
    setup();
    if not hill_climb(bluff);
    then bluff = 0.0;
    else printf("Hillclimb found a bluff of %f\n",bluff);
    Z >= bluff;
    if
        find_max: Z;
        subject_to
            search();
    then output();          // Left to reader
    else printf("No solution can be found\n");
}
```

The hill-climbing search is very successful and finds the optimal solution. This makes the work of the branch-and-bound code much simpler, since it reduces basically to verifying that the given solution is indeed optimal. An important thing about the bluff is that it equalizes the choice between the optimistic and pessimistic versions of the `decide` routine. If the pessimistic version is used without the bluff, some 22 solutions are found before the optimal one. If the optimistic version is used without the bluff, only two solutions are found before the optimal one. With the bluff, both find only the optimal solution.

<u>End of Model 11.3</u>

11.4 State space optimization

As discussed in Section 9.3, to solve the optimization form of an NP-Complete problem, both an NP problem and a co-NP problem must be solved. At this point in time the only general method for solving co-NP problems comes down to an enumeration of the search tree. The branch-and-bound technique is a natural step to take to prune this search space without losing the optimal solution. In the following chapter we treat a very important method for disjunctive linear programs that can often dramatically simplify things.

Exercises

11.3.1. Do the optimization form of Model 10.1, Corporate Paper, and apply the branch-and-bound method using `find_all`.

11.3.2. Do the non-preemptive version of the problem of Model 4.3, Batch Scheduling, as a disjunctive linear program. Compare the pruning power of the linear relaxation with that of Jackson's preemptive strategy in Exercise 10.2.3.

11.3.3. Find the optimal integer-valued solution to Model 2.2.

11.3.4. In a traditional programming language, write code to find the optimal solution to Model 11.2 as a discrete problem.

11.4 State space optimization

A state space search problem is a discrete application whose solution consists of a sequence of states that satisfy given local and global conditions. An example is the N-queens problem of Model 10.2, where it is a consistent sequence of rows in which to place the queens that yields the solution. Frequently there will an optimization dimension to a state space problem, namely to find a state sequence of minimum length that meets the necessary conditions. In this section we consider how to apply the branch-and-bound method to this kind of application.

The next model is a variation on a classic theme. It presents a situation where the generate-and-test paradigm is preferable to constrain-and-generate because of the strictly local nature of the connections in the problem. This means that the job that would be done by the machinery of continuous variables is better done by hand with variables of type `int` or `double`.

Model 11.4 Mathematicians and Physicists

Four mathematicians and four physicists, who are attending an international conference on mathematical physics at the Institut Henri Poincaré in Paris, have wandered from the conference site down to the left bank of the River Seine. They would like to cross the river to get to the right bank and in front of them is a small row boat. The row boat has a set of oars and can seat up to three people. Inspired by the motto of the city of Paris, *fluctuat nec mergitur*, they decide to cross the river by ferrying back and forth in the row boat. However, the mathematicians become hesitant because they fear that if physicists should outnumber mathematicians on either bank of the river during the ferrying, the conversation will turn to experimental matters. The physicists sense the mathematicians' discomfort and make them an offer they cannot refuse: The physicists will go along if the mathematicians can figure out a way of getting everyone to the other bank without the mathematicians' ever being outnumbered on either side of the river. The mathematicians realize that the problem is not all that simple, since the people in the boat can join in the conversation whenever the boat reaches a bank of the river. Can the mathematicians come up with a solution? If so, to save face, can they produce a solution that is guaranteed to have the minimum possible number of crossings?

What makes this problem different from ones for which continuous variables and the constrain-and-generate paradigm are the appropriate tools is that we are not presented with any global relations that connect the elements of the problem. Instead, all we have are strictly local relations: A move of the row boat across the river with certain passengers is possible if and only if the current situation permits it. What the mathematicians have to produce is a sequence of crossings that is similar to moves in a board game that change the state of the board. Moreover the moves must be *legal* in the sense that no more than three people can take the boat at any one time, and the moves must be *safe* in the sense that if there are mathematicians on a bank of the river, they cannot be outnumbered by the physicists on the same bank. Note that the number of mathematicians on a bank must include those in the boat. What is needed then is a sequence of states with the correct local properties leading to a state that is a solution, namely one that has everyone on the right bank.

First, let us address the problem of developing sound and complete code to find a solution. That done, we can wrap this code in a `find_all` and use branch-and-bound search to hunt for the optimal solution. As is often the case with applications involving search, it is best to place oneself in the middle of the process and to code the next step to take. The state of affairs as the boat is about to set out again from one bank to the other can be described by the location of the boat and the number of mathematicians and physicists on either bank. We can represent where the mathematicians and physicists are by means of four integers, the number of mathematicians and the number of physicists on the left bank, and the number of mathematicians and the number of physicists on the right bank. Since we will want to keep track of the sequence of moves back and forth across the river, we will keep a stack of the successive states that are generated.

11.4 State space optimization

To keep track of the branch-and-bound cutoff, we shall use a global variable

```
int best_so_far;
```

To help with the notation, let us introduce some symbolic constants. Two important constants in this model are the number of mathematicians and physicists and the capacity of the row boat:

```
#define ALL 4
#define SEATS 3
```

Furthermore, to denote the indices in the state vector, we have

```
#define LBM 0   // Index of mathematicians on left bank
#define LBP 1   // Index of physicists on left bank
#define RBM 2   // Index of mathematicians on right bank
#define RBP 3   // Index of physicists on right bank
```

The stack of states can be represented as a two dimensional array. If we let N be a symbolic constant for an upper bound on the number of consecutive safe states that can be generated, the following definition and declaration are in order:

```
#define N (ALL+1)*(ALL+1)*(ALL+1)*(ALL+1)   // Plenty big
int state[N][4];
```

Starting at 0, if a state is associated with an even integer, the boat is on the left bank, and if the state has an odd number for its index, the boat is on the right bank. The parity of the index of the state on the stack will tell us which bank the row boat is on. Thus we can store the number of mathematicians on the left bank after k crossings in state[k][LBM], the number of mathematicians on the right bank in state[k][RBM], and so on. Again, the index k is part of the state information in that if k is even the boat is on the left bank, and if k is odd the boat is on the right bank.

For simplicity, let us write state[k] to denote the kth row of the array state[N][4]. Note that the definition of N can be sharpened to 2*(ALL+1)*(ALL+1), based on the fact that the vector state[i] is completely determined by the pair of integers state[i][LBM] and state[i][LBP]. The factor 2 comes from the fact that the same combination of scientists can arise but with the boat on different banks of the river. We will have to take care to avoid having state[i] equal to state[j] for distinct i and j of the same parity because this will mean that the search is not progressing.

The initial state is given by the problem itself:

```
state[0][LBM] = ALL;
state[0][LBP] = ALL;
state[0][RBM] = 0;
state[0][RBP] = 0;
```

The ith row of the state array is a safe state if the following test succeeds:

```
if state[i][LBM] >= 1;
then state[i][LBP] <= state[i][LBM];
if state[i][RBM] >= 1;
then state[i][RBP] <= state[i][RBM];
```

If in the transition from the `state[i]` to `state[i+1]`, there are m mathematicians and p physicists going from the left bank to the right bank, we will have

```
state[i+1][LBP] = state[i][LBP] - p;
state[i+1][LBM] = state[i][LBM] - m;
state[i+1][RBP] = state[i][RBP] + p;
state[i+1][RBM] = state[i][RBM] + m;
```

The plus and minus signs are reversed if the boat is going from the right bank to the left bank.

The goal is to determine an integer n and a sequence of states `state[0],...,state[n]` with the following properties:

1. `state[i+1]` is a safe state that can result from `state[i]` by a legal crossing
2. `state[n][RBM] == ALL`
3. `state[n][RBP] == ALL`

Working top down, we will need a loop such as

```
and(int crss=0;crss<N-1;crss++) {  // Central loop
    try_crossings(crss);    // Generate next state
    if goal(crss+1);        // If solution found
    then break;             // then exit loop
}
```

The code for the goal procedure can also call an output routine:

```
goal(int next)
{
    next%2 == 1;              // Boat must be on the right bank
    state[next][RBM] == ALL;  // Mathematicians across
    state[next][RBP] == ALL;  // Physicists across
```

11.4 State space optimization

```
    output(next);    // To be written below
}
```

The `try_crossings` code must loop through all possible legal and safe ways of ferrying people across the river, and it must also generate `state[crss+1]` from `state[crss]` which is the current state. So the task of the `try_crossings` routine will be to read `state[crss]`, to generate a possible crossing of the river, to check that it leads to a safe state and, if so, to create this new state `state[crss+1]`:

```
try_crossings(int crss)
{
int lbm,lbp,rbm,rbp;
int maths,phys;

    // Enter current state in identifiers lbm,lbp,rbm,rbp
    read_current_state(crss,lbm,lbp,rbm,rbp);

    if crss%2 == 0;         // Boat on left bank
    then {
        maths = lbm;
        phys = lbp;
    }
    else {
        maths = rbm;
        phys = rbp;
    }
    // Select mathematicians
    or(int m=maths;m>=0;m--)
        // Select physicists
        or(int p=phys;p>=0;p--) {
            if crss%2==1;       // Boat on right bank ?
            then crss + 2 <= best_so_far;// Prune by bound
            m+p >= 1;           // Send at least one person
            m+p <= SEATS;       // but don't send too many
            if crss%2 == 0;     // Boat on left bank ?
            then create_new_state(crss+1,lbm-m,
                    lbp-p,rbm+m,rbp+p);
            else create_new_state(crss+1,lbm+m,
                    lbp+p,rbm-m,rbp-p);
        }
}
```

Notice that we have inserted the branch-and-bound trick of checking that it is still possible to equal the best solution found so far in the inner `or` loop. This way the bound check will be made when the code first enters the `or` loops and when the code backtracks into the `or` loops.

The code to read the current state can be

```
read_current_state(int crss, int & lbm, & lbp, & rbm, & rbp)
{
    lbm = state[crss][LBM];
    lbp = state[crss][LBP];
    rbm = state[crss][RBM];
    rbp = state[crss][RBP];
}
```

And the code to create the next state can be

```
create_new_state(int crssplus1,nextlbm,nextlbp,
                                     nextrbm,nextrbp)
{
    if nextlbm >= 1;           // LBP musn't outnumber
    then nextlbp <= nextlbm;   // LBM

    if nextrbm >= 1;           // RBP musn't outnumber
    then nextrbp <= nextrbm;   // RBM

    state[crssplus1][LBM] = nextlbm;
    state[crssplus1][LBP] = nextlbp;
    state[crssplus1][RBM] = nextrbm;
    state[crssplus1][RBP] = nextrbp;

    and(int i=crssplus1%2;i<crssplus1;i=i+2)
        c_or(int j=0;j<4;j++)   // Check state not repeated
            state[i][j] != state[crssplus1][j];
}
```

Let us note that this code is respecting the fairness condition because the only time create_new_state is called with the parameter crssplus1 at i+1 is when the loop variable is at crss=i.

To go back to the output routine, what we want is a listing of the states that lead to a successful strategy for the mathematicians. These are contained in the stack of states when a solution is reached.

```
output(int crss)
{
    printf("Solution found with %d crossings\n",crss);
        ...    // Detailed output left to reader
}
```

11.4 State space optimization

Summing up, the main procedure will wrap the central loop in the `find_all` construct:

```
2lp_main()
{
    best_so_far = N+2;      // Upper bound on crossings
    create_new_state(0,ALL,ALL,0,0);  // Initial state

    find_all {       // Keep searching for solutions
        and(int crss=0;crss < N-1;crss++) {// Central loop
            try_crossings(crss);    // Generate next state
            if goal(crss+1);
            then {
                best_so_far = crss + 1;   // Shortest trip
                break;  // Exit loop successfully
            }
        }   // Re-enter loop to find next solution
    }
}
```

The program finds that the mathematicians can have their way and that everyone can be safely on the right bank with 9 crossings in the row boat. With this code, many solutions are found with 11 crossings and then with 9 crossings. Note that we have pruned with the bound `crss + 2 <= best_so_far` when `crss` is odd; this is valid here because a better solution must improve on the previous one by at least two crossings. If it is strengthened to `crss + 2 < best_so_far`, this will guarantee that only one optimal solution will be found; in fact, this way one solution with 11 crossings and one with 9 are found. The analogous tricks for continuous models are discussed in Section 12.2.

End of Model 11.4

With state space search, depth-first methods run a serious risk of straying down long unsuccessful paths. To compensate for this, breadth-first methods can be used; in Section 14.1 we develop this kind of alternative for the above model using a depth-first/breath-first technique called *iterative deepening search*.

Exercises

11.4.1. Do Model 11.4 using recursion and conditional disjunction. Write it in a traditional programming language.

11.4.2. Change the values of ALL and SEATS in Model 11.1 and try to spot a pattern in the optimal solutions. The classic case of this problem sets these to 3 and 2, respectively.

12

The Injury Method

In this chapter we present an important technique, the *injury method*, for organizing the branching strategy in the search in a disjunctive linear program. This method takes advantage of the mathematical structure of disjunctive programs to reshape the search tree dynamically. The basic idea goes back to Land and Doig (1960). The terminology we use here is adapted from the branch of mathematical logic known as recursion theory.

12.1 The eureka effect

The gap between linear programs and disjunctive programs is due to the presence of logical requirements which cannot be modeled by linear constraints. These nonlinear conditions are captured procedurally by the search process of the program. In this chapter we investigate ways of making this process more intelligent. As noted often, there are two dimensions to the search of the tree: horizontal and vertical. Up till now we have been stolidly modeling the search by a simple and/or structure:

```
and(int i=0;<M;i++)       // Vertical
   or(int j=0;j<N;j++)    // Horizontal
      { ... }
```

We seek to exploit the freedom to change the vertical order of the search and the freedom at each node to change the horizontal order of the search. To that end, we can use information gathered during the search process and information provided at a node by the witness point of the linear relaxation. The advantages that will accrue are many. In particular, by dynamically reorganizing the search, we will be able to detect failure earlier and eliminate useless wanderings about the tree. We will also be able to reach good solutions much sooner in the process. This will dramatically enhance the pruning power of branch-and-bound search. An added advantage of a disjunctive linear program is that it is possible for the witness point of the linear relaxation to provide a solution to the disjunctive problem itself; when this happens it is called the *eureka effect*.

Let us look at a concrete example. In Model 11.2, Fast Food, the disjunctive requirements are that each menu item be ordered in an integral amount. This is naively modeled as

```
and(int i=0;i<ITEMS;i++)
    or(int k=0;k<=LIMIT;k++)
        Serv[i] == k;
```

If a global integer variable `cnt` is introduced, we can count the number of nodes visited by this code and encapsulate things in procedures:

```
naive_search()
{
    and(int i=0;i<ITEMS;i++)
        branch_on(i);
}

branch_on(int i)
{
    or(int k=0;k<=LIMIT;k++) {
        cnt = cnt+1;
        Serv[i] == k;
    }
}
```

The core of the model becomes

```
find_min: Z;
subject_to
    naive_search();
```

Printing the value of `cnt` at the end of the program, we find that 2832 nodes have been generated.

Let us first address the question of making the vertical search more imaginative so as to reduce the node count. This means that we want to find a more effective way of ordering the consideration of the menu items.

In the best possible situation, the linear relaxation would provide a solution in which all the values `wp(Serv[i])` were integral; this way we would have the eureka effect the node count would be 0. What is more, if this witness point is optimal for the linear relaxation, then we have a solution that is optimal for the disjunctive program as well. If this happens at the outset of the program, so much the better. If not, we can

12.1 The eureka effect

work to try to have it take effect locally at a node in the search. A variable Serv[i] for which wp(Serv[i]) is not integral is said to be *injured*. One way to formulate our search task is as a campaign to reduce the number of injured variables until it is zero. Let us develop a search strategy along these lines. The idea will be to concentrate on injured variables; if one is found the program must have a way to repair the injury before proceeding. When the situation is reached where there are no injured variables, the eureka effect has manifested itself and a solution has been found. Again, if the eureka effect takes place at a witness point that is the optimal solution to the linear relaxation at the node, then we have a solution that is optimal for the disjunctive program at that node. In other words, the node need not be developed further; no better solution can be found below it. In this case and in the case where the node is pruned, the node is said to be *fathomed*.

The following routine will hunt for injured variables. It will fail if there are none. If there are injured variables, the index i of the first injured variable will be returned in the storeback parameter bv:

```
#define fail 0==1
determine_first_injured(int & bv)
{
    and(int i=0;i<ITEMS;i++)
        if not integral(Serv[i]);
        then {
            bv = i;
            return;
        }
    fail;
}
```

Now the vertical dimension of the search strategy can be re-organized so as to be responsive to information that becomes available as the search proceeds. In particular, we will not branch on variables unless they are injured. Also we will check for injured variables after calling min: Cost to ensure that the witness point is optimal for the linear relaxation at this node in the search tree.

```
smart_search()
{
int bv;      // Storeback variable
    and(int i=0;i<ITEMS;i++) {
        min: Cost;
        if determine_first_injured(bv);
        then branch(bv);
        else break;     // Eureka effect
    }
}
```

With this new code, smart_search replaces naive_search and the optimal solution is found in 644 nodes, a significant improvement. Further progress can be made by working harder on the way the horizontal phase of the search is handled, as we will now see.

A commonly used strategy for selecting the branching variable is to take one that is "most fractional." Code for this technique can be written as follows:

```
determine_most_fractional(int & bv, n, continuous X[])
{
double fract;
int nearest;

    bv = n;        // n is not the index of an X[i]
    fract = 0.0;
    and(int i=0;i<n;i++) {
        if integral(X[i]);
        then continue;
        nearest = nint(X[i]);
        if fabs(X[i] - nearest) > fract;
        then {
            fract = fabs(X[i] - nearest);
            bv = i;      // Record injured variable
        }
    }
    bv < n;        // Succeed if injury found, else fail
}
```

The intuition here is that one wants to branch on a variable that will have as big an impact on the objective function as possible. This is important in branch-and-bound search. A node can be pruned if the objective function deteriorates to the point that the current bound cannot be met.

However, if we replace our simple determine_first_injured routine with this more sophisticated one, things only get worse, much worse, as is easily checked. The problem is that the way the branch routine treats the horizontal part of the search does not take advantage of information contained in the value wp(Serv[i]) when the variable is injured.

Intuitively, the place to start looking for the best integer value for injured Serv[i] is at floor(Serv[i]) or ceil(Serv[i]). So our next branching routine will start branching from the floor of the witness point of the continuous variable that is passed to it as a parameter. The routine will also be sent the lower and upper bounds on the continuous variable:

12.1 The eureka effect

```
better_branch(continuous X, int lo,hi)
{
int start;

    start = floor(X);   // Record integer below wp(X)
    either {
        or(int k=start;k>=lo;k--){ // If k==-1, sidetrack
            cnt = cnt+1;           // to lower or loop
            X == k;     // If successful, return to and loop
        }
    }
    or {
        or(int k=start+1;k<=hi;k++){//If k==hi+1, backtrack
            cnt = cnt+1;    // Still counting nodes
            X == k;     // If successful, return to and loop
        }
    }
}
```

This code divides the branching into two parts. If we are going to have to branch in two directions, one down and one up, the most fractional variable is one likely to have a big impact on the objective function for both directions of the branching. This too makes it a reasonable candidate for the branching variable.

Another important point about backtracking is illustrated with this branching routine. If backtracking into the disjunctive statements in this procedure occurs, the feasible region is restored to what it was when the persistent disjunction was first called. However, as emphasized in Section 9.1 and elsewhere, the witness point is determined by the procedural apparatus that supports the type continuous, so it is not guaranteed that the witness point will be the same as it had been. Therefore we have cached the value of the floor of the witness point's X coordinate in start so that when the second or loop is entered, the initial value of the control variable is sure to be the value we want.

Let us look yet more closely at the way we are handling the horizontal search at a node. At present the loop starts with k equal to start. The loop then continues down to k = lo; the second loop is begun with k equal to start+1 and continues up to k = hi. The range of possible values for the continuous variable X is a closed interval of the form [a,b], where b is the maximum value that X can have and a is its minimum value. What we know is that lo <= a and b <= hi, but in either case the equality might not hold. Consider the downward loop and the constraint X == k. Whenever this constraint itself fails, it must mean that now k < a. Similarly, when backtracking occurs and the second loop is entered, if X == k fails, it must mean that now k > b. Hence in both cases, when X == k fails, it is safe to exit the or loop with failure and backtrack. This can be done by means of the break statement. For and loops, `break` causes the loop to succeed and exit. For `or` loops the situation is dual and `break` causes the loop to fail and exit. Incorporating this analysis, we can change the branching routine to read

```
down_then_up(continuous X, int lo,hi) // Descriptive name
{
int start;  // Variable to store floor(X)

    start = floor(X);
    either {
        //X==start,...,X==k+1 do not lead to a solution
        or(int k=start;k>=lo;k--){//If k<lo, sidetrack
            cnt = cnt+1;
            c_either X == k;
            or break;    // Sidetrack to lower loop
        }
    }
    or {
        //X==start,...,X==k-1 do not lead to a solution
        or(int k=start+1;k<=hi;k++){// If k>hi, backtrack
            cnt = cnt+1;
            c_either X == k;
            or break;    // Backtrack
        }
    }
}
```

Progress is being made. The number of nodes is reduced to 1682 with this method of branching applied to the most fractional variable.

The injury method is working for us on several fronts. By finding and evaluating injuries, it is guiding the depth-first search toward a good solution. After good bounds for the objective function have been determined, it is pruning the tree by directing the branching in directions likely to reach a violation of the current bound quickly. By means of the eureka effect, it is often making it unnecessary to pursue the search to the bottom of the tree to reach a solution.

To monitor this last phenomenon, we can keep track of the depth in the search tree as solutions are found in the branch and bound search:

```
smarter_search()
{
int bv;     // Storeback variable

    and(int i=0;i<ITEMS;i++) {
        min: Cost;
        if determine_most_fractional(bv,ITEMS,Serv);
        then down_then_up(Serv[bv],0,LIMIT);
```

12.1 The eureka effect

```
        else {
            printf("Eureka: %f at depth %d\n",Cost,i+1);
            break;
        }
    }
}
```

Replacing `smart_search` with `smarter_search`, the output is

```
Eureka: 14.450000 at depth 12
Eureka: 14.250000 at depth 12
Eureka: 11.960000 at depth 14
Eureka: 11.760000 at depth 13
Eureka: 11.200000 at depth 14
Eureka: 11.000000 at depth 13
Eureka: 10.950000 at depth 13
Eureka: 9.800000 at depth 11
Eureka: 8.960000 at depth 13
Eureka: 8.760000 at depth 12
Eureka: 8.760000 at depth 11
Eureka: 8.710000 at depth 9
```

We see that to find the initial solution, the "smarter" search had to branch on 12 of the 15 variables. As the bound on the objective function reached 8.96, the trend was that the eureka effect occurred at higher levels in the tree, that is, levels closer to the root. All the solutions were found without requiring branching on all 15 variables.

There are still many other things to try. For example, we have been following a parsimonious strategy in deciding the number of servings of each item: The branch routine first tries to reduce the fractional number of servings prescribed by the linear relaxation, and continues toward 0, before backtracking to follow the upward path. What if we switch to a greedy strategy and try the upward path first? This makes sense in that the linear relaxation is suggesting that this item should appear more fully in the final diet. This simply means changing the branching code to

```
up_then_down(continuous X, int lo,hi)
{
int start;
    start = floor(X);
    either {
        //X==start,...,X==k-1 do not lead to a solution
        or(int k=start+1;k<=hi;k++){// If k>hi, sidetrack
            cnt = cnt+1;
            c_either X == k;
            or break;      // Sidetrack to lower loop
        }
    }
```

```
    or {
        //X==start,...,X==k+1 do not lead to a solution
        or(int k=start;k>=lo;k--){//If k<lo, backtrack
            cnt = cnt+1;
            c_either X == k;
            or break;   // Backtrack
        }
    }
}
```

This time the associated output is

Eureka: 8.760000 at depth 4
Eureka: 8.710000 at depth 9

and the node count is 138. The moral of the story is that it pays to play with a model to get a sense of how best to solve it.

The concern with the number of nodes generated is motivated by two things. The number of nodes can grow exponentially and, if not controlled, can make an application intractable. Clearly, if more analysis is performed at a node, the linear relaxation can be tightened, better evaluations can be gathered and the branching strategy enhanced. However, all this can take effort. For example, the most thorough way to evaluate the impact on the objective function of making variables integral would be to lookahead and compute the effect of making each nonintegral variable X equal to `floor(X)` and `ceil(X)`. However, in a large application, this could prove to be overwhelming. Instead, heuristic approximations are employed ranging from the easily computed "most fractional" to rather elaborate techniques known as *penalty methods* (see Salkin and Mathur 1989). In Section 13.4, we consider a simple form of this kind of computation.

Thus there is a trade-off between the amount of computation at a node and the number of nodes generated. In disjunctive linear programs, already the linear programming support is a measurable computational effort; so in these applications some node reduction effort is almost invariably worthwhile. However, there are situations, where it is preferable, or even necessary, to generate a larger number of inexpensive nodes to handle a problem; the optimization form of the quadratic assignment problem can be an example.

As we will see in many examples, the injury method applies to other disjunctive situations and not only to the one of having variables take integer values. It also applies to problems that do not involve optimization. Therefore let us formalize the method in the general setting of a disjunctive linear programming problem.

A disjunctive linear program consists of setup constraints followed by a list of disjunctive logical requirements $L_0,...,L_{M-1}$, where each of these requirements is expressed as a disjunction of conjunctions of con-

12.1 The eureka effect

straints. Each requirement L_i corresponds to a disjunctive set $\cup F_{ij}$. Branching is usually needed to force these requirements to be satisfied by selecting one of the F_{ij} for each i. This situation can be modeled by means of an and/or loop

```
and(int i=0;i<M;i++)          // For each i
    or(int j=0;j<N;j++)       // find j to satisfy Lᵢ
        cc(i,j);              // by means of Fᵢⱼ
```

where `cc(i,j)` is a procedure call for generating a conjunction of constraints that meets the requirement L_i, namely the one that generates the polyhedron F_{ij}.

The goal is to reach a node where the solution to the linear relaxation to the problem also satisfies the logical requirements; this is the eureka effect. When the eureka effect occurs, no additional branching is needed, and a solution has been found.

If the witness point does not satisfy L_i, we say that this requirement is *injured*. The following template outlines the *injury method* for a general disjunctive linear programming problem:

```
and(int k=0;k<M;k++)
    if <there is an injured requirement L_injured>
    then
        or(int j=0;j<N;j++)      // Satisfy L_injured
            cc(injured,j);       // by means of F_injured j
    else break; // Eureka effect
```

This strategy is opportunistic in that the code will exit once a witness point is reached where no requirements are injured. In this way unnecessary branching can hopefully be avoided.

Annotating this code with loop invariants, we have

```
// Distinct integers i₀,...,i_{k-1}
// Integers j₀,...,j_{k-1}
// Successful calls to cc(i₀,j₀),...,cc(i_{k-1},j_{k-1})
and(int k=0;k<M;k++)
    if <there is an injured requirement L_injured>
    then
        // cc(injured,0),...,cc(injured,j-1) do not
        // lead to a solution
        or(int j=0;j<N;j++)// If j==N, decrement k
            cc(injured,j); // If successful, increment k
    else break;
```

Clearly the code satisfies the vertical conditions of consistency and monotonicity. If the program segment terminates, a solution has been found, and so it is sound. We are supposing that the or loop cycles through all feasible ways of repairing the injury to requirement i, which makes the search enumerative. The search is fair, since we are dealing with a disjunctive linear program. But for completeness, we cannot simply say that the code will find the first solution in lexicographic order as we did in Chapter 11, since the order in which the requirements are addressed is determined dynamically. In other words, we have to guarantee that all necessary levels of the search tree are accounted for. However, permuting the order of the requirements does not change the underlying solution set for the disjunctive program. So, as in Section 10.1, the argument for completeness can be made by induction on the number of requirements M. If there are no injuries when the body of the and loop is first entered, the code terminates successfully. If the first injury to be repaired is to requirement r, say, then the following loop is entered:

```
or(int j=0;j<N;j++)
    cc(r,j);
```

If there is a solution, the loop on j will reach a j_0 such that the constraints generated by cc(r,j_0) can be extended to a solution. At this point, we are in the case of M-1 requirements and the induction hypothesis applies.

The injury method is especially powerful for optimization problems. Again, it is critical that the check for injuries be made when the witness point is optimal for the linear relaxation. This way we are applying the principle that if the linear optimum satisfies the logical requirements, then it is optimal among all solutions satisfying the logical requirements.

When the eureka effect takes place in the optimization form of a disjunctive program, what happens is that both an NP-Hard problem and a co-NP-Hard problem have been merged into a linear programming problem. That is, the optimal solution of the linear relaxation provides both a solution to the NP-Hard problem of solving the disjunctive program and a solution to the co-NP-Hard problem of verifying optimality.

The next model is a classic going back to World War II and is narrated in the historical present. Here the injury method will be used and the logical requirements will be somewhat more complex than integrality requirements.

Model 12.1 Military History

A sector of the North Atlantic is divided into 30 quadrants, and air surveillance must be provided to try to protect convoys passing through this area from prowling submarines. Planes specially fitted out with new radio detecting and ranging equipment (RADAR) are based at three airports, one in England, one in Scotland, and one in Wales. Each quadrant must be covered by a squadron of planes from one of the

12.1 The eureka effect

TABLE 12.1 Available aircraft

Base 0	Base 1	Base 2
41	40	48

bases. The 12-hour stretch for patrolling from 6 AM to 6 PM is broken into 3 periods of 4 hours each. A schedule is needed for the next patrol period. Table 12.1 lists the number of planes available from each base for this patrol; all planes are to be used.

During the patrol, each quadrant must be covered by a squadron composed of a certain minimum number of aircraft and can be gainfully covered by up to a certain maximum number of aircraft. Table 12.2 gives these upper and lower bounds. Table 12.3 lists the distance in nautical miles from each base to each quadrant. The task is to find an assignment of planes from the different bases to the quadrants so as to meet minimum surveillance and not exceed maximum surveillance requirements. It would also be good to try to keep the total of the distances traveled by the aircraft in going out to the quadrants and back as small as possible.

Thus for each quadrant we must determine which airbase is to patrol it and how many aircraft from this base should be assigned to the quadrant. Let us define some constants:

TABLE 12.2 Minimum and Maximum Number of Planes per Quadrant

Quad	Minimum Cover	Maximum Cover	Quad	Minimum Cover	Maximum Cover
0	3	4	15	2	4
1	2	6	16	3	6
2	4	5	17	4	5
3	4	9	18	3	9
4	1	8	19	5	8
5	1	6	20	4	6
6	1	9	21	5	9
7	1	5	22	2	5
8	1	3	23	1	3
9	2	4	24	3	4
10	6	9	25	5	9
11	3	6	26	3	6
12	2	7	27	4	7
13	5	8	28	3	8
14	2	8	29	4	8

TABLE 12.3 Nautical Miles from Bases to Quadrants

Quad	Base 0	Base 1	Base 2	Quad	Base 0	Base 1	Base 2
0	475	200	575	15	750	400	400
1	550	225	475	16	850	475	325
2	650	250	400	17	950	500	300
3	750	300	300	18	300	425	825
4	350	225	700	19	375	400	750
5	450	275	600	20	475	400	700
6	550	275	500	21	575	400	600
7	650	300	425	22	675	400	525
8	750	350	325	23	750	475	475
9	850	400	275	24	850	500	425
10	250	375	800	25	950	550	400
11	350	350	725	26	500	475	725
12	450	325	475	27	600	475	650
13	550	400	525	28	700	475	600
14	650	400	475	29	800	500	525

```
#define BASES 3
#define QUADS 30
```

Next, let us introduce a two dimensional array of continuous variables and notation for the objective function

```
continuous BtoQ[BASES][QUADS];
continuous TravelDist;
```

The continuous variable `BtoQ[i][j]` will represent the number of planes from base i sent to quadrant j; the variable `TravelDist` will represent the total distance traveled to the quadrants by the aircraft from the three bases.

What makes this model challenging is the two-part logical requirement on each quadrant j: A solution must have only one of `BtoQ[0][j]`, `BtoQ[1][j]` and `BtoQ[2][j]` be nonzero, and that variable must be integer valued.

We can enter the data in integer arrays

```
int aircraft[BASES];
int lo[QUADS], hi[QUADS];
int knots[BASES][QUADS];
int cnt;         // If nodes are to be counted
```

12.1 The eureka effect

Then the linear constraints of the application can be put together in setup routines such as

```
assign_all_aircraft()
{
    and(int i=0;i<BASES;i++)    // From each base
        sigma(int j=0;j<QUADS;j++) BtoQ[i][j]
        == aircraft[i];         // Send all aircraft
}

stay_within_limits()
{
    and(int j=0;j<QUADS;j++)
        sigma(int i=0;i<BASES;i++) BtoQ[i][j] >= lo[j];
    and(int j=0;j<QUADS;j++)
        sigma(int i=0;i<BASES;i++) BtoQ[i][j] <= hi[j];

    and(int j=0;j<QUADS;j++)          // Redundant bounding
        and(int i=0;i<BASES;i++)      // constraints to help
            BtoQ[i][j] <= hi[j];      // tighten model
}
```

To keep track of the total distance that must be traveled by the aircraft in going to and from the quadrants under surveillance, we introduce the objective function:

```
set_travel_distance()
{
    TravelDist ==
        sigma(int i=0;i<BASES;i++)
            sigma(int j=0;j<QUADS;j++)
                knots[i][j]*BtoQ[i][j];
}
```

The main procedure can follow the template of the injury method by coding procedures to determine if there is an injured quadrant and if so to repair the injury. Repairing the injury to quadrant j can be done by trying to make each base i the one that serves the quadrant and then branching on BtoQ[i][j] to make the number of planes assigned to j from i take an integral value between lo[j] and hi[j]. This procedure for repairing an injury to a quadrant can use the routine up_then_down and be written as follows:

```
branch_on_bases(int j)
{
    or(int i=0;i<BASES;i++) {  // Select base i
        and(int t=0;t<BASES;t++)
            if t != i;           // Other bases t cannot send
            then BtoQ[t][j] == 0;  // planes to quadrant j
        up_then_down(BtoQ[i][j],lo[j],hi[j]); // Branch on
    }                                          // planes
}
```

The tasks remaining are to determine if there is an injured quadrant and, if so, which quadrant to branch on. We can do both jobs in one procedure by passing an identifier, say, quad, of type int to the following routine:

```
determine_quad(int & quad)
{
    and(int j=0;j<QUADS;j++)
        if not ok(j);
        then {
            quad = j;    // Record injured quadrant
            return;      // Return with success
        }
    0==1;        // Return with failure
}
```

where the ok test is

```
ok(int j)
{
    c_or(int i=0;i<BASES;i++){ // Is there a base i sending
        integral(BtoQ[i][j]);  // an integral number of
                                // planes to quadrant j
        and(int t=0;t<BASES;t++)   // such that all others
            if t != i;              // send none
            then wp(BtoQ[t][j]) == 0;
    }
}
```

Then if the determine_quad routine returns successfully, quad is set to the index of an injured quadrant.

We can organize the search procedure according to the injury method template:

```
assign_bases_to_quadrants()
{
int quad;    // Storeback variable
```

12.1 The eureka effect

```
    // A sequence of successful calls to
    // branch_on_bases(quad₀),...,branch_on_bases(quad_{k-1})
    // where quad₀,...,quad_{k-1} are distinct quadrants
    and(;;) {       // Loop variable plays no role
        min: TravelDist;
        if determine_quad(quad);       // Injured quadrant ?
        then branch_on_bases(quad);    // Repair injury
        else break;      // Eureka effect
    }
}
```

The loop invariant asserts that at the top of the loop a sequence of distinct quadrants has been taken care of. Each turn through the loop will guarantee that one more quadrant is taken care of if necessary.

As pointed out before, the injury method is especially effective in optimization problems. If we want the best possible solution, we can wrap the search for a solution in the find_min construct. Thus for this model, to find the solution that minimizes TravelDist, we envelop the assign_bases_to_quadrants procedure with the find_min construct:

```
21p_main()
{
    data();
    stay_within_limits();
    assign_all_aircraft();
    set_travel_distance();

    find_min: TravelDist;
    subject_to
        assign_bases_to_quadrants();

    output();
}
```

This code finds several solutions, the best placing TravelDist at 48,075. Again, what makes the method work here is the fact that the objective function is optimized before injuries are looked for. Thus if the witness point is one at which there are no injuries, that witness point yields a new incumbent solution in the branch-and-bound search.

End of Model 12.1

The injury method, while far from a panacea, is remarkably successful at inducing the eureka effect and reducing branching in search problems. One reason is that a vertex will tend to place a continuous variable at either its lower or upper bound. If the logical requirements are integrality requirements, threshold conditions, or combinations of these, frequently a vertex will satisfy them. This is an additional reason why it is good programming practice to state bounding constraints explicitly even if they are mathematically redundant. This is also an additional reason for being careful in devising the branching strategy, for branching adds constraints that re-shape the feasible region and that can create vertices for the eureka effect. In the next chapter, we will consider the underlying geometry and discuss some of the many techniques that have been developed for adding constraints and otherwise changing the model so as to coax the eureka effect into taking place.

The injury method is widely used both in its depth-first form and in its breadth-first form. We consider its role in the breadth-first search form of the branch-and-bound technique and in other search methods in Chapter 14.

A point to note here is that our analysis so far of search and branching has been unfailingly local in the sense that all decisions are made with information that is generated at the current node; no use is being made of information that might be gathered from the history of the process that led to this node, except for the bound on the objective function. An antidote that is commonly used in branch-and-bound search is the method of *pseudocosts*, (Land and Powell 1979). A piece of information that is very useful in choosing among requirements is the effect that repairing an injured requirement has on the objective function in the linear relaxation. If repairing a requirement has a significant impact on this value, it is an especially good candidate for branching during the co-NP phase of the process where optimality is being verified and the bound on the objective function is the principal pruning device. Although the information might be gathered from faraway nodes in the search tree, the hope is that a requirement that has an impact at one node is likely to have an impact at other nodes.

Exploitation of historical information is done in local search algorithms by metaheuristics such as tabu search. This is one of the strengths of this kind of algorithm.

Exercises

12.1.1. Explain why the injury method is not compatible with the `find_all` construct.

12.1.2. Into the code for Model 12.1, introduce a mechanism for counting nodes and for computing the depth of the search. Analyze the effectiveness of the injury method.

12.1.3. The injury method works nicely with progressive roundoff. Do Models 2.3 and 3.5 this way. One idea: Pick the injured variable that is closest to its ceiling.

12.2 Gaps

12.1.4. Write a branching strategy that alternates between going up and going down. For example, if wp(X) is 5.5, the code should try X==5, X==6, X==4, X==7, ...

12.2 Gaps

In order to push a branch-and-bound search along more quickly, it is often helpful to increment the bound more aggressively. Suppose that the last value of the objective function found at a solution for a maximization problem was bound; then we can replace bound with a higher value, say, bound + cheat where cheat is greater than 0.

If the find_max construct is used, then this can be carried out by a call to the built-in routine absgap. The code will read something like

```
find_max: Z;
subject_to {
    absgap(cheat);
    search();
}
```

The call to absgap(cheat) will have the effect of making the gap between successive solutions at least cheat in value. This is called an *absolute gap* because the increments are constant. The same command works for the find_min constructs and makes the successive decrements be at least cheat in value. In both cases the value of the parameter cheat must be nonnegative.

If one wants the gap to be computed dynamically as a function of the value of the best solution found so far, one can call the absgap routine each time the search has found a solution and pass it a newly computed value. For example, let us look at a formula which is used to push the objective along in this way (e.g. see Cplex 1995). Suppose that x is set to be a number strictly between 0 and 1; the following code has the effect of making the increments about 100*x percent of the value of wp(Z1-Z2):

```
find_max: Z1 - Z2;
subject_to {
    search();
    max: Z1 - Z2;
    absgap((x + fabs(Z1 - Z2))/(1 - x) - fabs(Z1 - Z2));
}
```

The formula is somewhat abstruse but has the virtue of making the increment nonnegligible when wp(Z1-Z2) is at or near 0.0.

Naturally in the above code we optimized the objective function before resetting the gap. The objective function is automatically optimized by the `find_max/subject_to` construct, but this takes place at the end of the execution of the body of code wrapped by the construct. If the value of the optimized objective function is needed during execution of this code, the optimizing call should be made. It is often helpful, for example, to print the value of the objective function as successive solutions to the logical requirements of the application are found. For this, the code segment would be something like

```
find_max: Z1 - Z2;
subject_to {
    search();
    max: Z1 - Z2;   // In case Z1-Z2 is not at optimum
    printf("Best value so far is %f\n",Z1-Z2);
}
```

Gaps are especially useful in an application where there are many solutions to the logical requirements that yield the same value for the objective function, a situation we will encounter in the next model. The model also illustrates the situation where the objective function must increase by a fixed gap in order to be meaningful as the incumbent solution evolves in the branch-and-bound search.

Model 12.2 More Fast Food

Let us take another look at the situation in Model 11.2, Fast Food. The day's regimen found there was not without all culinary merit, but decidedly lacking in variety. To vary the menu, so to speak, we can try to continue on to find a more satisfying solution to the student's problem, one where a greater number of different menu items form part of the daily diet. Rethinking the whole problem and starting from scratch, let us first find the minimum cost of a diet that meets the student's nutritional requirements; then let us find a way to maximize the number of menu items while staying within a certain range of the minimal possible cost, say, 50%. We relax the requirement of spending the least amount possible, because this requirement will in all likelihood restrict choices too greatly to afford much variety in the diet. This then is a kind of goal program. The first goal is to stay within 50% of the minimum cost, and the second goal is to maximize variety in the diet.

In order to keep track of which items will appear in the diet, we can introduce continuous variables `Y[i]` to serve as fuzzy counters and a variable `Variety` to add up the counters:

```
continuous Variety, Y[ITEMS];
```

In order to make the `Y[i]` do the counting job we want, we need some extra constraints that relate `Y[i]` to `Serv[i]` and to `Variety`:

12.2 Gaps

```
extra_setup()
{
    and(int i=0;i<ITEMS;i++) {
        Y[i] <= 1.0;
        Y[i] <= Serv[i];
    }
    Variety == sigma(int i=0;i<ITEMS;i++) Y[i];
}
```

When it comes to optimizing Variety, we will use a gap of 1.0, since more variety means at least one more item in the diet.

The main routine will first compute the least cost solution as in Model 11.2 or as in the previous section; then the result of this will be recorded in a variable cache of type double. That done, the feasible region will be restored to the one initially created by the setup routine of Model 11.2. Then the extra constraints on the Y[i] will be added and the second goal will be pursued with an upper bound of 1.5*cache on the cost of the daily regimen.

```
2lp_main()
{
double cache;

    data();
    setup();
    if not {
        find_least_expensive_meal();   // As before
        cache = wp(Cost);
        printf("The best possible cost is %.2f\n",cache);
    }
    then {
        printf("There is no solution\n");
        return;
    }
    extra_setup();
    Cost <= 1.5*cache;
    find_greatest_variety();
    output();
}
```

It remains to develop the routine find_greatest_variety. For this we can use the injury method. In the situation where Variety has been maximized, if there are no fractional items Serv[i], then we must have an integral solution to this phase of the problem. To see this, if wp(Serv[i]) is integral and greater than or equal to 1.0, then because of the optimization of Variety, the witness point value wp(Y[i]) must be 1.0. On the other hand, if wp(Serv[i]) is 0.0, then wp(Y[i]) must be `0.0`. So we can restrict ourselves to branching on injured `Serv[i]`. For the sake of simplicity let us use the routine `determine_first_injured` of the previous section to find an injured menu item. For the sake of variety, in the branching process let us record the current fractional value of `wp(Serv[i])` in a local variable `cwp`, and let us react to the injury by splitting the problem into two subproblems, the one with `Serv[i] >= ceil(cwp)`, the other with `Serv[i] <= floor(cwp)`:

```
split_in_two_halves(continuous X)
{
double cwp;

    cwp = wp(X);
    either X <= floor(cwp);
    or     X >= ceil(cwp);
}
```

This splitting of the problem into two subsets of the feasible region does not guarantee that henceforth `wp(Serv[i])` will be integral; rather, it repairs the injury to `Serv[i]` of lying strictly between `floor(cwp)` and `ceil(cwp)`. It is this situation that is rectified by the call to the routine `split_in_two_halves`. It is possible for the same variable to be injured again after this call, but that will take place in a different interval and eventually, if necessary, this splitting will force the continuous variable to a single integer value. This manner of handling injuries is classic (see Dakin 1965), and is the one usually presented in describing the branch-and-bound technique for mixed integer programs.

All the ingredients for the `find_greatest_variety` procedure are now accounted for:

```
find_greatest_variety()
{
    find_max: Variety;
    subject_to {
        absgap(1.0);// Fractional improvements impossible
        and(;;) {
            max: Variety;
            if determine_first_injured(bv);
            then split_in_two_halves(Serv[bv]);
            else break;     // Eureka effect
        }
    }
}
```

12.3 Branching and duality

This above code shows that a palatable diet can be had with 12 different items at a daily cost of $12.55.

End of Model 12.2

In Chapter 6 we discussed the role of randomization in local search algorithms. In Chapter 10 we briefly touched on using it in a backtracking search to scramble the order in which the horizontal and vertical dimensions of the tree search were carried out. Without going into detail, let us point out a way to use randomization in conjunction with branch-and-bound search and the injury method.

Suppose that we have a disjunctive program with a list of logical requirements $L_0,...,L_{M-1}$. For simplicity suppose that L_i reduces to making X[i] take an integer value. Let us consider the situation where we can solve this problem for K << M variables to optimality but not for M. Let us also suppose that through a quick-and-dirty routine we know that there is a solution, given by the M vector $e_0,...,e_{M-1}$. Then randomly select a subset of [0,...,M-1] of length M-K; using negation-as-failure, fix X[i] to e_i for i in this subset and solve for the optimal solution in the remaining K variables. Record this new solution in $e_0,...,e_{M-1}$ and repeat the process. This method is an example of a *shuffle*. For an application to job shop scheduling, see Applegate and Cook (1991).

Exercises

12.2.1. Do Model 12.2 with discrete code in a traditional programming language.

12.3 Branching and duality

Model 3.4, The Committee, was an example of a set covering problem: A collection of subsets of a base set is given, and one wants a cover of the base set by means of a small or even minimum number of subsets. In that code and in the code of Model 6.2, The Committee Realigned, the strategy was to branch on subsets to determine which ones to put in the cover. The next model is based on a related mathematical problem known as *set partitioning*. Again, we have to cover a set with subsets, but this time the subsets have to be pairwise disjoint. It is one of those instances where a crisp mathematical problem corresponds exactly to an important practical problem. In this model the number of sets that are candidates to be in the partition is large compared to the number of elements in the base set. Therefore we will approach the branching issue from the dual point of view and branch on elements rather than on subsets. The model addresses a classic application known as *airline crew scheduling*.

Model 12.3 Yet Another Crew Compilation

A small regional airline has 32 scheduled short flights each day and has to assign crews to them. A list of tours is drawn up; these tours consist of the set of flights that a potential crew could service. For example, a single tour could be formed to provide the service on flights 0, 1 and 24. Changing viewpoint, with each flight is associated a set of possible crew tours, namely, the tours that include this flight. Thus Flight 0 can be served by tours 34, 4, 50, 40, 22, 33 and 61. This information is given in Table 12.4. The number of different tours that can service each flight is given in Table 12.5. The task is to assign a tour to each flight. Naturally one objective is to minimize the number of tours needed. However, not all tours are equal in that in some tours at some points during the work day, the crew is in the wrong city. This requires that the crew take up space as passengers on a flight to their next city. This costly practice is known as *deadheading*. The airline can cost out the impact of deadheading, and in Table 12.6 the cost associated with each of the tours is given. These costs account for such elements as crew size and deadheading. Thus the challenge is to compile a consistent assignment of tours to flights so as to minimize the associated cost.

TABLE 12.4 The Tours for Each Flight

Flt	Tours	Flt	Tours	Flt	Tours
0	34,4,50,40,22,33,61	1	8,62,21,19,32,34,1	2	51,3,63,13,25,35,2
3	31,5,14,29,36,6,NA	4	10,41,27,37,5,NA,NA	5	11,56,20,23,38,4,NA
6	12,14,52,42,39,32,NA	7	18,1,43,28,40,9,NA,	8	21,13,44,28,41,6,NA
9	12,4,62,53,45,42,7	10	54,55,21,19,46,43,38	11	21,3,26,13,25,44,41
12	4,5,56,29,45,42,NA	13	47,55,27,46,44,NA,NA	14	11,31,20,23,47,41,NA
15	57,54,7,22,48,49,NA	16	48,32,16,28,49,6,NA	17	17,33,53,50,40,NA,NA
18	19,30,26,59,51,33,NA	19	15,44,16,49,52,46,NA	20	17,22,26,60,53,12,NA
21	58,14,64,8,2,54,36	22	51,11,21,9,2,55,50	23	27,23,26,3,5,56,37
24	14,15,4,9,57,34,NA	25	0,8,17,58,25,NA,NA	26	1,19,10,13,59,0,NA
27	2,24,7,12,60,39,NA	28	8,0,16,18,61,22,NA	29	7,13,6,28,62,45,NA
30	2,24,17,8,12,63,50,	31	28,31,26,9,49,64,60		

TABLE 12.5 Number of Tours That Can Service Flight

Flights	Number of Tours										
0-10	7	7	7	6	5	6	6	6	6	7	7
11-21	7	6	5	6	6	6	5	6	6	6	7
21-32	7	7	6	5	6	6	6	6	7	7	

12.3 Branching and duality

TABLE 12.6 Costs for Each Tour

Tours	Costs in thousands of dollars												
0-12	7	3	7	2	5	4	6	4	6	2	2	3	4
13-25	5	3	3	3	2	3	3	3	2	2	2	3	8
26-39	8	5	6	9	10	11	7	2	2	6	3	3	6
40-52	3	6	7	2	7	2	3	3	6	3	2	4	3
53-65	3	2	2	3	2	4	3	3	3	6	2	2	4

What must be determined is whether or not each tour is to be included among the tours to be employed. We will introduce a continuous variable `Tour[t]` for each tour `t`. In a solution this variable will have witness point value `1.0` if the tour is included and `0.0` if not.

```
#define TOURS 65
#define FLIGHTS 32

continuous Z,Tour[TOURS];
```

The data of Table 12.5 and 12.6 can be stored in one-dimensional arrays:

```
int card[FLIGHTS]; // Cardinality of set of tours for flight
double cost[TOURS];// Cost of tour
```

To avoid ragged arrays, we will associate a vector of length MAXNUM with each flight, where MAXNUM is the greatest number of tours associated with any flight. The data of Table 12.4 will be stored in a two-dimensional array:

```
#define MAXNUM 7
int set[FLIGHTS][MAXNUM];
```

The routine to enter the data will record the information of Tables 12.4 through 12.6.

```
data()
{
#define NA -1
    set = {
        34,4,50,40,22,33,61,
        8,62,21,19,32,34,1,
        51,3,63,13,25,35,2,
        31,5,14,29,36,6,NA,
        10,41,27,37,5,NA,NA,
```

			...
		2,24,17,8,12,63,50,
		28,31,26,9,49,64,60
	};
#undef NA

	card = { 7,7,7, ... ,7,7 };

	cost = { 7,3,7, ... ,2,4 };
}
```

The procedure to set up the basic constraints of the application will bound the continuous variables, assert the condition that each flight be covered by exactly one tour in fuzzy logic, and define the objective function:

```
setup()
{
    and(int i=0;i<TOURS;i++)
        Tour[i] <= 1;   // Each tour can be used at most once

    and(int flt=0;flt<FLIGHTS;flt++)  // Cover each flight
      sigma(int i=0;i<card[flt];i++)Tour[set[flt][i]] == 1;

    // Objective function
    Z == sigma(int k=0;k<TOURS;k++) cost[k]*Tour[k];
}
```

In this application, we can consider the continuous variables `Tour[k]` as sets and the flights as elements. The flights are represented by the left-hand sides of the constraints laid down in the `setup` routine, the constraints that state that each flight must be covered exactly once:

```
    sigma(int i=0;i<card[flt];i++)Tour[set[flt][i]] == 1;
```

Following the lead of Models 2.3 and 3.5, we would branch on the continuous variables `Tour[k]` to force each to a 0-1 value. However, in this model the number of tours is considerably larger than the number of flights, and the variables outnumber the constraints. Therefore rather than branching on tours, we take the dual point of view and branch on flights. In other words, rather than going through subsets to put in the partition, let us go through elements and make sure each element is covered.

A flight is injured if no set `Tour[j]` covering it has witness point value `1.0`. To select the injured flight, we will take the one with the smallest number of tours that can serve it. This is an example of the first fail strategy: Address the most constrained situation first:
```

12.3 Branching and duality

```
injured_flight(int & be)
{
int nots;    // Number of tours
    nots = MAXNUM+1;
    be = FLIGHTS;
    and(int flt=0;flt<FLIGHTS;flt++)// Loop through flights
        if (
            and(int i=0;i<card[flt];i++)   // Injured ?
                wp(Tour[set[flt][i]]) < 1.0;
            card[flt] < nots;        // Worse than previous ?
        }
        then {                  // Worst injury found so far
            be = flt;           // This is element for branching
            nots = card[flt];   // Update number of tours
        }
    be < FLIGHTS;   // Injury found ?
}
```

Having found an injured element, how do we repair the injury? We use a *divide-and-conquer* strategy. We simply divide the collection of tours that can service the flight into two groups, the first half and the second half; it is here that branching takes place.

The constraint representing the fact that a flight must have exactly one tour assigned to it has the form

$$X[0] + \ldots + X[n-1] == 1;$$

and exactly one `X[i]` must have witness point value 1.0 in the solution. A constraint of this kind is called a *generalized upper bound* or *gub*. This terminology originated with work on special handling of these constraints in the simplex method itself (Dantzig and Van Slyke 1967). From the point of view of logic, what we have is an n-ary exclusive-or.

To process a gub, first constraints `X[i] == 0.0` are added for $i=(n-1)/2,\ldots,n-1$ to the effect that a member of second half cannot go into the partition. Upon backtracking, these constraints are undone, and zeroing constraints asserting that a member of the first half cannot go into the partition are added instead. Several bookkeeping issues must be dealt with. During the search, with each flight `flt`, we will associate a closed interval and a set of tours. To start, this interval will be `[0,card[flt]-1]`, and the set will be all the tours that can be assigned to the flight, `set[flt][0],...,set[flt][card[flt]-1]`. When the element `flt` is the selected injured element, we will split the current interval `[m,n]` into the two halves, `[m,mid]` and `[mid+1,n]`, with `mid = (n-m)/2`. The corresponding sets of tours will be

```
        set[flt][m],...,set[flt][mid]
        set[flt][mid+1],...,set[flt][n]
```

The first branch in the search will try the constraints

```
and(int i=mid+1;i<=n;i++)
    Tour[set[flt][i]] == 0;
```

and update the interval associated with flt to [m,mid]. The other branch will add the constraints

```
and(int i=m;i<=mid;i++)
    Tour[set[flt][i]] == 0;
```

and update the interval associated with flt to [mid+1,n].

To keep track of the progress of the branch-and-bound search, we will introduce some auxiliary arrays and employ the trailing method of Model 10.2; to count nodes, we will introduce a node counter. The following declarations of global variables are made at the top of the program:

```
int trail[3*FLIGHTS];  // > (log base 2 (MAXNUM))* FLIGHTS
int top;               // Top of trail stack
#define N 10           // > (log base 2 (TOURS))
int cnt;               // Node counter
int itinerary[FLIGHTS][N]; // The vector itinerary[f] is
                       // stack for flight f
int depth[FLIGHTS];    // depth[f]-1 is top of stack for f
```

The trailing mechanism will be used to keep track of the intervals $[m, n]$ associated with each flight. If both alternatives of a choice point fail, the trail must be undone, and the interval reset to its previous state. The vector itinerary[flt] will serve as a stack, and the integer depth[flt] will keep track of the top of this stack. The integer depth[flt] will be incremented each time flt is selected as the injured element. When the branching goes to the left, we will set itinerary[depth[flt]]-1 equal to LEFT; when it goes right, we set it to RIGHT. This way when the flight is re-injured, we can easily find the current interval $[m, n]$ by means of a binary search. When all alternatives fail, the state of the array itinerary can be restored by decrementing depth[flt]; in turn decrementing depth[flt] will be done by popping the stack trail. The length of this binary search process to retrieve the interval $[m, n]$ is bounded by the number of times the original interval can be split in two, which is certainly no more than the $\log_2$ of the size TOURS of the collection of all tours.

To keep track of how many decrements of depth[flt] must be made when backtracking occurs, we use a global variable top of type int. To keep track of the current depth in the tree, we can make the central and/or loop of the depth-first search have a control variable k for the and loop. When things are moving forward, top will be set equal

12.3 Branching and duality

to k. When deep backtracking takes place, the loop control variable k will drop back to its value at the last open choice point. All increments made in the interval from k+1 to top must then be undone to get back on track, and top must be reset to k.

The code to compute the interval associated with a flight is the following:

```
#define LEFT 0
#define RIGHT 1

compute_interval(int flt, int & m, & mid, & n)
{
    m = 0; n = card[flt]-1;            // Original interval
    and(int i=0;i<depth[flt];i++)      // Binary path
        if itinerary[flt][i] == LEFT;  // of branches
        then n = m + (n - m)/2;        // left and right
        else m = m + (n - m)/2+1;

    mid = m+(n-m)/2;                   // New midpoint
    depth[flt] = depth[flt]+1;         // Increment depth
}
```

This code to recompute the interval associated with the injured element returns the values m, mid and n. The candidate intervals are now [m,mid] and [mid+1,n], which present two branching choices: one can branch first to the left and then to the right, and *vice versa*. In the following routine to take care of an injured element, the choice is made first to branch in the direction which the linear relaxation is giving more weight to:

```
tend_to_flight(int flt,k)
{
int m,mid,n;      // Storeback variables

    compute_interval(flt,m,mid,n);    // Find m,mid,n
    top = k;                          // Increment top of stack
    trail[top] = flt;  // Push flt onto trail stack
    // Bias search towards heavier half
    if wp(sigma(int i=m;i<=mid;i++) Tour[set[flt][i]]) >
        wp(sigma(int i=mid+1;i<=n;i++) Tour[set[flt][i]]);
    then left_then_right(flt,m,mid,n,k);
    else right_then_left(flt,m,mid,n,k);
}
```

Let us develop the code for first branching to the left. This requires a persistent disjunction to go first to the left and then, upon backtracking, to the right. If backtracking or sidetracking to the second alternative occurs, the top of the `itinerary[flt]` stack must be reset. This is accomplished by cleaning up the trail stack before entering the or part of the `either/or`.

To go left and then to sway right, we have

```
left_then_right(int flt,m,mid,n,k)
{
    either left(flt,m,mid,n);   // top is equal to k
    or {
        if top > k;       // Deep backtracking has occurred
        then {    // Untrail required
            and(int t=top;t>k;t--)
                depth[trail[t]] = depth[trail[t]] - 1;
            top = k;    // Reset
        }
        right(flt,m,mid,n);
    }
}
```

The code to go left can be

```
left(int flt,m,mid,n)
{
    cnt = cnt+1;                          // Count new node
    itinerary[flt][depth[flt]-1] = LEFT;
    shrink_sets(flt,m,mid,n,LEFT);    // Branch left
}
```

while the code to go right can be

```
right(int flt,m,mid,n)
{
    cnt = cnt+1;                          // Count new node
    itinerary[flt][depth[flt]-1] = RIGHT;
    shrink_sets(flt,m,mid,n,RIGHT);   // Branch right
}
```

The codes for going left and right update the itinerary and call a routine to zero out the appropriate part of the gub. When this is done, the objective function is re-optimized for the sake of the injury method:

```
shrink_sets(int flt,a,b,c,dir)
{
    if dir == LEFT;
    then and(int i=b+1;i<=c;i++)
        Tour[set[flt][i]] == 0;     // Annihilate right
```

12.3 Branching and duality

```
        else and(int i=a;i<=b;i++)
            Tour[set[flt][i]] == 0;    // Annihilate left
        min: Z;                // Re-optimize for injury method
}
```

For completeness, to go back and forth in the other direction, we have

```
right_then_left(int r,m,mid,n,int k)
{
    either right(r,m,mid,n);
    or {
        and(int t=top;t>k;t--)      // Untrail if needed
            depth[trail[t]] = depth[trail[t]] - 1;
        top = k;                    // Re-enforce if needed
        left(r,m,mid,n);
    }
}
```

With all this machinery in place we can turn to the main routine:

```
2lp_main()
{
int flight;

    data();
    setup();
    top = -1;       // Empty stack
    optimize();
}

optimize()
{
int flight;

    min: Z;
    find_min: Z;
    subject_to
        and(int k=0;;k++){ // k records depth of search
            absgap(1.0);    // Each solution strictly better
            if injured_flight(flight);
            then tend_to_flight(flight,k);
            else break;     // Eureka effect
        }
    output();
}
```

```
output()
{
    printf("There will be %.0f tours needed\n",
        sigma(int k=0;k<TOURS;k++) Tour[k]);
    printf("At a cost of %f\n",Z);
    printf("There were %d choice nodes generated\n",cnt);
}
```

This codes finds a schedule with only 12 tours at a cost of $39,000. In all 72 nodes were generated. This result is better in terms of performance than that obtained by branching on tours, as is easily checked.

In practice, this kind of application proves to be a computationally difficult problem. The number of flights for an airline is usually very substantial; the number of potential tours is very, very large. In fact, establishing a list of tours is a time consuming process itself. This means that even the linear relaxation can be difficult to solve. For developments in this field (see Hoffman and Padberg 1991).

End of Model 12.3

The method of treating a block of continuous variables as a unit is used in MIP solvers to treat special constraint structure. In this last model we have the example of the gub: a constraint summing to 1.0, where only one variable can be nonzero in the solution. Another term for gub is *specially ordered set of type* 1. There are also the *specially ordered sets of type* 2. In this situation the sum of a sequence of continuous variables is constrained to be equal to 1.0, but no more than two of the variables can be nonzero and they must be successive in the list. Specially ordered sets of this type are used to work with piecewise linear approximations to nonlinear terms (see Williams 1992).

Exercises

12.3.1. Do the minimum cover problem with a branching strategy modeled on that of Model 12.3. Try different strategies for picking the injured element.

12.3.2. Use a specially ordered set of type 2 to handle a linear program with a term $... + x^2 + ...$ in the objective function.

12.3.3. Change the code of the crew scheduling model so that if after being tended to, a flight is still injured, the code continues to work on that flight without returning to the main loop.

13

Tightening the Linear Relaxation

The linear relaxation is a powerful heuristic in disjunctive linear programming. It is used to guide the search and to prune the search space. In this chapter we consider several methods for making the declarative knowledge represented by the linear relaxation assist more in the procedural search process.

13.1 Fuzzy booleans

We have seen many applications where the logical requirements were to force continuous variables to have 0-1 witness point values. In the next example, we start with a straightforward disjunctive program and then deliberately rewrite the logical requirements of the model to express every persistent disjunction in the form of a continuous variable's being 0 or 1. Thus, in effect, the original disjunctive program is transformed into the special kind of problem known as a Mixed Integer Program, or MIP. In the case we consider next, the transformation will serve both to simplify the model formulation and to speed up the solution process. The model is a classic known as the *capital budgeting problem*.

Model 13.1 The Capitol University Budget

The board of trustees of Capitol University, an important state school, has been given a budget for capital improvements over the next 7 years. In all, 20 projects are competing for funding and the ground rules are that a project must be funded fully for the 7 year period or not funded at all. The amount of capital required by each project in each year is given in Table 13.1. The total amount of capital available each year is given in Table 13.2.

These projects will be underwritten by the state dormitory authority, which issues bonds. The price of these bonds is determined in large

TABLE 13.1 Project Requirements

Project	Year 0	Year 1	Year 2	Year 3	Year 4	Year 5	Year 6
Project 0	50	50	50	50	50	50	50
Project 1	100	100	0	0	30	30	0
Project 2	90	100	100	70	60	50	50
Project 3	90	80	70	60	50	40	30
Project 4	50	100	100	100	0	0	0
Project 5	60	60	90	20	20	30	30
Project 6	120	40	20	10	10	0	50
Project 7	180	30	20	0	0	0	0
Project 8	55	44	53	22	10	10	10
Project 9	77	55	68	11	0	0	0
Project 10	40	95	80	35	50	44	20
Project 11	30	90	70	0	20	20	0
Project 12	30	90	70	50	50	39	10
Project 13	90	60	70	65	60	22	10
Project 14	30	70	60	90	20	8	0
Project 15	70	100	50	30	0	0	20
Project 16	10	45	20	20	15	0	40
Project 17	80	40	80	33	10	0	0
Project 18	65	34	100	0	0	10	7
Project 19	57	125	40	31	9	0	0

TABLE 13.2 Budgeted Capital

Year	0	1	2	3	4	5	6
Available	640	430	390	290	350	290	320

part by the value of the facilities that will be built. So with each project is associated a measure in current dollars of its worth in seven years. Using classical accounting terminology, this number is called the *net present value* of the project.

The goal of the trustees is to maximize the sum of the net present values of the funded projects. For starters, let us introduce the symbolic constants:

```
#define YRS 7
#define PJS 20
```

and the data arrays:

```
double capital[YRS],req[PJS][YRS],npv[PJS];
```

13.1 Fuzzy booleans

TABLE 13.3 Net Present Value

Project	NPV	Project	NPV
Project 0	551	Project 10	651
Project 1	497	Project 11	222
Project 2	319	Project 12	529
Project 3	270	Project 13	150
Project 4	430	Project 14	630
Project 5	653	Project 15	530
Project 6	365	Project 16	665
Project 7	300	Project 17	240
Project 8	275	Project 18	175
Project 9	330	Project 19	280

To encode this application as a straightforward disjunctive program, we start with continuous variables `Invest[i][j]` for projects i and years j. Each year's budget determines a resource constraint

```
sigma(int i=0;i<PJS;i++) Invest[i][j] <= capital[j];
```

A search procedure is needed that has to branch on i and decide whether or not to fund project i. The contribution to the sum of the net present values of a project can be represented by a continuous variable `NetPresVal[i]`; the branch-and-bound code can be something like

```
find_max: sigma(int i=0;i<PJS;i++) NetPresVal[i];
subject_to
    and(int i=0;i<PJS;i++)
        either {                    // Fund
            and(int j=0;j<YRS;j++)
                Invest[i][j] == req[i][j];
            NetPresVal[i] == npv[i];
        }
        or {                        // Do not fund
            and(int j=0;j<YRS;j++)
                Invest[i][j] == 0.0;
            NetPresVal[i] == 0.0;
        }
```

This is a straightforward formulation of the model, as a disjunctive program. For the record, this code will visit 195,778 nodes and find 10 solutions in all, the last yielding the optimal solution of a net present value of 3968.

Clearly the treatment of this application can be greatly simplified by the observation that the fate of each project is a YES/NO decision. This decision completely determines the amounts to be invested in it each year as well as its contribution to the sum of the net present values of the funded projects.

Since for each project there is a single unknown, namely, whether or not the project is to be funded, what is needed here is a kind of boolean, a type that only takes the values 0 and 1. So let us try to use continuous variables to simulate booleans. Introducing continuous variables Y[i] to stand for the fate of project i, we can bound these variables to range between 0.0 and 1.0, and we can use search and branching to force each Y[i] to have witness point value 0.0 or 1.0.

```
continuous Y[PJS];
```

Knowing that at every solution all of these variables will be either 0.0 or 1.0, we can proceed with the model and use the Y[i] in constraints while *thinking of them* as booleans! Since booleans themselves are discrete, the continuous variables Y[i] will function therefore as *fuzzy* booleans in that at various points in the program the witness value of Y[i] may well lie strictly between 0.0 and 1.0.

With this approach, the resource constraints become

```
sigma(int i=0;i<PJS;i++) req[i][j]*Y[i] <= capital[j];
```

and the objective function becomes

```
Z == sigma(int i=0;i<PJS;i++) npv[i]*Y[i];
```

Thus the variables Invest[i][j] and Npv[i] have simply disappeared. This means that basic setup is changed to

```
setup()     // New version
{
    and(int i=0;i<PJS;i++)
        Y[i] <= 1;
    and(int j=0;j<YRS;j++)
        sigma(int i=0;i<PJS;i++) req[i][j]*Y[i]
        <= capital[j];
    Z == sigma(int i=0;i<PJS;i++) npv[i]*Y[i];
}
```

We have moved much of the program into setting up the initial constraints by using fuzzy logic. That is, we have expanded the constraint phase of the model and will be therefore able to simplify the generate phase: At this point the only requirements left to the generate phase of the model are the 0-1 conditions on the Y[i]. So the only persistent dis-

13.1 Fuzzy booleans

junctions needed in the model concern these variables: This comes down to a routine such as

```
decide_project(continuous Y)
{
    either Y == 1;      // Fund
    or     Y == 0;      // Do not fund
}
```

and a straightforward branch-and-bound will use a search such as

```
naive_search()
{
    and(int i=0;i<PJS;i++)
        decide_project(Y[i]);
}
```

This approach finds and verifies the optimal solution with a node count of 764, a considerable improvement over 199,404.

But even in the search phase of the program, we can continue to use the intuition behind the fuzzy logic to help guide the search by means of the injury method. With fuzzy logic we can restrict the branching to injured Y[i]. As usual, there is a wide range of strategies available for choosing the branching variable. This variable is chosen in the injury method after a call that optimizes the objective function. Here let us follow a greedy strategy and branch on the variable with current witness point value closest to 1: the intuition is that wp(Y[i]) is a measure of the likelihood that Y[i] should be 1.0 at an optimal solution. This way of looking for a branching variable can be coded as follows:

```
determine_most_likely(int & bv, continuous Y[], int p)
{
double gauge;

    gauge = 0.0;
    bv = p;
    and(int i=0;i<p;i++)
        if {
            not integral(Y[i]);// Injured variable ?
            gauge < wp(Y[i]);  // Closer to 1.0 ?
        }
        then {
            bv = i;            // New candidate for branching
            gauge = wp(Y[i]);  // Record witness point value
        }
    bv < p;                    // Injury found ?
}
```

So the structure of the main routine is simply

```
21p_main()
{
    data();             // Initialize arrays
    setup();            // Resource constraints

    Z == sigma(int i=0;i<PJS;i++) npv[i]*Y[i];

    find_max: Z;
    subject_to
        search();       // Use injury method
}
```

To code the search routine with the strategy of branching on the most likely injured variable, we have

```
search()
{
int bv;

    and(;;) {           // Standard injury method code
        max: Z;
        if determine_most_likely(bv,Y,PJS);
        then decide_project(Y[bv]);
        else break;
    }
}
```

Now the solution is found in 156 nodes. The injury method performs far better than the strategy of simply branching on each project in order from 0 to PJS-1, a technique that yielded a node count of 764.

In fact, the fuzzy booleans also suggest another improvement to the code. Since the constraints of the application are basically all resource constraints, a progressive roundoff can be made as a quick-and-dirty way of getting an initial solution to use as a bluff for the branch-and-bound search. A call to the following routine will determine a solution without recourse to backtracking by proceeding through the projects, optimizing the objective function and fixing the variable Y[i] at 1.0 only if the witness point is currently at 1.0. This way feasibility is maintained and the code is sure to reach a solution. In order to undo this solution and to restore the feasible region, the code will use a double negation-as-failure.

13.1 Fuzzy booleans

```
quick_and_dirty()
{
double cache;
    not not {                // Progressive round down
        and(int i=0;i<PJS;i++){
            max: Z;
            Y[i] == floor(Y[i]);
        }
        cache = wp(Z);       // Record the bluff
        printf("The bluff is %.2f\n",Z);
    }
    Z >= cache;              // Enforce the bluff
}
```

This routine can be called before entering the find_max construct. With this addition the code finds a bluff of 3815 as compared to the optimal solution of 3968; this brings the node count down to 108.

To simplify the discussion, we did not use a gap to force the solutions to get better and better at a faster clip, another reasonable ploy to improve performance. Other techniques that might also be used are randomization to get better quick-and-dirty solutions, pseudocosts to help with the choice of branching variable, and shuffle strategies in case the problem size is formidable.

End of Model 13.1

The trick of using continuous variables as fuzzy booleans is classic. In fact, any disjunctive linear program can be coded using fuzzy booleans so that all the logical requirements reduce to making these variables 0 or 1. For example, the persistent disjunction

```
either X >= 20;
or X <= 10;
```

can be recoded as

```
X >= 20*Y;
X <= 10 + bigM*Y;
either Y == 0;
or Y == 1;
```

where Y is a new continuous variable that serves as a fuzzy boolean and bigM is a known upper bound on X. In the literature, constraints of the form X <= bigM*Y, which force X to be 0 when a fuzzy boolean Y takes the value 0 and do nothing when Y takes the value 1, are known as *big M constraints*.

Coding a disjunction of equality constraints can be done by treating the constraints as an opposing pair of inequality constraints. For example, the disjunction

```
either X + Y - Z == 20;
or     X - Y + Z == 10;
```

can be recoded using two large constants `bigM` and `bigMM` and a fuzzy boolean variable `U` as follows:

```
X + Y - Z <= 20 + bigM*U;
X + Y - Z >= 20*(1-U) - bigMM*U;
X - Y + Z <= 10 + bigM*(1-U);
X - Y + Z >= 10*U - bigMM*(1-U);
```

The upshot is that every disjunctive linear program can be given the form of a MIP problem, one where all the disjunctive requirements reduce to integrality requirements. This approach has the advantage of standardization; the models can be formulated independently of the strategy for solving them, and solvers can be written for generic MIP problems. Thus, for example, modeling languages such as GAMS and AMPL generate matrices that can be passed to LP and MIP solvers. These languages also facilitate the handling of sparse data and AMPL supports network formulations of models. The basic references are Brooke, Kendrick and Meerhaus (1988) for GAMS and Fourer, Gay, and Kernighan (1993) for AMPL.

On the solver side, an advantage to the standardized MIP formulation is that the injury method is easy to apply, since all injuries reduce to variables' not being integral.

In Section 10.2 we noted that applications where the set of values a variable can take becomes punctured in the course of the search process are often best treated with discrete rather than continuous methods. However, let us see how we can model this kind of behavior with MIP methods. In fact let us look at a classic puzzle analyzed by Newell and Simon (1972). The challenge is to find all solutions to the crypto-arithmetic equation GERALD + DONALD = ROBERT. The letters represent distinct digits. Since there are ten letters, this means that they must correspond to a permutation of 0,...,9. For the MIP representation of the puzzle, we can introduce continuous variables

```
#define TEN 10
#define G Letter[0]
...
#define T Letter[9]
continuous Letter[TEN];
continuous X[TEN][TEN];
```

13.1 Fuzzy booleans

The variable X[i][j] will be a fuzzy boolean, which will become 1.0 if the ith letter is mapped to the jth digit. The setup code is

```
layout()
{
    and(int i=0;i<TEN;i++)
        and(int j=0;j<TEN;j++)
            X[i][j] <= 1.0;
    and(int i=0;i<TEN;i++)
        sigma(int j=0;j<TEN;j++) X[i][j] == 1.0;
    and(int j=0;j<TEN;j++)
            sigma(int i=0;i<TEN;i++) X[i][j] == 1.0;
    and(int i=0;i<TEN;i++)
        Letter[i] == sigma(int j=0;j<TEN;j++) j*X[i][j];
}
```

The equation is

```
equation()
{
    100000*G + 10000*E + 1000*R + 100*A + 10*L + 1*D
+   100000*D + 10000*O + 1000*N + 100*A + 10*L + 1*D
/*  ---------------------------------------------- */
==  100000*R + 10000*O + 1000*B + 100*E + 10*R + 1*T    ;
}
```

To put this together as a MIP application we have

```
21p_main()
{
    layout();
    equation();
    search_for_all_solutions();// Make all X[i][j] integral
}
```

We leave it to the reader to formulate an effective search routine to find all solutions. For a MIP solver one usually has to concoct an objective function. The fact is that MIP solvers find this a surprisingly difficult problem. On the other hand, people are able to solve this puzzle easily enough although not without some effort. In the work of Newell and Simon, subjects were given the value of one of the letters at the start.

What human beings typically do is to employ a strategy known as *constraint propagation*. With this strategy one systematically reduces the possible values that each letter can take. Reducing the range of one letter impacts the ranges of the others; one continues iteratively until, hopefully, each variable converges to a single value. If the variables do not all converge to a single value, then branching can be resorted to and the process relaunched. To give an idea of how one would write code for this kind of approach, we have the following:

```
2lp_main()
{
    equation();
    vanilla_constraint_propagation();
}

vanilla_constraint_propagation()
{
int cnt;                    // To count nodes
    cnt = 0;
    and(int k=0;k<TEN;k++)  // Constrain variables
        Letter[k] <= 9;
    find_all                // Obtain all solutions
        and(int k=0;k<TEN;k++) {
            propagate(k);   // Propagate constraints
            generate(Letter[k],k,cnt);     // Branch
        }
    output(cnt);            // Print number of nodes
}
```

A simple way to propagate the constraints is to enforce the best possible upper and lower integral bounds on each variable Letter[i].

```
propagate(int k)
{
int flag;
    flag = 1;
    and(;flag==1;) {        // While flag is on, propagate
        flag = 0;           // Turn flag off
        and(int i=k;i<TEN;i++) {
            max: Letter[i];     // Upper linear range
            if wp(Letter[i]) < ub(Letter[i]);
            then {          // Make upper bound integral
                flag = 1;       // Turn flag on
                Letter[i] <= floor(Letter[i]);
            }
            min: Letter[i];     // Lower linear range
            if wp(Letter[i]) > lb(Letter[i]);
            then {          // Make lower bound integral
                flag = 1;       // Turn flag on
                Letter[i] >= ceil(Letter[i]);
            }
        }
    }
}
```

13.1 Fuzzy booleans

In generating values at which to fix the continuous variables Letter[i] we have to take care not to use a value that has already been taken by a previous variable; the propagate routine does not account for "holes" in the range of a variable.

```
generate(continuous X, int k, int & cnt)
{
    or(int i=lb(X);i<=ub(X);i++) {// Loop over range
        and(int j=0;j<k;j++)         // Check for puncture
            i != wp(Letter[j]);
        cnt = cnt+1;                 // Still counting nodes
        X == i;                      // Fix letter's value
    }
}
```

Typically in a constraint propagation approach to a problem like this, one will try to apply a first-fail strategy and branch on the remaining variable with the smallest range. This simply requires maintaining a list of the indices of the continuous variables and ordering this list accordingly. Another thing that people will do in solving this problem is to introduce carry bits and to add the constraints that express the column by column relations such as $D + D == 10*Carry + T$. This idea is the topic of Exercise 13.2.5 below. Let us also note that for this small crypto-arithmetic puzzle, the impatient programmer will resort to a nested sequence of loops in C or other traditional programming language and solve the problem by introducing carries and by generating an exceedingly high number of inexpensive nodes. This is not very sporting since it drowns the problem in computing power, although all's fair in love and modeling.

In fact LP and MIP solvers themselves use constraint propagation under the name of *pre-solving* at the start of the solving process. The classic reference for linear programming is Brearley, Mitra, and Williams (1975). For a different approach to the analysis of LPs, see Greenberg (1993a,b). For MIP, references are Williams (1994) and Savelsbergh (1994). In the pre-solve phase of the solution of a MIP model, a pass is made, and variable bounds are tightened iteratively to reflect as far as possible the fact that the certain variables will have to be integer valued in all solutions. This tactic can of course be inserted into a 2LP program to solve a MIP model. It can also be reiterated during the search as we just did with the puzzle.

We have been discussing a situation where MIP is not the most effective way of formulating a search problem. Before that we looked at the case of Model 13.1, where fuzzy booleans and MIP were very much the thing to do. The analysis of situations where the technique of using fuzzy booleans is preferable to direct coding of the associated logic in the disjunctive program style is very much "application sensitive."

Exercises

13.1.1. How does the solution in Model 13.1 change if any money not used in a given year's budget is added to the amount for next year.

13.1.2. Program the optimization form of Model 9.1 with fuzzy booleans.

13.1.3. Program Model 13.1 as a discrete application. Then add complications such as restricting funding to at most n projects or requiring mutual exclusions among various sets of projects.

13.1.4. Code the optimization form of the graph coloring problem of Model 5.1 as a MIP model.

13.1.5. Write code to find all solutions to the crypto-arithmetic puzzle

$$\begin{array}{r} L\ Y\ N\ D\ O\ N \\ *\ b \\ \hline N\ O\ S\ N\ H\ O\ J \end{array}$$

13.1.6. Find all solutions to the puzzles SEND + MORE = MONEY and CROSS + ROADS = DANGER.

13.1.7. Formulate Model 8.5 as an Integer Program.

13.2 Valid cuts

What makes disjunctive programs difficult to solve in general is the need for branching to smoke out a solution. In the simplest case the disjunctive program is a linear program, and then the problem is straightforward.

Redundant constraints that do not change the solution set of a disjunctive program can serve to make the linear relaxation tighter. An example of such a constraint is illustrated in Figure 13.1. Before going on, let us make a formal definition.

Let a disjunctive program be given by an initial feasible region intersected with an intersection of disjunctive sets:

$$F \cap \bigcap_{i<M} \bigcup_{j<N} F_{ij}$$

Let c be a constraint, and let C be the set of points that satisfies this constraint. Then c is a *valid cut* if and only the following equality holds:

$$F \cap \bigcap_{i<M} \bigcup_{j<N} F_{ij} = C \cap F \cap \bigcap_{i<M} \bigcup_{j<N} F_{ij}$$

13.2 Valid cuts

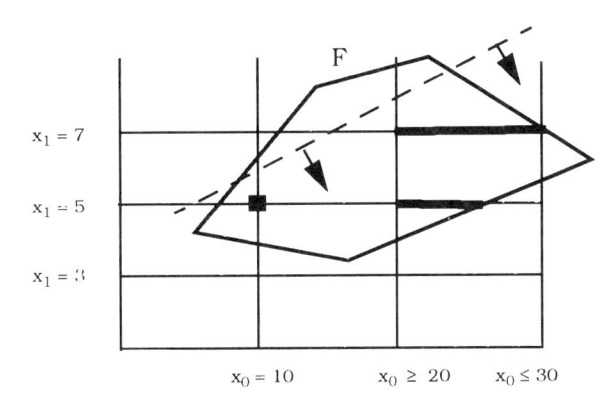

FIGURE 13.1 A disjunctive program with initial feasible region. The dashed line represents a cut.

A valid cut is a constraint that is satisfied by the witness point whenever a solution to the constraints and logical requirements of the disjunctive program is found. When the context makes it clear, we will often simply call a valid cut a *cut*. If $C \cap F$ is strictly included in F, the cut is called a *proper valid cut*. The principal function of valid cuts is to tighten the linear relaxation.

When a proper valid cut is given by an inequality constraint, the equality form of the constraint defines a hyperplane that cuts the feasible region of the linear relaxation into two parts, one of which can be ignored. For this reason, cuts are historically called *cutting planes*.

In a program the valid cuts can be added to the setup constraints that define the initial feasible region:

```
setup();     // Constraints for initial feasible region
valid_cuts();   // Add cuts
and(int i=0;i<M;i++)    // Disjunctive program
    or(int j=0;j<N;j++)
        cc(i,j);
```

To reiterate, valid cuts are constraints that become redundant when a feasible region that is a solution to the logical requirements of a disjunctive program is reached. Although a constraint is known to be redundant in this way, it does not mean that the constraint will be redundant during the process of reaching that solution. In fact, we have already seen this phenomenon in a constrain-and-generate application, namely Model 8.5, Who Has the Zebra. There the redundant constraints were a fuzzy approximation to the requirement that the solution yield a permutation of each category.

A set E in n-dimensional space is *convex* if for any pair of points p, q in E, the straight line segment from p to q is included in E; see Figure

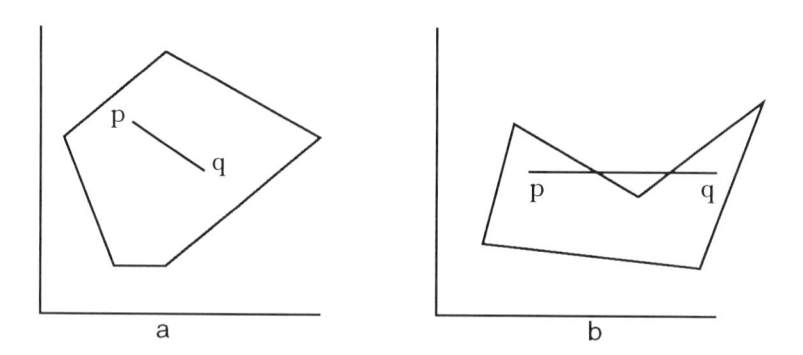

FIGURE 13.2 (a) A convex region; (b) a non convex region.

13.2. The entire space R^n is convex of course, and half spaces are convex. The intersection of two convex sets is also convex. It follows that feasible regions are convex, since they can always be written as the intersection of a collection of half spaces. It also follows that every subset of n-dimensional space is contained in a smallest convex set, namely the intersection of all convex sets containing it; this set is called its *convex closure* or *convex hull*.

As noted in Chapter 9, by putting a disjunctive program in DNF, one sees that the solution set of a disjunctive program is itself a disjunctive set and so has the form $\cup F_i$, where the F_i are feasible regions. In general, disjunctive sets are not convex. The mathematical result that underwrites the search for valid cuts is that the convex closure of a disjunctive set is itself a polyhedral set and that the vertices of this polyhedral set are among the vertices of the components F_i of the disjunctive set. In Figure 13.3 we have an illustration of how this works. This result is proved in Chapter 15.

To make the linear relaxation as tight as possible, one would like the feasible region of the relaxation to be the convex closure of the solution set of the disjunctive program. Ideally one would like to find cuts that yield this convex closure; this special situation is illustrate in Fig-

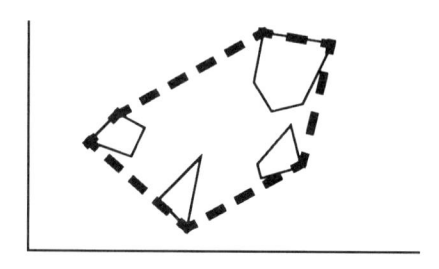

FIGURE 13.3 The convex closure of a disjunctive set.

13.2 Valid cuts

ure 13.4. This will guarantee the eureka effect. Although in practice it is rare to be able to find such a family of cuts, an important strategy is to seek cuts that generate vertices of the convex closure of the solution set. This is the geometric interpretation; in our examples, the cuts will be found by considering the semantics of the problem statement.

To illustrate the power of cuts in a disjunctive program, let us look at a classical challenge problem for propositional theorem provers, the pigeonhole principle: It is impossible to place $n + 1$ pigeons in n pigeonholes without placing more than one pigeon in at least one of the holes. To formulate this in propositional logic, one models this impossible situation with propositional formulas and then shows the formulas to be unsatisfiable. While easy enough for humans armed with induction or an insight into the symmetry of the situation, this problem is difficult for automated systems for propositional logic.

Letting p_{ij} represent the case where the i th pigeon goes to the j th pigeonhole, we can express the requirement that each pigeon have a hole by means of a conjunction of clauses

$$\bigwedge_i \bigvee_j p_{ij}$$

To ensure that no hole is occupied by more than one pigeon, we have

$$\bigwedge_j \bigwedge_i \left(p_{ij} \to \bigwedge_{k \neq i} \sim p_{kj} \right)$$

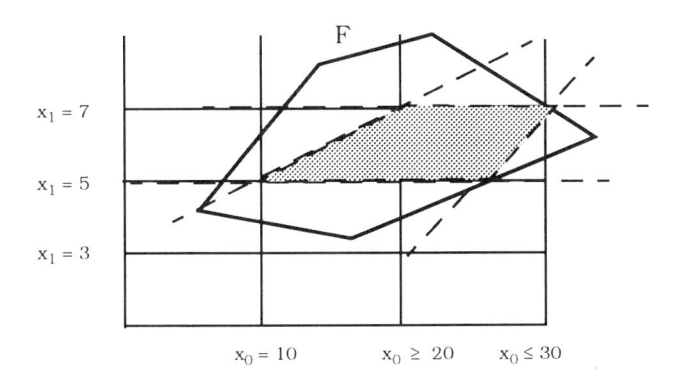

FIGURE 13.4 A disjunctive program with initial feasible region. The shaded region represents the convex closure of the set of solutions.

which becomes a collection of clauses

$$\sim p_{ij} \vee \sim p_{kj}, \quad k \neq i$$

which, in turn, is equivalent to

$$\sim p_{ij} \vee \sim p_{kj}, \quad k < i$$

The formulation is now in CNF. To translate these clauses into constraints is easy enough. One declares variables

```
continuous P[N+1][N];
```

and the constraints are

```
pigeon_hole_principle()
{
    and(int i=0;i<N+1;i++)     // Bound the propositional
        and(int j=0;j<N;j++)   // variables
            P[i][j] <= 1;

    and(int i=0;i<N+1;i++)     // For each pigeon
        sigma(int j=0;j<N;j++) P[i][j] >= 1;  // a hole

    and(int j=0;j<N;j++)                  // For each hole
        and(int i=0;i<N+1;i++)            // for each pair
            and(int k=0;k<i;k++)          // of pigeons
                (1-P[i][j]) + (1-P[k][j]) >= 1;// at most 1
                                             // in hole
}
```

As linear constraints, these are consistent. Using the algorithm of Model 9.2, even as enhanced by the injury method of Section 11.1, to show there is no 0-1 solution will take an inordinate amount of time even for small values of N. The trouble is that the linear relaxation will simply place half of every pigeon in hole 0 and half in hole 1. This is consistent with the fact that the conditions excluding pigeons from the same hole apply to pigeons two at a time. This came about because in the translation of the statement

$$\bigwedge_j \bigwedge_i \left(p_{ij} \to \bigwedge_{k \neq i} \sim p_{kj} \right)$$

to the clauses $\sim p_{ij} \vee \sim p_{kj}$, there is nothing to assert globally that a total of at most 1 pigeon out of N+1 can go into each hole. In other words,

13.2 Valid cuts

what is needed is a new logical connective, one that expresses "At most 1 of n propositions is true." The cuts that serve as a silver bullet here are those that model this connective in a fuzzy way:

```
and(int j=0;j<N;j++)              // For each hole
    sigma(int i=0;i<N+1;i++) P[i][j] <= 1;// at most 1
                                          // pigeon
```

In fact we can consider the single constraint

```
sigma(int j=0;j<N;j++)            // For N holes
    sigma(int i=0;i<N+1;i++) P[i][j]   // at most N
<= N;                                   // pigeons
```

Adding this redundant constraint makes the linear relaxation infeasible and so detects the unsatisfiability of the formulas at once. In other words, the following code detects the impossibility of placing too many pigeons in the holes:

```
#define N ...
continuous P[N+1][N];

2lp_main()
{
    if not {
        pigeon_hole_principle();  // Clauses as constraints
        sigma(int j=0;j<N;j++)    // Valid cut
            sigma(int i=0;i<N+1;i++) P[i][j]
        <= N;
    } then printf("N+1 pigeons can't fit in N holes\n");
}
```

As noted in Section 10.4, the pigeonhole principle is a difficult challenge problem for resolution based theorem provers and Davis-Putnam methods. What we have seen here is an application of Cook, Coullard and Turán (1987) of the cutting planes proof method of Chvátal (1973). This method also has unavoidable exponential behavior on a infinite class of problem instances, as has been shown in Pudlák (1996). For work on automatic generation of cutting planes for problems like this, see Barth (1996).

Naturally cuts play an important role if one seeks a solution to the disjunctive program that optimizes a linear objective function. If the feasible region for the linear relaxation is the convex closure of the solution set of the disjunctive program, then the optimal solution of the linear relaxation will satisfy the disjunctive requirements and be the optimal solution to the disjunctive program. In fact, much less is needed than to have the convex closure of the solution set; it suffices that the optimal

TABLE 13.4 Capacities and Costs

Supplier	0	1	2	3	4	5	6	7	8	9
Capacity	700	1000	1000	1000	400	500	4700	9930	3440	7550
Unit Cost	18.0	18.5	20.0	20.0	14.0	17.5	22.0	27.5	24.5	23.8

solution to the linear relaxation satisfy the disjunctive requirements and so lie in the solution set.

Let us treat a classic model known as the *fixed charge problem*. Good valid cuts for the model will be particularly simple and easy to find.

Model 13.2 The Jobber's Fee

A children's toy store is desperately seeking an additional 5000 units of a sought-after item for the upcoming holiday season. The store has contacted a jobber who in turn will negotiate potential deals with his suppliers. Each supplier can make a bid on the item in question and at a certain price per unit. This price includes shipping, handling, and the like. Of course each supplier only has a limited number of these items on hand. In addition to the price per unit, the jobber adds to each supplier's bid his own fee, which is $3000. This fee is a fixed charge that is only paid if the supplier is hired to furnish merchandise. The store's goal is to stock its shelves at the least possible cost to itself.

What makes this application special is the that the fixed charge of the jobber's fee is only paid if the supplier is signed up to deliver some merchandise. The data are given in Table 13.4

Let us introduce symbolic constants and arrays for the data:

```
#define N 10
#define DEMAND 5000
#define FEE 3000.0

double cap[N],cost[N];
```

Let us put aside the issue of the fixed charge for the moment. With this simplification, if we are asked to decide how to meet demand most economically, we have a linear programming problem. The unknowns are the number of items ordered from each supplier, and so the declarations of the continuous variables are

```
continuous X[N],Z;
```

There are the bounding constraints and the store's demand constraint

13.2 Valid cuts

```
basic_constraints()
{
    and(int j=0;j<N;j++)        // Bounding constraints
        X[j] <= cap[j];

    // Store's demand constraint
    sigma(int j=0;j<N;j++) X[j] >= DEMAND;
}
```

The objective function is simply

```
    Z == sigma(int j=0;j<N;j++) cost[j]*X[j];
```

However, with each supplier is associated that fixed charge of the jobber's fee. The most direct way of getting a solution to this more complex version of the problem is to take the solution to the linear programming problem just considered and, for every supplier used, to add to the cost the full fee charged by the jobber. In other words, this is a version of the simple roundoff technique discussed in Section 2.2. The code for this step is

```
quick_and_dirty()
{
double scratch;

    min: Z;
    scratch = wp(Z);           // Cost without fees
    and(int i=0;i<N;i++)
        if wp(X[i]) > 0;       // Supplier used ?
        then scratch = scratch + FEE;  // then pay fee
    Z <= scratch;    // A bluff for branch-and-bound to come
    printf("quick_and_dirty gets a cost of %f\n",scratch);
}
```

This simple heuristic is not guaranteed to yield the best possible solution; in fact, it will perform rather badly if there are resources that have low proportional cost `cost[i]` but are in short supply. Moreover there is no way to take advantage of a progressive roundoff technique for a quick solution, since re-optimizing the objective function in the body of the and loop will not help; the method will find the same quick-and-dirty solution.

To account more fully for the added complication of the fixed charge in the model, we can introduce further global continuous variables

```
continuous FixedCost[N];
```

and make the objective function be

```
        Z ==
        sigma(int i=0;i<N;i++) (cost[i]*X[i] + FixedCost[i]);
```

Despite these changes, the `quick_and_dirty` routine does not have to be changed, nor does the `basic_constraints` routine.

The all-or-nothing nature of the fixed charge introduces the logical requirement that either `X[i]` be fixed at `0.0` or `FixedCost[i]` be fixed at `FEE`. Code to enforce this requirement is

```
enforce(int i)
{
    either FixedCost[i] == FEE;
    or {
        X[i] == 0.0;
        FixedCost[i] == 0.0;
    }
}
```

Now the basic solution strategy can be coded as follows:

```
21p_main()
{
    data();
    basic_constraints();
    Z ==                          // Define objective function
    sigma(int i=0.i<N;i++) (cost[i]*X[i] + FixedCost[i]);
    quick_and_dirty();            // Compute and enforce bluff

    find_min: Z;
    subject_to
        and(int i=0;i<N;i++)
            enforce(i);
}
```

This code visits 324 nodes in the task of finding the optimal solution and finds 8 solutions to the problem.

For aesthetic purposes and to emphasize the fact that there is a YES/NO decision involved in dealing with fixed costs, let us replace the variables `FixedCost[i]` by variables `Y[i]` and make the logical requirements reduce to having `Y[i]` be at 0 or 1. This changes the objective function to

```
        Z == sigma(int i=0;i<N;i++) (cost[i]*X[i] + FEE*Y[i]);
```

and the `enforce` procedure to

13.2 Valid cuts

```
enforce(int i)           // New version
{
    either Y[i] == 1.0;
    or {
        X[i] == 0;
        Y[i] == 0.0;
    }
}
```

The main routine does not change, but the `basic_constraints` routine now should include the bounding constraints on the `Y[i]`:

```
and(int i=0;i<N;i++)     // Fuzzy logic bounds
    Y[i] <= 1.0;
```

The `quick_and_dirty` routine can also stay the same.

So this version of the code uses the continuous variables `Y[i]` as fuzzy boolean variables. But this has only been a cosmetic change, and the code still visits 324 nodes.

Let us now try the injury method to make things better. In this situation an injury occurs when we have `wp(X[i]) > 0.0` but `wp(Y[i]) < 1.0`. In other words, there is an injury when supplier i is being used but the fuzzy logical variable `Y[i]` is not at `1.0`. So the code to determine a branching variable could be

```
#define fail 0==1
determine_first_pair(int & bv)
{
    and(int i=0;i<N;i++)
        if {
            wp(X[i]) > 0.0;
            wp(Y[i]) < 1.0;
        }
        then {
            bv = i;
            return;
        }
    fail;
}
```

The code to repair the injury would be the new version of the `enforce` routine above.

Incorporating these changes, the `find_min` code reads as follows:

```
find_min: Z;
subject_to
    search();
```

where the `search` routine is

```
search()
{
int bv;

    and(;;){
        min: Z;
        if determine_first_pair(bv);
        then enforce(bv);        // New version
        else break;
    }
}
```

This effort brings the node count down to 264 but still generates 8 solutions to the problem. The drawback of this code is that after the call to `min: Z`, if the variable `Y[i]` has not already been branched on and fixed at `1.0`, the optimization will always send the witness point at `Y[i]` to `0.0`. So the only case where the injury method will avoid a branch point is when this optimization call also makes the witness point at `X[i]` equal to `0.0`.

When the witness point at `X[i]` is not at `0.0`, what we need is a way to force the witness point at `Y[i]` to reflect the fact that the witness point at `X[i]` is nonzero. Now, if the fixed fee did not have an all-or-nothing character but instead was proportional to the amount of available resource used, we could express this by means of the constraint

```
            X[i] <= cap[i]*Y[i];
```

and not require `Y[i]` to be `0.0` or `1.0`. This way we would once again have a linear programming problem rather than a disjunctive programming problem to deal with. However, for the disjunctive program, we claim that the constraint `X[i] <= cap[i]*Y[i]` is a valid cut. To see this, note that if `wp(Y[i])` is `1.0`, then the constraint reduces to the bounding constraint `X[i] <= cap[i]`, which is already in place. On the other hand, if `wp(Y[i])` is `0.0`, then it reduces to `X[i] == 0.0`, which must hold if `Y[i] == 0.0` holds because of the `or` branch in the above `enforce` routine:

```
    or {
        X[i] == 0.0;
        Y[i] == 0.0;
    }
```

So at some point before the code enters the branch-and-bound search, let us add a call to a procedure to adjoin these cuts to the constraints

13.2 Valid cuts

```
add_cuts()
{
    and(int i=0;i<N;i++)          // Valid cuts
        X[i] <= cap[i]*Y[i];
}
```

Let us make a remark about the `quick_and_dirty` routine in the presence of cuts. If this routine is invoked after the call to the `add_cuts` routine, then it must be changed to reflect the presence of these cuts. The cuts force the objective function to take into account the products `cap[i]*Y[i]` when Z is minimized in the linear relaxation. For all that, the adjustment in the code is easy to make:

```
quick_and_dirty()            // After cuts
{
double scratch;

    min: Z;
    // Initialize scratch to be wp(Z) less the FEE*Y[i]
    scratch = wp(sigma(int i=0;i<N;i++) cost[i]*X[i]);
    and(int i=0;i<N;i++)
        if wp(X[i]) > 0;                     // Supplier used ?
        then scratch = scratch + FEE; // then pay fee
    Z <= scratch;   // A bluff for branch-and-bound
    printf("quick_and_dirty yields a cost of %f",scratch);
}
```

It is to be expected that this will yield a bluff at least as good and probably somewhat better than that obtained when the cuts are not present.

Putting things together, we have the main routine:

```
21p_main()
{
    data();
    basic_constraints();
    add_cuts();
    Z == sigma(int i=0;i<N;i++) (cost[i]*X[i] + FEE*Y[i]);
    quick_and_dirty();      // Compute and enforce bluff

    find_min: Z;
    subject_to
        search();           // Use injury method
}
```

Now the node count is 2 and the optimal solution is the first one found. A closer analysis of the search tree will reveal that the cuts help in two ways. First, if after the call minimizing the objective function the witness point at X[i] is equal to cap[i], then there is no injury at i and Y[i] need not be branched upon. Also in the branch-and-bound search, the contribution of the term FEE*Y[i] in the objective function will be more realistic and more nodes are pruned by the bound on the objective function. Thus the combination of the cuts and the injury method produces a striking improvement over the earlier code.

Moreover the presence of the cuts improves the performance of the quick_and_dirty routine. In fact, the simple roundoff yields the excellent solution of 112,500, while the branch-and-bound search finds the optimal solution to be 112,300.

End of Model 13.2

The goal of bringing the feasible region of the linear relaxation of a disjunctive program close to the convex hull of the solution set can sometimes be furthered by following a method introduced by Jeroslow and Lowe (1984). Suppose that we have to find a solution that lies in one of the half spaces defined by the m constraints

$$a_{11}x_1 + \ldots + a_{1n}x_n \leq b_1$$
$$\vdots$$
$$a_{m1}x_1 + \ldots + a_{mn}x_n \leq b_m$$

We split the x_j into components on each of the m half spaces,

$$x_j = x_{j1} + \ldots + x_{jm}$$

New fuzzy boolean variables $y_1,\ldots,y_m$ are introduced and the constraints that one works with are

$$y_1 + \ldots + y_m = 1$$
$$a_{11}x_1 + \ldots + a_{1n}x_n \leq b_1 y_1$$
$$\vdots$$
$$a_{m1}x_1 + \ldots + a_{mn}x_n \leq b_m y_m$$

This transformation can be very effective, even though it greatly increases the number of continuous variables needed to model the application. It can make the linear relaxation much tighter in situations where this can otherwise be hard to achieve. For work in which its power is nicely illustrated, see Hanson and Martin (1990).

Let us conclude this section by citing some theoretical work on cutting planes. Gomory (1958, 1963) gave a fundamental algorithm for integer programs that yields a sequence of cuts that lead to an integer point that is a vertex of the solution set of the problem. For work on the relation between cutting plane algorithms for IP and resolution methods for propositional logic, see Hooker (1992). For disjunctive programs there has been the work of Jeroslow (1980) who showed that there is a converging sequence of cuts for the class of *facial disjunctive programs*, a class that includes the 0-1 mixed integer programs. More recently, Lovász and Schrijver (1991) and Balas, Ceria, and Cornuéjols (1993) have developed the *lift-and-project* method for generating cuts for MIP problems which has computational promise.

A very different approach to MIP is taken in Cook, Rutherford, Scarf, and Shallcross (1993). In this work an algorithm is given based on H. W. Lenstra's beautiful theorem that in fixed dimension, IP is in P (see Lenstra 1983).

Exercises

13.2.1. Run the algorithms of Model 9.2 on the pigeonhole principle with and without cuts with 7 pigeons and 6 pigeonholes.

13.2.2. Show that the valid cuts for the pigeonhole principle detect the inconsistency without even formulating the pairwise incompatibility constraints $\sim p_{ij} \vee \sim p_{kj}$.

13.2.3. Add carry equations such as `D + D == T + 10*Carry[1]` as cuts to `GERALD + DONALD == ROBERT` and the crypto-arithmetic puzzles of the previous exercises. Comment on how these cuts affect node count. Also in `GERALD + DONALD == ROBERT`, one can add the cut which expresses the fact that the letters must sum to `sigma(int i=0;i<TEN;i++) i`.

13.3 Branch-and-cut

In this section we develop a simple example of a very powerful technique. In the previous section we considered the notion of a valid cut. There the valid cuts which were added to the constraints were all adjoined at the outset of the program. In many situations, the number of cuts that can be added this way is too large to be manageable. It will also happen that good cuts will emerge as the search process proceeds. In this section we blend the injury method with the method of valid cuts to add cuts dynamically as they are needed, a technique known as *branch-and-cut*. We illustrate these ideas with a classic application known as the *capacitated warehouse location problem*. This application is similar to the previous model in that it requires a mixed integer program rather than a linear program because of the presence of a fixed charge.

TABLE 13.5 Monthly Demand for Each of the 30 Supermarkets

Stores 0-9	2146	187	672	133	2131	559	2370	1089	2133	2132
Stores 10-19	2495	904	1466	2143	3615	564	226	3016	253	2195
Stores 20-29	2138	807	551	304	814	3337	4368	577	482	495

Model 13.3 New York Supermarkets

Thirty supermarkets that are part of a chain of stores in New York City have monthly demands for products to be supplied from the warehouses operated by the chain. Since the merchandise required by the stores will be very much the same, the total monthly demand of each store can be expressed as a single number, which is given in Table 13.5.

To serve these supermarkets, 6 new warehouse sites have been prospected. Data has been gathered on the capacity of each warehouse and the cost of supplying the different supermarkets from each of these sites. The capacity of each warehouse and the fixed cost of running, maintaining, and paying down the financing of each warehouse for each month are given in the Table 13.6.

In Table 13.7 the costs of meeting all of a supermarket's requirements from each warehouse are given. This includes among other things the cost of trucking the products from the warehouse to the supermarket. The cost of supplying part of the supermarket's demand from a warehouse is then proportional to the cost of supplying all the demand.

The objective is to minimize monthly cost while meeting the needs of the supermarkets. In other words, the problem is to determine which warehouses to use so as to minimize cost while meeting demand and to provide a schedule of how much of each supermarket's demand should be supplied from each warehouse.

If the issue of the fixed charge did not arise, then this application would be an example of the transportation problem of Section 4.4. For we can consider each warehouse and each supermarket to be a node of a network with an arc from each warehouse to each supermarket. The flow into each supermarket must equal its demand, and the flow out of each warehouse cannot exceed its capacity. However, the presence of the fixed cost leads us to formulate it as a MIP problem.

The data for this model can be entered in arrays of type double, namely

```
#define M 6              // Number of warehouses
#define N 30             // Number of supermarkets

double c[M][N];          // Objective function coefficients
double fc[M];            // Fixed cost of running warehouse i
double kap[M];           // Capacity of warehouse i
double d[N];             // Demand of supermarket j
```

13.3 Branch-and-cut

TABLE 13.6 Warehouse Capacities and Fixed Costs

	Capacity	Fixed Cost
Warehouse 0	19000	10500
Warehouse 1	27000	19500
Warehouse 2	29000	30500
Warehouse 3	39000	20500
Warehouse 4	23000	17500
Warehouse 5	44000	24500

TABLE 13.7 Cost of Supplying All the Demand of Each of the 30 Supermarkets from Each Warehouse

	6739	3204	4914	32372	1715	6421	81972	33391	2020	1459
0	141015	17684	38207	1953	17181	25640	7031	78453	9452	8597
	1581	23170	12087	4883	24063	4124	281463	11056	8585	12480
	3727	4673	13451	372672	9745	12055	97602	60774	54470	7146
1	38011	39723	16111	16981	168663	57109	15576	2542	34056	7095
	10355	5457	26409	29982	2152	23701	28499	26544	2480	1995
	205925	32069	42477	5044	36054	35602	10492	92515	12441	14113
2	2030	48702	19877	12851	39682	12148	406770	22113	22449	25455
	11116	13346	35106	229188	18070	18181	73603	63568	65177	8618
	70728	52917	20714	32575	210766	66703	18481	3928	34221	11999
3	7650	3845	19622	21024	1577	16197	43134	6370	1869	1402
	104130	15322	15319	4089	25399	25154	6305	36644	7754	10500
	1326	36072	9670	10822	24603	8180	325852	11424	14122	22151
4	8229	7880	25927	203364	12049	11400	59561	27330	52117	6428
	39587	32225	15620	23312	169251	53124	14368	3020	24448	7886
	5219	2396	13876	29681	1061	10383	65767	16770	1324	869
5	12638	8429	15832	3428	16297	15763	2542	27445	3542	7254
	693	26166	3801	8930	11050	5611	253234	5582	7458	19069

We will model this application by using continuous variables `X[i][j]` to represent the flow of goods from warehouse `i` to supermarket `j`. Since the data provided give the cost of supplying all of store `j`'s demand from warehouse `i`, we let `X[i][j]` be the fraction of supermarket `j`'s demand that is provided by warehouse `i`; this means that the flow from `i` to `j` is in fact `d[j]*X[i][j]`. To deal with the fixed costs, we introduce continuous `Y[i]` to represent warehouse `i`; the idea is to use `Y[i]` as a fuzzy boolean variable, which will be `1.0` if the warehouse is to be used and `0.0` otherwise. The declarations of the needed continuous variables are

```
continuous X[M][N];       // Fraction of j's demand met by i
continuous Y[M];          // Fuzzy boolean for warehouse i
continuous Cost;          // The objective function
```

We can express the constraint that all of supermarket j's demand be met by stipulating that the sum of the X[i][j] over all warehouses i be equal to 1.0:

```
and(int j=0;j<N;j++)      // Supermarket constraints
    sigma(int i=0;i<M;i++) X[i][j] == 1.0;
```

Conversely, to express the fact that amount shipped from each warehouse cannot exceed its capacity, we stipulate

```
and(int i=0;i<M;i++)      // Warehouse constraints
   sigma(int j=0;j<N;j++) d[j]*X[i][j] <= kap[i];
```

Note that we can tighten these capacity constraints by replacing them with what would otherwise be added as valid cuts:

```
and(int i=0;i<M;i++)      // Tight warehouse constraints
   sigma(int j=0;j<N;j++) d[j]*X[i][j] <= kap[i]*Y[i];
```

The cost of the supply that goes from warehouse i to supermarket j is c[i][j]*X[i][j], and so the objective function can be written as

```
Cost ==
    sigma(int i=0;i<M;i++) fc[i]*Y[i]  // Fixed costs
    +
    sigma(int i=0;i<M;i++)      // Proportional costs
        sigma(int j=0;j<N;j++) c[i][j]*X[i][j];
```

The objective function, the supermarket constraints, and the tight warehouse constraints can be loaded by means of a setup routine that also includes bounding constraints on the Y[i] and X[i][j] that set their upper bounds at 1.0.

Once the routines to initialize the data arrays and to set up the basic constraints of the model are called, a quick-and-dirty solution can be found much as in the previous model:

```
quick_and_dirty() // Compute and enforce bluff
{
double scratch;
    min: Cost;
    // Add up proportional costs
    scratch = sigma(int i=0;i<M;i++)
                sigma(int j=0;j<N;j++) c[i][j]*wp(X[i][j]);
```

13.3 Branch-and-cut

```
    // Add full fixed costs for any partially open warehouse
    and(int i=0;i<M;i++)
        scratch = scratch + fc[i]*ceil(Y[i]);

    Cost <= scratch;          // A bluff

    printf("Quick and dirty solution is %f\n",scratch);
}
```

A branch-and-bound search now can be made to find the optimal solution. The decision whether or not to open a warehouse can be reduced to the simple branch:

```
either Y[i] == 1.0;  // To open
or     Y[i] == 0.0;  // or not to open
```

To see that this suffices, note that when the second branch is chosen and Y[i] is fixed at 0.0, the constraint

```
sigma(int j=0;j<N;j++) d[j]*X[i][j] <= kap[i]*Y[i];
```

forces all X[i][j] also to be 0.0, thus ensuring that no supermarkets are supplied from warehouse i. However, just as it is good programming practice to place upper and lower bounds on variables as tightly as possible, so in the course of a model, if a continuous variable is forced by the constraints to one possible value, then that should be made explicit. In the case at hand, when it is decided that a warehouse is not to be opened, then Y[i] is explicitly fixed at 0.0 and the X[i][j] are implicitly fixed at 0.0. The code to fix a variable Y[i] should therefore be written so as to zero out the associated vector X[i]:

```
fix(continuous Y, X[])     // Called as fix(Y[i],X[i])
{
    either
        Y == 1.0;           // Open the warehouse
    or {
        Y == 0.0;           // Do not open the warehouse
        and(int j=0;j<N;j++)
            X[j] == 0.0;   // Good programming practice
    }
}
```

As in the previous model the main routine takes the form

```
2lp_main()
{
    data();
    setup();
    quick_and_dirty();      // Compute and enforce bluff

    find_min: Cost;
    subject_to
        search();           // Yet to be written

    and(int i=0;i<M;i++)
        if wp(Y[i]) == 1;
        then printf("\nUse warehouse %d",i);
}
```

So far we have used fuzzy booleans and valid cuts in a static fashion. Now we will add additional cuts dynamically in order to further prune the search tree.

Let us look more closely at the link between X[i][j] and Y[i]. In any solution if wp(X[i][j]) is nonzero, then wp(Y[i]) must be 1.0. In the linear relaxation the constraints we have in place so far do ensure that if wp(X[i][j]) is greater than 0.0, then wp(Y[i]) must also be greater than 0.0. However, this link is muffled by the factors d[j] and kap[i]. This means that wp(X[i][j]) might be as great as 1.0 and Y[i] might still be close to 0.0. The constraints X[i][j] <= Y[i] are all valid cuts because they must hold at every solution where Y[i] is either 0.0 or 1.0. However, until Y[i] is forced either to 0.0 or 1.0, these cuts can be violated. The simplest thing would be to add all these cuts as part of the setup constraints. If the problem size is large, however, adding M*N constraints could well slow things down unacceptably. The trick in branch-and-cut is only to add a cut X[i][j] <= Y[i] that is violated by the current witness point after the objective function has been optimized. In other words, we apply a form of the injury method in determining which cuts to add. There are many ways the details of this strategy can be managed. The following code will add cuts in a loop that ends when there are no more violated cuts. The strategy in this code to add cuts dynamically merges nicely with the injury method, and after the cuts are added, the witness point is a vertex that is optimal for the linear relaxation.

```
add_cuts()
{
int flag;

    flag = 1;                   // Turn flag on
    and(;flag != 0;){           // Loop while flag is on
        flag = 0;               // Turn flag off
```

13.3 Branch-and-cut

```
        and(int i=0;i<M;i++)
            and(int j=0;j<N;j++)
                if wp(X[i][j]) > wp(Y[i]);  // Violated cut ?
                then {
                    X[i][j] <= Y[i];    // Add cut
                    flag = 1;           // Turn flag on
                    min: Cost;          // Re-optimize
                }
    }
}
```

In the `add_cuts` routine we have re-optimized the objective function after adding a cut and before seeking another violated cut; this is akin to the progressive roundoff technique of incorporating changes incrementally. Let us see how to work this into the branch-and-bound code. What we want is to add the cuts before determining whether or not there are injuries to contend with. So we can simply insert the `add_cuts` routine before the check for injuries:

```
search()
{
int bv;

    and(;;) {
        min: Cost;              // Optimize linear relaxation
        add_cuts();             // Add the cuts
        if determine_branching_variable(bv);
        then fix(Y[bv],X[bv]);  // Repair injured variable
        else break;             // Exit the loop
    }
}
```

A very important part of this and other search programs is the choice of the branching variable; in this code we have called this the `determine_branching_variable` routine. As usual there are many possible strategies for choosing this variable. Several such strategies are discussed in the exercises. However, with the data from the above tables, the program finds the optimal solution without needing to carry out a branch-and-bound search at all. It is not unusual for the branch-and-cut method to work very well on this kind of application.

End of Model 13.3

The branch-and-cut method is an important tool for attacking difficult problems in combinatorial optimization. In this last model the cuts suggested themselves. In other situations it can be far more difficult to smoke out the cuts to be added dynamically. A program of automating the process is undertaken in the MINTO system of Nemhauser, Savelsbergh, and Sigismondi (1994).

In early work, including such classic papers as Dantzig, Fulkerson, and Johnson (1954, 1959) and Gomory (1958, 1963), the emphasis was on finding a sequence of cuts that lead to the solution of an integer programming problem without requiring branching. Work blending branching with cut generation was done by Glover (1965), Geoffrion (1969), and Balas (1971).

Branch-and-cut gained impetus with the work of Crowder and Padberg (1980) and Crowder, Johnson, and Padberg (1983) on especially large problems. The branch-and-cut technique is discussed in theory and in detail in Schrijver (1986) and in Nemhauser and Wolsey (1988).

For discrete models the geometric picture of the cutting plane cannot be invoked. In this field the analogous method is to introduce redundant constraints that are deduced from the current constraint set; the impact of these new constraints is captured by the process of constraint propagation. For important applications such as job shop scheduling, this method can prove very powerful (see Carlier and Pinson, 1989).

In the exercises below we explore the importance of the `determine_branching_variable` routine and the `add_cuts` routine. Data sets for the capacitated warehouse problem and for many other problems can be obtained from the OR-Library via anonymous `ftp` to `mscmga.ms.ic.ac.uk` or, alternatively, at the web page with address `http://mscmga.ms.ic.ac.uk/`.

Exercises

13.3.1. Write a `determine` routine for Model 13.3 that selects the injured warehouse that has the greatest number of users.

13.3.2. Write a `determine` routine for Model 13.3 that selects the warehouse that has the largest percentage of its capacity used by the supermarkets in the linear relaxation.

13.3.3. In the code for Model 13.3, add a variable that counts nodes visited in the `fix` routine. Then, using capacitated warehouse data from the OR-Library or other source, compare the number of nodes visited with different `determine` routines with and without using the `add_cuts` routine.

13.3.4. Grains are stored in silos, but each silo can only hold one kind of grain. A farm needs to store the following quantities (in tons) of 4 different grains: 80, 48, 28, 78. There are 7 silos available with the following capacities: 31, 24, 39, 55, 82, 100, 95. To open each of these silos generates a fixed cost; these are 100, 130, 130, 100, 180, 90, 80. After that, storage is charged by the ton. The cost of storing a ton of grain j in silo i is given in Table 13.8.

13.4 Penalties

TABLE 13.8 The Cost of Storing Grain

	Grain 0	Grain 1	Grain 2	Grain 3
Silo 0	75	100	100	80
Silo 1	150	150	100	80
Silo 2	150	150	75	160
Silo 3	225	150	50	160
Silo 4	300	50	25	240
Silo 5	375	250	125	400
Silo 6	375	250	125	400

Find a least cost solution to this storage problem. This is an example of a segregated storage problem. The capacitated warehouse location problem is a relaxation of this problem. Experiment with using the capacitated warehouse location problem as a lookahead in the branch-and-bound search to solve this new problem.

13.4 Penalties

In this section we present a most classic technique that uses sensitivity analysis to tighten the model during a branch-and-bound search. Suppose that we are dealing with a maximization problem and that the continuous variable X has lower bound a and upper bound b. Let us also suppose that the threshold condition either X == a or X >= c is among the logical requirements of the model, where a < c <= b. Suppose that cache contains the current lower bound on the objective function for the branch-and-bound search. Let us place ourselves in the context of the injury method. Suppose the objective function is optimized in the linear relaxation, suppose that X is at its lower bound a, and suppose that r is the reduced cost of X. Then if X is to leave its current spot at its lower bound a and migrate to the interval [c,b], the objective function will deteriorate by at least r*(c-a). This can be seen from Figure 7.7; the reduced cost will either be a true evaluation or an underestimate of how fast the optimum function will decrease as the lower bound on the variable is increased. The quantity r*(c-a) is called the *penalty* associated with moving X from its current position. If paying this penalty would make the objective value lower than the current bound of the branch-and-bound search, we can freeze the variable X at a. Similarly, if X is at its upper bound b, and if r*(b-a) + cache is greater than the current optimal value of the objective function in the linear relaxation, we can impose the bounding constraint X >= c, since paying the penalty r*(b-a) would be too costly. Here the story is told in Figure 7.3. If variable X is at neither of its bounds, the reduced cost is zero and provides no usable information.

To illustrate this, let us return once more to Model 3.2, British Cooking, the problem of blending cooking oils. We now address the full version of the problem as given in Williams (1994).

Model 13.4 British Cooking Refined

Management is not able to use the plan determined by the linear program of Model 3.2 because several key elements of the situation were not fully explained to the group that developed the linear programming model. It turns out that there are some logical requirements on each month's production, which can be summarized as follows:

1. No more than 3 oils can be used in each month's production run.
2. If an oil is used in a monthly production run, then at least 20 tons of it must be used.
3. If either sesame or sunflower oil is used in the blend, then corn oil must also be used.

Requirements 1 and 3 reflect the culinary insight and skill that the company has acquired over the years, while requirement 2 comes from the mechanics of the production process itself. These requirements change a linear program into a disjunctive program. Williams (1994) notes that these changes make the model "difficult to solve."

This is still a six-month planning problem but now each month also has a set of logical requirements. For each month and for each oil, it must be determined whether that oil is to be included in the product blend, and these determinations must satisfy requirements 1, 2 and 3 for each month.

The logical requirements are imposed on each month's production, so we can take things a month at a time. For each oil it must be decided whether or not to include that oil in the blend. The corn oil plays a special role. If it is not included, then the last two oils, sesame and sunflower, cannot be included.

To start, we need the definitions, declarations and setup routines of Model 3.2. Let us add some definitions, a utility variable, and some auxiliary routines:

```
#define CORN 0
#define PEANUT 1
#define SAFFLOWER 2
#define SESAME 3
#define SUNFLOWER 4
double cache; // Utility variable to store best value found
```

13.4 Penalties

To make the decision not to use one of the oils in the blend, we will pass the following routine the integer corresponding to the type of oil and the vector Plan[m][PROC]:

```
dont_use_oil(int k, continuous Oils[])
{
    Oils[k] == 0;
    if k==CORN;
    then {  // Excluding CORN means excluding the last two
        Oils[SESAME] == 0;
        Oils[SUNFLOWER] == 0;
    }
}
```

It will be necessary to monitor the requirement that at most 3 oils figure in the blend each month. So it will be useful to have a routine to check that at least i of the next j oils among Oils[m],...,Oils[KINDS-1] have zero witness point values:

```
#define EPSILON .000001
zero(int i,j,m, continuous Oils[])
{
    if i == 0;        // Enough zeroes accounted for
    then return;
    j >= i;           // Check room left to find enough zeroes
    if wp(Oils[m]) < EPSILON;      // Another zero found ?
    then zero(i-1,j-1,m+1,Oils);   // A recursion
    else zero(i,j-1,m+1,Oils);     // Ditto
}
```

This last routine can be used to check whether a new oil can be used in the blend for a given month. The following procedure is passed a vector Oils and an integer k such that Oils[0],...,Oils[k-1] have been dealt with:

```
#define THRESHOLD 20
use_oil(int k,continuous Oils[])
{
    Oils[k] >= THRESHOLD;      // Use Oil[k]
    if {
        k==2;   // SAFFLOWER
        not zero(1,2,0,Oils);//All previous oils nonzero ?
    }
    then {      // Last two oils cannot be used
        Oils[k+1]==0;
        Oils[k+2]==0;
    }
```

```
    if {
        k==3;      // SESAME
        not zero(2,3,0,Oils);// Two previous oils nonzero?
    }
    then Oils[k+1]==0;     // Last oil cannot be used
}
```

Already with these routines we can put together a hillclimbing strategy to try to find a quick-and-dirty solution:

```
hill_climb()
{
    if not {
        and(int i=0;i<MONTHS;i++)
            and(int k=0;k<KINDS;k++)
                steeper_gradient(i,k,Plan[i][PROC]);
        max: Revenue;
        cache = wp(Revenue);
    }
    then {
        printf("The hillclimb failed\n");
    }
    else Revenue >= cache; // Bluff for branch-and-bound
}
```

The `steeper_gradient` routine will look into the future to choose between two alternatives:

```
steeper_gradient(int i,k,continuous Oils[])
{
double u, uu;
    if not {
        use_oil(k,Oils);
        max: Revenue;
        u = wp(Revenue);
    }
    then u = 0;
    if not {
        dont_use_oil(k,Oils);
        max: Revenue;
        uu = wp(Revenue);
    }
    then uu = 0;
    if u >= uu;
    then use_oil(k,Oils);
    else dont_use_oil(k,Oils);
}
```

13.4 Penalties

This is an example where the injury method will require some thought to develop. But first, let us write down the main routine.

```
21p_main()
{
    data(price,hardness);              // As in Model 3.2
    ...                                // Ditto
    find_revenue(Plan,price,Revenue);  // Ditto

    hill_climb();

    find_max: Revenue;
    subject_to
        smart_search();
}
```

The basic unit of this model is the month of production. Therefore we will organize the injury method around the months. A month is injured if one of the requirements 1, 2, 3 is violated. A simple way to structure an injury is by means of the oils themselves, starting with corn oil. The following routine is passed a months production Plan[i][PROC] and an integer j; the integer j will keep track of which oils have already been repaired.

```
a_ok(int j, continuous Oils[])
{
    if j == KINDS;                     // All repairs done
    then return;

    and(int i=j;i<KINDS;i++)
        if wp(Oils[i]) > EPSILON;      // Condition 2
        then wp(Oils[i]) >= THRESHOLD - EPSILON;

    if {
        j==0;                          // Condition 3
        wp(Oils[j]) < EPSILON;
    }
    then and(int k=B;k<KINDS;k++) wp(Oils[k]) < EPSILON;
    else zero(2,KINDS,0,Oils);         // Condition 1
}
```

We will use the trailing method of Model 10.2, much in the same way as in Model 12.3. An array `depth` will mark how far into each month the repairs have been made; the stack `trail` will record the changes that have been made to depth. When backtracking takes place, the `trail` stack will be popped and `depth` will be restored to its proper state. Furthermore, to make the code search for injuries among the months in a more equitable way, we will make the check for injuries circular, treating the 6 months as a kind of 6 hour clock.

As the search progresses, penalty computations will be made and continuous variables that can be forced to stay at 0.0 or above the threshold of 20.0 will be detected.

```
#define GAP 1.0
#define ZERO 0.0

smart_search()
{
int top;                        // Top of trail stack
int trail[MONTHS*KINDS];        // Trail stack
int depth[MONTHS];              // Depth of injury repair
int current_month;              // For circular loop
int eye;                        // Storeback variable

    top = -1;           // Empty stack
    and(int i=0;i<MONTHS;i++)   // Initialize to first oil
        depth[i] = 0;
    current_month = 0;  // Initialize to first month
    absgap(GAP);        // Avoid virtually duplicate solutions
    and(int m=0;;m++) {
        penalty_computation(depth,GAP);// To tighten model
        if find_injury(eye,current_month,depth);
        then fix(m,eye,Plan[eye][PROC],depth,trail,top);
        else break;
    }
    printf("Solution at %f\n",Revenue);
    cache = wp(Revenue);
}
```

If a month i is injured, depth[i] records the first place at which injury to oils from this month have not already been repaired; so new repairs start at this point:

```
fix(int m,i,continuous Oils[],int depth[],trail[],& top)
{
int j;

    j = depth[i];
    depth[i] = depth[i]+1;      // Push stacks
    top = top+1;                // Same as top = m
    trail[top] = i;
    if wp(Oils[j]) >= THRESHOLD;
    then {
        either use_oil(j,Oils);
```

13.4 Penalties

```
        or {
            and(int t=top;t>m;t--) // Clean up stacks
                depth[trail[t]] = depth[trail[t]]-1;
            top = m;
            dont_use_oil(j,Oils);
        }
    }
    else {
        either dont_use_oil(j,Oils);
        or {
            and(int t=top;t>m;t--) // Clean up stacks
                depth[trail[t]] = depth[trail[t]]-1;
            top = m;
            use_oil(j,Oils);
        }
     }
}
```

To distribute the search for injuries more evenly among the months, we follow a circular loop among the months `0,...,MONTHS-1` in the hunt for an injured month. The variable `current_month` will keep track of the month among `0,...,MONTHS-1` to start with. In essence this needs an *until* loop, one with exit condition `current_month==i%MONTHS` but which must be executed at least once. We can get this effect by running the test `current_month != (i+1)%MONTHS` inside an and loop.

```
find_injury(int & eye, & current_month, int depth[])
{
    and(int i=current_month; ;i=(i+1)%MONTHS) {
            if not a_ok(depth[i],Plan[i][PROC]);
            then {
                eye = i;
                current_month=(i+1)%MONTHS;// New month
                return;
            }
         current_month != (i+1)%MONTHS;// Circled round ?
    }
}
```

Finally, we get to the penalty computations. But first we need an auxiliary routine to detect if a variable is not in position to be assessed a penalty.

```
ineligible(continuous Oil)
{
    c_either lb(Oil) != 0.0;    // Oil already used ?
    or ub(Oil) == 0.0;           // Oil already eliminated ?
}
```

Then we use the reduced cost function `rc` to compute the penalties:

```
penalty_computation(int depth[], double gap)
{
double cwp,src,crev;
int eye[MONTHS*KINDS];
int jay[MONTHS*KINDS];
int dir[MONTHS*KINDS];
int apex;
    apex = 0;
    max: Revenue;   // Linear optimization for reduced costs
    crev = wp(Revenue);    // Record linear optimum
    and(int i=0;i<MONTHS;i++)
        and(int j=depth[i];j<KINDS;j++) {
            if ineligible(Plan[i][PROC][j]);
            then continue;    // Move on if ineligible
            cwp = wp(Plan[i][PROC][j]);    // Current value
            src = rc(Plan[i][PROC][j]);    // Reduced cost
            if crev + THRESHOLD*src < cache + gap; // Bingo
            then { // Record variable
                eye[apex] = i;
                jay[apex] = j;
                dir[apex] = 0;     // To fix at lower bound
                apex = apex + 1;
            }
            else {
                if crev - src*cwp < cache + GAP;   // Bingo
                then { // Record variable
                    eye[apex] = i;
                    jay[apex] = j;
                    dir[apex] = 1; // To fix threshold bound
                    apex = apex + 1;
                }
            }
        }
    and(int k=0;k<apex;k++)    // Use results
        if dir[k] == 0;
        then Plan[eye[k]][PROC][jay[k]] == ZERO;
        else Plan[eye[k]][PROC][jay[k]] >= THRESHOLD;

    max: Revenue;      // For injury method
}
```

With this code the optimal solution of 100278.70 is found and verified. The reader can check that the call to `penalty_computation` bounds as yet unbound variables a significant number of times and that this reduces the node count. One can also find it interesting to experi-

13.4 Penalties

ment with heuristics for choosing the initial and subsequent values of the parameter `current_month`. This model is an example of a disjunctive linear program which is not tightened by translating the disjunctive requirements into 0-1 conditions and making it a MIP application.

End of Model 13.4

Penalty computations are among the earliest techniques developed for branch-and-bound search, see Driebeek (1966) and Tomlin (1971). The version employed in the above model is relatively simple and has small computational overhead. More elaborate penalty computations can be made by extracting more information from the simplex method. These more elaborate methods mean greater computational overhead but can apply to variables that are currently between their lower and upper bounds. A source is Salkin and Mathur (1989).

Exercises

13.4.1. Formulate the problem of Model 13.4 as a MIP model. Compare this version with the disjunctive programming version of Model 13.4.

13.4.2. Apply the penalty method of this section to Model 13.1. Comment on its effectiveness or ineffectiveness.

14

Further Search Methods

With local search the idea is to start with a initial solution or partial solution to the application at hand and then to improve this state iteratively by means of a hill-climbing process. If necessary and if possible, the process can be applied to many different initial states so as to achieve a broad penetration of the search space. The signal advantage of local search is that branching is avoided or at least kept to a minimum. This is especially important if the search space is large or if the space includes many "equivalent" states, which lead to useless searches of "equivalent" subtrees when an enumeration-based approach is used. Local search methods typically provide sound but not complete algorithms.

For enumeration based methods, the advantages of depth-first search are that it is efficient in its use of storage space and it is relatively easy to keep track of the states generated by the search. With the depth-first technique, the search goes from parent node to child node or backtracks from child up to parent, so the passage from node to node is local and can be implemented with a stack. Moreover, since depth-first search is a stack-based technique, it merges nicely with the stack of a programming language interpreter. And this is true whether one is using the built in disjunctive constructs of a logic-based language like 2LP or using recursion and conditional disjunction to capture persistent disjunction.

In a *breadth-first search* of a tree, exploration starts at the root node at level 0, then examines all nodes at level 1, followed by all nodes at level 2, and so on. For the tree in Figure 14.1, a breath-first search will visit the nodes in the order of their labels: 0,1,2,... . When the tree to be searched is part of modeling an application, for a breadth-first search of the tree to be sound and complete, the horizontal and vertical conditions of Chapter 10 must still be observed.

Depth-first and breadth-first searches are dual to one another in that depth-first search is organized vertically and breadth-first search is organized horizontally. One of the advantages of breadth-first search is that a goal node that is close to the root node of a tree will be found early. This can be especially important in state space optimization where the length of the path in the tree is often what is to be minimized. Another advantage of breadth-first search is that it will avoid the folly of plunging down a fruitless path while disregarding other alternatives, something depth-first search has trouble avoiding. As its basic data structure breadth-first search employs a queue rather than a stack. A disadvantage of breadth-first search is that its queues can require unavailable amounts of space on intermediate or large problems.

In practice, hybrid methods are used that meld features of these different basic models. In fact, in the following section, we treat a depth-first method for doing breadth-first search.

14.1 Iterative deepening

Some of the applications we have considered require that a successful search reach the deepest level in the tree; the N-queens is an example of this. With the injury method, on the other hand, the eureka effect can limit the depth to which search has to proceed. In Model 11.4, Mathematicians and Physicists, we have a situation where the task itself is to minimize the depth the search has to reach in the tree; with this example, we were able to use branch-and-bound depth-first search. In this section we develop a different method for dealing with discrete applications that require the length of the path to a successful solution to be minimized.

One way to avoid the complications of breadth-first search is to make repeated use of depth-first search. An *iterative deepening search* is a depth-first search that is only permitted to go to a specified depth, and this depth is increased with each iteration until a solution is found. The

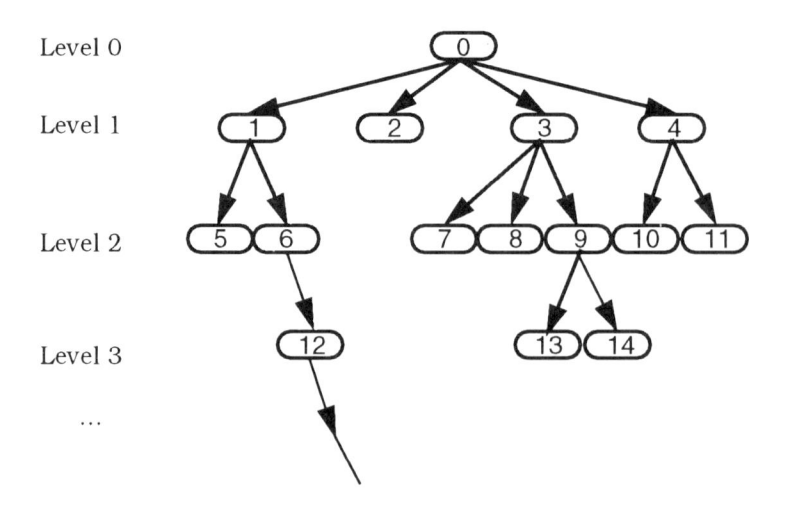

FIGURE 14.1 A basic breadth-first search.

14.1 Iterative deepening

idea is to simulate a breadth-first search without maintaining a queue. For the tree of Figure 14.1, if the node numbered 10 is a goal node, an iterative deepening search will visit the nodes 0, 1, 2, 3, 4 when the depth of the search is 1 and nodes 0, 1, 5, 6, 2, 3, 7, 8, 9, 4, 10 when the depth is 2. Note that nodes 0, 1, 2, 3, and 4 are visited twice. Note also that a depth-first search would visit all the nodes below 12 before coming back to find the goal node at 10; this is avoided by the iterative deepening search.

To illustrate this technique, we return to Model 11.4, Mathematicians and Physicists. In this model, the principal way of getting people back and forth across the River Seine is the call to the routine try_crossing(crss); this call pushes a state with index crss+1 onto the stack of states state[N][4]. The code for finding a single solution by means of depth-first search is

```
create_new_state(0,ALL,ALL,0,0);

and(int crss=0;crss<N-1;crss++) {
    try_crossings(crss);   // Generate next state
    if goal(crss+1);
    then break;            // Exit loop
}
```

In and of itself, this code does not solve the minimization problem because the leftmost solution found by this tree search is not necessarily the shortest solution.

To do an iterative deepening search to find the minimal solution, we need only to modify slightly the code for a single solution. Note that in the code listed backtracking occurs in the routine try_crossings where all possible next states are examined. The symbolic constant N is a generous upper bound on the length of successful paths to solutions. Since the loop control variable crss is bounded by N-2 and since N-1 is certainly larger than the total number of legal states, the and loop can only exit successfully if a solution is indeed found. In other words, the only way for the and loop to be exited successfully is through the break statement which is triggered if a solution is found.

With iterative deepening the idea is to start with a sound and complete method for a depth-first search. Then successive limits to the level to which the depth-first search can go are set. Since the search algorithm is sound and complete, this will force an enumeration of the entire tree down to the level given. If a solution exists in this limited subtree, the leftmost one will be found; if not, the limit is increased, and the entire process begins again. Eventually the *minimum* level at which a solution resides will be reached.

Let us modify the code for Model 11.4 for iterative deepening search. In the new 21p_main below, the central and loop is preceded by a c_or loop whose control variable is called depth; this variable restricts the depth of the search. All paths of length depth or less will be searched before incrementing depth; this way once a solution is found at a particular level, the program will succeed and will print out the directions for getting all the mathematicians and physicists across the river. Furthermore, once a solution is found at a particular level, we know that it is a solution at minimum distance in the tree from the root. Since we seek to minimize the number of crossings, this solution will be optimal.

```
2lp_main()    // Model 11.4 modified for iterative deepening
{
    create_new_state(0,ALL,ALL,0,0);

    c_or(int depth=1;depth<N-1;depth=depth+2) {
        and(int crss=0;crss<depth;crss++)
            try_crossings(crss);   // Generate state crss+1
        goal(depth);               // Is this a goal node ?
    }
}
```

It should be pointed out that in this new version of the code, the and loop can exit successfully without reaching a state that provides a solution; at that point, however, depth is equal to crss+1 and the call to goal will provoke the required backtracking. Let us also note that the increment is depth=depth+2, since all solutions require an odd number of crossings.

Recall that in Model 11.4, there are 4 mathematicians, 4 physicists, and a boat seating 3 people. The depth-first branch-and-bound search found two solutions, the first with 11 crossings and the second with 9. The iterative deepening search will find the optimal solution with 9 crossings directly.

In Chapter 10 we observed that transforming the depth-first search code for a discrete application into C, C++, or other procedural language is straightforward; the way to do this is to simulate persistent disjunction by means of recursion and conditional disjunction. In turn, such procedural code can then be easily adapted to the iterative deepening paradigm by mapping the c_or construct into a for or while loop.

Next we introduce a family of puzzles which we will solve in this chapter using four different search techniques: iterative deepening, breadth-first search, a breadth-first variant of branch-and-bound search known as *A* search*, and a hybrid technique known as *iterative deepening A* search*. These simple puzzles enable us to emphasize the programming techniques for search and to keep modeling questions at a minimum, while still providing for some nontrivial mathematical observations.

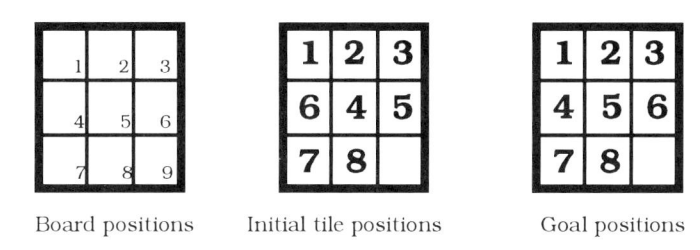

FIGURE 14.2 Board positions for the 8 puzzle.

Model 14.1 The S^2-1 Puzzle

In this family of puzzles, one is given an S-by-S board of squares containing S^2-1 tiles labeled $1,...,S^2$-1 in some order. In the case where S is equal to 3, the board is 3x3 and distinct tiles labeled from 1 through 8 occupy different squares while one square is left blank. The basic move is to move a numbered tile into the empty square. This can be viewed as exchanging the empty square with one of its surrounding tiles. The object is to transform the board from its initial state to a given goal state. Typical examples of initial and goal states are illustrated in Figure 14.2. In the initial state of the figure, two moves are possible: Tile 5 or 8 can be moved to the empty square.

In the late 1870s the American public was fascinated by the 15 puzzle, and two articles on the puzzle appeared in 1879 in the second volume of the *American Journal of Mathematics*. We have the following editorial comment from the article by Story (1879):

> The "15" puzzle for the last few weeks has been prominently before the American public, and may safely be said to have engaged the attention of nine out of ten persons of both sexes and of all ages and conditions of the community. But this would not have weighed with the editors to induce them to insert articles upon such a subject in the American Journal of Mathematics, but for the fact that the principle of the game has its root in what all mathematicians of the present day are aware constitutes the most subtle and characteristic conception of modern algebra, viz: the law of dichotomy applicable to the separation of the terms of every complete system of permutations into two natural and indefeasible groups, a law of the inner world of thought, which may be said to prefigure the polar relation of left and right-handed screws, or of objects in space and their reflexions in a mirror. Accordingly the editors have thought that they would be doing no disservice to their science, but rather promoting its interests by exhibiting this a priori polar law under a concrete form, through the medium of a game which has taken so strong a hold upon the thought of the country that it may almost be fairly be said to have risen to the importance of a national institution. Whoever has made himself master of it may fairly be said to have taken his first lesson in the theory of determinants.

With these puzzles, three different questions arise: (1) Can the puzzle be solved from a given initial position, (2) if it can be solved, can you find a sequence of moves to do it, and (3) if it can be solved, can you find a solution with a minimal sequence of moves? The first question is a decision problem; the other two require that a sequence of moves be produced.

Question 1 turns out to be an interesting problem in permutation groups; in the exercises, a proof is developed that yields a simple test whether an instance of the puzzle is solvable. Question 2 is one that is readily solvable by people as many school children will attest. There are also fast programs to find solutions; a recursive approach is indicated in the exercises. On the other hand, question 3 is computationally difficult for $S \geq 4$. It is the optimization form of an NP-Complete problem, so is NP-Hard and co-NP-Hard, (Ratner and Warmuth 1990). The challenge is to develop a program to find a sequence of moves leading from the initial state to the goal state using the least possible number of moves. Let us put our minds to it then.

In the 8 puzzle, the board positions are numbered 1 through 9 as in Figure 14.2. In analogy with the Mathematicians and Physicists model, in order to find a minimal correct sequence of moves, we will use a state space search. For this problem, we can represent a state as a vector. The goal position is given by the vector [1,2,3,4,5,6,7,8,0] and the initial position is given by [1,2,3,6,4,5,7,8,0]. We will need to keep a stack of these vectors as we did for the states in the Mathematicians and Physicists model. For this purpose we declare a two-dimensional array to stack vectors of length 9. The stack will be of length 32, since 31 is known to be the maximum number of moves needed to find a solution to the 8 Puzzle from a solvable position. We will also require a vector to store the goal state. We have the declarations

```
#define S 3                      // Length of a side
#define STACKSIZE 32             // Big enough
int nsquared[STACKSIZE][S*S];    // A stack of states
int goal_node[S*S];              // The goal state
```

The initial state and the goal state are given to us:

```
init_first_node()
{
    nsquared[0] = { 1,2,3,6,4,5,7,8,0 } ;
}

initialize_goal()
{
    goal_node = { 1,2,3,4,5,6,7,8,0 } ;
}
```

14.1 Iterative deepening

To use iterative deepening, we will follow the structure of the code we just developed for the problem of the Mathematicians and Physicists; there the states were generated by the call to `try_crossings` which cycled through all possible next trips across the river. In the present situation, we need a procedure `try_moves` for cycling through all possible next moves. The structure of the puzzle dictates the form of this routine. This time we try to add a state for each of the (at most) four possible moves. Note that a move is equivalent to interchanging the empty square with the square above it, below it, or to the right or left of it. Some of these moves might be impossible should they place the empty square off the S-by-S board. If a move is legal and does not lead back to a previous state, it can be pushed onto the stack of moves.

```
try_moves(int lte)
{   // lte is the level to examine in the search
    // First free position on the state stack is lte + 1

int emp_pos;    // Storeback parameter

    // Find empty position; emp_pos is a storeback parameter
    find_empty_pos(emp_pos,nsquared[lte]);

    either {// The empty square can be moved to the right
        // if its current position is not divisible by S.
        (emp_pos+1)%S != 0;
        add_new_state(lte,lte+1,emp_pos,1);
    }
    or {// The empty square can be moved to the left
        // if its current position does not have
        // remainder 1 when divided by S.
        (emp_pos + 1)%S != 1;
        add_new_state(lte,lte+1,emp_pos,-1);
    }
    or {// The empty square can be moved down if its
        // current position is less than or equal to S*S-S
        emp_pos + 1 <= S*S - S;
        add_new_state(lte,lte+1,emp_pos,S);
    }
    or {// The empty square can be moved up if its
        // current position is greater than S
        emp_pos + 1 > S;
        add_new_state(lte,lte+1,emp_pos,-S);
    }
}
```

The `find_empty_pos` routine is as follows:

```
find_empty_pos(int & emp_pos, int vector[S*S])
{
    and(int i=0;i<S*S;i++)
    {
        if vector[i] == 0;
        then
        {
            emp_pos = i;
            return;
        }
    }
}
```

Note that the first statement in each clause of the `either/or` statement checks that the move is legal. If so, `add_new_state` is called, and the new position becomes the top of the stack provided that `add_new_state` succeeds.

```
add_new_state(int lte,newstate,emp_pos,k)
{   // lte is the index of the state being examined
    // newstate is the index of the new state
    // emp_pos is the position of the empty square
    // k is the change in position of the empty square

    // Copy all squares
    and(int j=0;j<S*S;j++)
        nsquared[newstate][j] = nsquared[lte][j];

    // Put empty square in proper place
    nsquared[newstate][emp_pos+k] = 0;

    // Put new square where empty square was
    nsquared[newstate][emp_pos] = nsquared[lte][emp_pos+k];

    // Verify state is new
    check_not_there(newstate);
}
```

Note that since this is a stack-based search, the parameter `new_state` is always one more than the parameter `lte`, the index of the state that is currently at the top of the state stack. The procedure `add_new_state` will succeed only if the `check_not_there` call succeeds. The procedure `check_not_there` makes sure that the new state is not already on the stack: If it is on the stack, adding it again would create an unnecessary cycle in the solution of the puzzle, and we are looking for the shortest solution.

14.1 Iterative deepening

```
check_not_there(int newstate)
{
    and (int sta=0;sta<newstate;sta++)
        not and(int j=0;j<S*S;j++)
            nsquared[sta][j] == nsquared[newstate][j];
}
```

Let us put together a procedure that tests for a goal node and one to output the results:

```
goal(int depth)
{
    and(int j=0;j<S*S;j++)
        nsquared[depth][j] == goal_node[j];
    output(depth);
}

output(int depth)
{
int emp_pos;
    and(int j=0;j<=depth;j++)
    {
        // Find empty position
        find_empty_pos(emp_pos, nsquared[j]);
        if j == 0;
        then
        printf("\n The empty tile moves from position %d",
                                                emp_pos+1);
        else
        printf("\n to position %d ", emp_pos+1);
    }
}
```

Finally, we have the main procedure. Note that each time through the iterative deepening search we increase the depth by 2: We can do this because the empty square is the same position in both the initial and final configurations. This means that the total number of moves must be even.

```
21p_main()      // An iterative deepening search
{
    startup(); // Initialize positions
    c_or(int depth=0;depth<STACKSIZE;depth=depth+2){
        and(int lte=0;lte<depth;lte++)    // Loop down levels
            try_moves(lte);    // Generate new state
        goal(depth);           // Check if goal reached
    }
}
```

```
startup()
{
    init_first_node();
    initialize_goal();
    if not check_reachability(); // See exercises
    then {
        printf("Unreachable position\n" );
        exit();
    }
}
```

This code does not use continuous variables; the persistent disjunction in the `try_moves` routine can be readily simulated by recursion and conditional disjunction in a traditional programming language. In fact, the only reason 2LP is used is not to switch programming languages at this point.

End of Model 14.1

Iterative deepening yields a simple way of transforming a depth-first algorithm for finding a path to a goal in a state space application into one for finding a path of shortest length. Generally speaking, it avoids the potential pitfall of depth-first search of pursuing long paths that lead far away from solutions that can be found higher up in the tree.

In Korf (1985) it is remarked that iterative deepening has probably been rediscovered many times but that its first appearance in the literature is in Slate and Atkin's *CHESS 4.5* program (1977). Iterative deepening search also made its mark in distributed computing because it avoids bottlenecks associated with parallelizing breadth-first search, (see Rao, Kumar and Ramesh 1987).

Exercises

14.1.1. Run the mathematicians and physicists code with 3 mathematicians, 4 physicists, and a boat holding 2 people with both a depth-first search and an iterative deepening search. Compare the solutions.

14.1.2. Run the 8 puzzle without using the routine `check_not_there`. What happens? What happens when a straight depth-first search is used?

14.1.3. You are given two glasses with integer capacities; the first has capacity C1 and the second has capacity C2, where C2 is larger than C1. Six different operations are allowed.

14.1 Iterative deepening

1. Fill C1 from the tap.
2. Fill C2 from the tap.
3. Pour the contents of C1 into C2.
4. Pour the contents of C2 into C1.
5. Empty C1 into the sink.
6. Empty C2 into the sink

Write an iterative deepening search to determine how to obtain a given integer quantity of water Q in either glass. Note that Q must be less than C2. The iterative deepening search should give you the shortest possible way of obtaining quantity Q if it exists. If C1 and C2 are relatively prime, all possible quantities can be obtained by your search.

14.1.4. A permutation is a one-to-one and onto function on a finite set. It can be represented in terms of its cycle structure. Consider any configuration of the S^2-1-puzzle with the empty square in the bottom right corner. We can consider this configuration to be the result of a permutation of the numbers $(1,...,S^2-1)$. For example, the permutation (1 2 3 6)(4 5 8 7) gives rise to the middle board in Figure 14.2. Note the distinction being made between the mapping that is the permutation and the board that results from the permutation. Now consider the situation where the empty tile is in the bottom right square for the initial configuration and the goal configuration is the standard configuration given in the third board of Figure 14.2. This way the goal configuration corresponds to the identity permutation. With these conventions, prove that the configuration corresponding to the identity permutation can be reached if and only if the permutation that generates the initial configuration can be written as the product of an even number of transpositions. A sketch of a proof follows.

Outline of Proof: First recall that an *even* permutation of $(1,2,...,S^2-1)$ is one that can be written as the product of an even number of transpositions, where a transposition is simply the interchange of any two symbols. It is well know that the even permutations form a subgroup of index 2 in the group of all permutations. This subgroup is called the alternating group on n symbols and is usually denoted A_n.

As the board is transformed from the initial position to the goal position, the empty square moves around the board in a possibly complex path and a permutation of the tiles is determined by the path of the empty square.

Now define a *simple closed loop* as one that moves the empty square around the board so that the path of the empty square never crosses itself. The paths in Figure 14.3 (b) and (c) are examples of simple closed loops.

Claim 1: Any path of the empty square and hence any reachable permutation can be obtained as a succession of simple closed loops.

Claim 2: A simple closed loop always has an even number of moves.

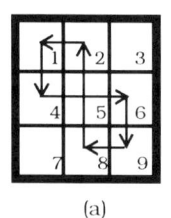

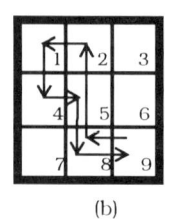

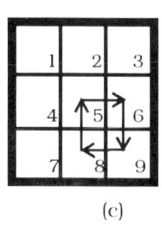

FIGURE 14.3 (a) A path of the empty square that corresponds to the permutation (2 4 1)(5 8 6) of the board positions, (b) a simple closed path resulting in the cycle (2 4 1), and (c) a simple closed path resulting in the cycle (5 8 6).

Claim 3: A simple closed loop resulting from n moves of the empty square produces a cyclic permutation of length $n - 1$.

Claim 4: Cyclic permutations of odd length are even permutations.

14.1.5. Now show that all even permutations of $(1,2,...,S^2-1)$ can be reached from the identity permutation when $S = 3$ and $S = 4$. To do this, one can use the fact that the alternating group on n symbols can be generated by the following permutations (see Coxeter and Moser 1980):

$$s = (3\ 4\ ...\ n),\ t = (1\ 2\ 3),\ (n\ \text{odd})$$

$$s = (1\ 2)(3\ 4\ ...\ n),\ t = (1\ 2\ 3),\ (n\ \text{even})$$

14.1.6. Show that if the empty square is not at the bottom right position on the board then it can be moved into that position in fewer than 2*S moves. Conclude that to determine if an arbitrary configuration of the puzzle is reachable, one can transform it into a configuration with the empty tile at the bottom right and then determine if the goal position can be reached from that permutation.

14.1.7. Write code to determine the cycle structure of the permutation that generates an initial position with the blank tile at the lower right hand corner. Then write a routine that determines if it is possible to reach the goal node from this configuration by determining from the cycle structure if the permutation is even or odd.

14.1.8. To determine whether a permutation is even or odd, one can sort the corresponding configuration using the bubble sort of Exercise 10.1.4 and count the number of switches of adjacent elements. Show that this amounts to describing the inverse permutation as a product of transpositions. Use this for a simple test for solvability: the number of pairs of (not necessarily adjacent) elements that are in the wrong order is equal to the number of switches in the usual bubble sort.

14.1.9. Code the following recursive strategy for the 8, 15, and 24 puzzles: First arrange the first row and first column correctly; then solve the resulting inner square.

14.2 Breadth-first search

As can be seen from the code in the previous section, it is straightforward to program an iterative deepening search if a depth-first search has already been developed. However, there are situations where a breadth-first search or some of its more efficient variants are more appropriate than iterative deepening search.

Breadth-first search requires a queue or priority queue rather than a stack, because this kind of search can jump from a node in one part of the tree to a node with very different ancestors. The queue explicitly lists the states or nodes in the order in which they are to be visited. As a result the programmer cannot use the 2LP interpreter's stack in the same way as with depth-first search, and space can become an issue as the queues grow. In short, a breadth-first search will prove somewhat more complicated to program.

To program a breadth-first search, the queues will list the nodes of the search tree in the order in which they should be visited and explored. The next node to be visited is taken from the front of the queue. When a node is visited, its children are generated and placed at the end of the queue. The search stops when a goal is found or there are no additional nodes to place on the queue.

In LISP and Prolog, lists are built in data structures that can readily serve to implement queues. In C, C++, and Pascal, a linked list can be built using structures with one of the entries in the structure being a pointer to the next element in the list. In FORTRAN, linked lists and queues are implemented by means of arrays; this strategy is often used by people in mathematical programming even when working in C. Perforce, a 2LP program will take this approach too.

In this section we set up the basic machinery in 2LP code for the simplest form of breadth-first search; in Sections 3 and 4 of this chapter more sophisticated variations are considered.

In this form of search, backtracking is not used to move about the tree, and all the alternatives to examine are kept explicitly on a queue. Therefore the loop structure for controlling the search will require only an and loop as opposed to an and/or combination. For continuity we will continue to treat the example of Model 14.1, The S^2-1 Puzzle. In this section we will consider only the most elementary kind of breadth-first search. We will use the array `nsquared` as before to contain the states; however, here the array will represent a queue instead of a stack. The global variable `ste` will represent the state to examine; the global variable `end` will keep track of where to add new states. Since the variable end keeps track of how many positions of the array nsquared have been used, it can be used to make sure that we do not go out of bounds.

Stacking states in depth-first search serves an additional purpose: It makes it possible to read off the path from the root node to the goal node. Here we will use the array nsquared to hold the queue and we will use a new array predecessor to store a node's immediate ancestor, since this no longer will be the element before it on a stack.

Below we have the 2lp_main routine for a simple breadth-first search. The key new element is that the routine try_moves is replaced by a deterministic routine bf_try_moves. The additional declarations we will need are the predecessor array and the two global variables ste and end of type int to keep track of the next state to examine and to keep track of the end of the queue. The array nsquared must be redeclared to be of length QUEUESIZE rather than STACKSIZE.

```
#define QUEUESIZE 10000// The maximum length of the queue
#define ALTERNATIVES 4 // At most 4 moves available

int nsquared[QUEUESIZE][S*S];
int predecessor[QUEUESIZE];
int ste;     // The state to examine
int end;     // The end of the queue

2lp_main()
{
    // Initializations
    startup();
    ste = 0;        // First state to examine is root node
    end = 1;        // First empty slot on queue

    and(;end < QUEUESIZE - ALTERNATIVES;) {
        if goal(ste);
        then return;
        bf_try_moves(ste);    // Breadth-first try_moves
        ste = ste + 1;        // Update front of queue
        end > ste;     // Fail if no progress is being made
    }
    printf("Out of space\n");
}
```

The queue is mined out if ste reaches the end of the queue, that is, if ste is equal to end. The test end > ste checks this. The queue has grown too large if end is too close to QUEUESIZE. The test end < QUEUESIZE - ALTERNATIVES checks for this.

In bf_try_moves, the persistent disjunctions of the either/or construct in try_moves are replaced by if/then statements. When the children of the state being examined are generated, they are added

14.2 Breadth-first search

to the end of the queue by the routine `add_new_state`; this routine, however, needs no changes from the preceding section. The variables end and the array `predecessor` must be updated in `bf_try_moves`.

```
bf_try_moves(int ste)
{
int emp_pos; // emp_pos is a storeback parameter

    find_empty_pos(emp_pos,nsquared[ste]);

    if {// The empty square can be moved to the right
        // if its current position is not divisible by S.
        (emp_pos+1)%S != 0;
        add_new_state(ste,end,emp_pos,1);
    }
    then {
        predecessor[end] = ste;      // Ancestor of new node
        end = end + 1;               // New end of queue
    }

    .
    .    //Similar adaptation of other three cases
    .
}
```

It should be noted that the queue is divided into two parts: the part that has already been examined and the part that has not. The variable `ste` always denotes the first state on the unexamined part of the queue; if it is not a goal node, it is expanded, and its children are placed at end of the queue.

The procedure `check_not_there`, which is called from the procedure `add_new_state`, is the same as in the previous section. It checks both parts of the queue when a new node is generated in order to prevent redundant nodes from appearing on the queue. Since we want the shortest solution to the puzzle, it is only necessary to keep on the queue one node for a given position of the tiles, even though there may be many ways of reaching that node or position. In the current situation the first copy of the state to be placed on the queue will always be the one with the shortest path through it. This is an example of the *dynamic programming principle* which states that the best path through an intermediate node H to a goal node G is the best path to H from the root node followed by the best path from H to G.

All the other routines with the exception of `output` are identical to those employed in the iterative deepening approach to the problem. With iterative deepening, we were able simply to pop the stack of states to read off the path to the solution, since the predecessor to each state lay before it on the stack. Now, in the `output` routine, we have to use the additional array `predecessor`; the entry in `predecessor[i]` is the ancestor of the node in the `i`th position on the breadth-first search queue. This information can now be used for printing the path from the initial position to the goal node. The printout is the reverse of that in the iterative deepening case.

```
output(int ste)      // Breadth-first version
{
int emp_pos;

    predecessor[0] = -1;    // To prevent underflow
    and(int i=ste;i>=0;i=predecessor[i]) {
        find_empty_pos(emp_pos,nsquared[i]);
        if i == ste;
        then
        printf("\nThe empty tile's final position is %d\n",
                                          emp_pos+1);
        else
        printf("The previous position is %d\n",
                                          emp_pos+1);
    }
}
```

While this code represents a solution to the problem posed, it is very naive as the algorithm reduces to a brute force enumeration of the tree. Even on small examples such as that of Figure 14.2, this code is significantly slower than the iterative deepening code. The difference is best studied when the code is written in a language such as C or FORTRAN with optimized compilation. In the following sections we study ways of adapting the basic mechanism of breadth-first search to make for more intelligent exploration of the search tree.

14.3 A* search and IDA* search

In many instances the application will have a heuristic that allows one to restructure the basic breadth-first search so that a goal node can be reached more quickly. For example, with the S^2-1 puzzle, we might assume that a state in which most of the tiles were in their goal positions is closer to the goal node than a state in which most of the tiles were not in their goal positions. Therefore the goal node might be found faster if states with more tiles in their goal positions were expanded first. What is needed is a quantitative estimate of how close the root node is to a goal via a given path through a node. Traditionally, in state space search, this quantitative estimate for a node n is denoted $f(n)$, and $f(n)$ is itself written as the sum of two values: $g(n)$ and $h(n)$. The function $g(n)$

14.3 A* search and IDA* search

measures the distance from the root to node n and the function $h(n)$ is a heuristic estimate of the remaining distance to a goal node. In the case of the S^2-1 puzzle, $g(n)$ is simply the number of moves already made; for $h(n)$, one simple heuristic is the number of tiles that are not in their goal positions. If the heuristic $h(n)$ never overestimates the distance from node n to the closest goal node, $h(n)$ is called an *admissible heuristic*, and the search is called an A* *search*. The notion of an admissible heuristic was introduced in the context of depth-first search in Section 11.2.

To transform the no-frills breadth-first search into an A* search, the code needs some changes. Essentially we must be able to reorder the nodes so that they are examined in a prioritized order. This means our data structure becomes a priority queue. However, it is not necessary physically to sort this queue to reflect these priorities. Rather we reset links going from node to node on the queue. To that end, we add a declaration of an array, n_lnk[QUEUESIZE], for maintaining links to the next state to examine, and a call to a new procedure insert_new_states to set up the linking so that the states will be examined according to the order of priority to be specified by the A* search.

The new declarations needed are

```
#define NIL -99     // The NULL pointer
int n_lnk[QUEUESIZE];
double f[QUEUESIZE],g[QUEUESIZE],h[QUEUESIZE];
```

An additional bit of initialization is required to set up the priority queue:

```
init_queue_vars()
{
    ste = 0;          // First state to examine is root node
    end = 1;          // First empty slot on priority queue
    n_lnk[0] = NIL;  // First element has no ancestor
}
```

The compute_f_g_h procedure computes the evaluation of the node. For this we can use the simple heuristic mentioned above, namely the number of tiles not in their goal positions.

```
compute_f_g_h(int ste, end)
{
    // g is the length of the path so far
    g[end] = g[ste] + 1;
    h[end] = S*S;  // Make h equal to its maximum value
    and (int j=0;j<S*S;j++)
        if nsquared[end][j] != 0;  // Not the empty square ?
```

```
        then {
            if nsquared[end][j] == goal_node[j];
            // The heuristic estimate is the number of
            // tiles that are not in their goal positions.
            then h[end] = h[end]-1;
        }
    f[end] = g[end] + h[end];
}
```

Note that the above version of compute_f_g_h will find the value for f[end] in all cases. However, for any case except the first, the change in f can be computed simply by noting what happened to the tile that was moved. This is addressed in the exercises. Also, a better heuristic based on the Manhattan distance is discussed in the exercises.

The main procedure is the same as for basic breadth-first search except that adding nodes to the priority queue must be done more carefully.

```
21p_main()      // A* search
{
    // Initializations
    startup();
    init_queue_vars();

    and(int i=end;i<QUEUESIZE-ALTERNATIVES;i=end) {
        if goal(ste);  // Goal has reached front of queue
        then {
            output(ste);
            return;     // Exit main procedure
        }
        bf_try_moves(ste);          // Breadth-first version
        insert_states(i,ste,end);   // Reorder priority queue
        ste = n_lnk[ste];   // Update front of priority queue
        ste != NIL;         // Fail if priority queue exhausted
    }
    printf("Out of space\n");
}
```

The insert_states routine to be developed must maintain the list of states as a priority queue. The queue must be reordered so that the states are examined in the order specified by the function value f[n]. The states are not physically reordered; instead, a link is maintained to the next state that should be examined. This is equivalent to having the priority queue sorted in order of increasing f values and is illustrated in Figure 14.4. In this situation the array-based representation of the linked list (as opposed to one using pointers) will prove very convenient when it comes to checking whether a node has already been

14.3 A* search and IDA* search

	0	1	2	3	4
f array	f[0]=3	f[1]=7	f[2]=6	f[3]=9	f[4]=8

	0	1	2	3	4
n_lnk array	2	4	1	NIL	3

(a) The n_lnk representation

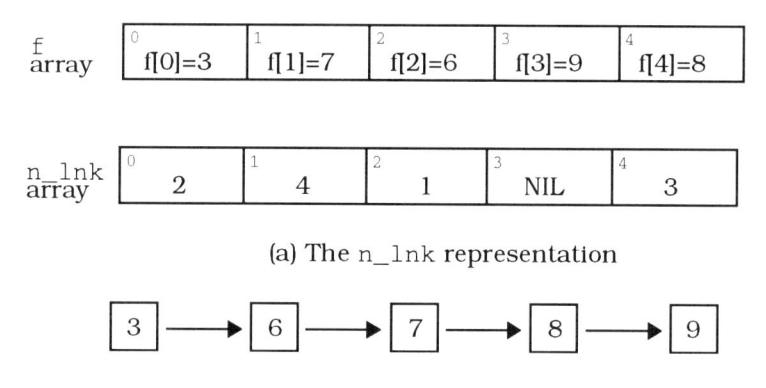

3 → 6 → 7 → 8 → 9

(b) The equivalent linked list representation

FIGURE 14.4 Heuristically reordered priority queues.

placed on the priority queue; this kind of checking will be done to avoid redundancy on the queue.

The job of insert_states is to find the proper position in the priority queue for these new states based on the f value of the new states. Since each of the new states must have an f value greater than or equal to the f value of the state that generated them, namely the state ste, we can start looking for the place to insert the new states after state ste on the priority queue.

The key step will be inserting a node in the priority queue or, more precisely, splicing the index of a node into the queue. To insert node j between the adjacent nodes pre and post, we can call

```
splice(int j,pre,post)
{
            n_lnk[pre] = j;
            n_lnk[j] = post;
}
```

If there are no states currently on the priority queue with values that are greater than f[ste], then n_lnk[ste] will be NIL indicating the end of the heuristically ordered queue. Hence we simply insert the new state j at the end of the queue with the call

```
splice(j,ste,NIL);
```

Now in the general case, we must follow the priority queue using n_lnk until we find the proper insertion place. This is done using the local integer variables pre_ptr and ptr. The variable pre_ptr is always one position behind ptr in the heuristically reordered queue. At the end of the and loop below, `pre_ptr` will point to the position before the position in which the new state should be inserted, and `ptr` will point to the position after. To search for the point at which to insert node j, we can employ the following routine:

```
find_and_fill_slot(int j)
{
int ptr,pre_ptr;

    pre_ptr = ste;
    ptr = n_lnk[ste];       // Never insert before ste
    and(;n_lnk[pre_ptr] != NIL;){
        if f[j] < f[ptr];
        then break;         // Insertion spot found
        pre_ptr = ptr;      // Move pre_ptr up to ptr
        ptr = n_lnk[ptr];   // Move ptr up to its successor
    }
    splice(j,pre_ptr,ptr); // Insert the new state
}
```

The next task is the insertion routine.

```
insert_states(int i,ste,end)
{
    // end-i states to be added to priority queue
    and(int j=i;j<end;j++) {
        if n_lnk[ste] == NIL;
        then splice(j,ste,NIL);
        else find_and_fill_slot(j);
    }
}
```

At the end of the routine `add_new_state`, `compute_f_g_h` must be called. Also, the routine `check_not_there` must now be changed to `check_not_better_g`. In addition to checking if a node is there, the revised routine must check if the new node has been previously discovered but with a worse g value. In this case the new node should be added to the priority queue and the old node abandoned.

```
add_new_state(int ste,newstate,emp_pos,k)// Modified for A*
{
    // ste is the index of the state being examined
    // newstate is the index of the new state
    // emp_pos is the position of the empty square
    // k is the change in position of the empty square
    ...
    // Copy remaining squares
```

14.3 A* search and IDA* search

```
    ...
    // Find square to move
    ...
    check_not_better_g(newstate);  // Node is new or node
                                   // has a better g value
    compute_f_g_h(ste,newstate);   // Evaluation of node
}
check_not_better_g(int newstate)
{
    and (int sta = 0; sta < newstate; sta++)
        if and(int j=0;j<S*S;j++)
            nsquared[sta][j] == nsquared[newstate][j];
        then g[sta] > g[ste] + 1;
}
```

The computation of the values `g[n]`, `h[n]` and `f[n]` provides us with a simpler test for a goal node, since node n is a goal node if `h[n]` is equal to 0.

```
goal(int n)      // For A* search
{
    n != 0;      // Obviate by initializing h[0] = -1
    h[n] == 0;
}
```

Note that instead of maintaining a sorted queue, the priority queue could be maintained as a heap. This means that the cost of inserting a new element into the queue would be $O(\log n)$ instead of $O(n)$. For large applications this can prove a more effective data structure. For the above code, one would only need to change the routine `insert_states` in order to change representations.

This finishes the development of the routines needed for a vanilla A* search. Let us now consider a blend of A* search with a depth-first method by combining it with iterative deepening search. The idea simply is to measure "depth" in the iterative deepening search by the heuristic estimate $f[n] = g[n] + h[n]$. If the heuristic is admissible, this will again find the shortest path to the solution. This search technique is denoted *IDA** or *iterative deepening A* search*.

For the S^2-1 puzzles, the basic idea in IDA* search is that we launch the search by only expanding the tree down to nodes whose f evaluation is no greater than the initial f estimate of the number of moves from the beginning position to the goal. Then we try again from the initial starting position, but this time we proceed down the tree to the next f level. Note that if the empty square for the goal node is in the same board position as in the initial state, then the number of moves must be even. This is because the empty tile must move the same number of squares up as down and left as right. Hence we can increment the depth of the search to by 2 if a goal node is not found. The code for 21p_main is as follows:

```
21p_main() // Iterative Deepening A* Search
{
int begin;

    startup();  // Initializations as in ID search

    g[0] = -1;  // To rectify the count
    compute_f_g_h(0,0);     // Evaluate root node
    begin = 2*(f[0]+1)/2); // Nearest even integer >= f[0]

    c_or(int fb=begin;fb<STACKSIZE;fb=fb+2){// fb is bound
        and(int lte=0; ;lte++) {
            try_moves(lte);// As in depth-first search
            compute_f_g_h(lte,lte+1);  // Evaluate node
            f[lte+1] <= fb;            // Observe bound
            if h[lte+1] == 0;          // Goal node ?
            then {
                output(lte+1);
                break;
            }
        }
    }
}
```

As noted above, a goal node is reached if h[n] == 0. Since we are trying to minimize the length of the path to a goal and each move counts as one, this situation is reached if the total heuristic estimate f[n] = g[n] + h[n] is in fact equal to n. The use of the heuristic has the effect of altering the geometry of the search tree in that paths through nodes with poor f values are considered "long" and paths through nodes with small f values are considered "short"; an optimal path to the goal is one where the f value at the end of the path coincides with the g value.

For the 15 puzzle, optimal solutions are known to take up to 81 moves and 88 moves is a proven upper bound (see Gasser 1995). Solving a puzzle of this size would simply not be doable with a breadth-first search, since the queue could reach a length greater than 2^{88}. In fact, the best published results on finding optimal solutions for the 15 and 24 Puzzles use IDA* search. For more on this topic, see Korf (1985), Culberson and Schaeffer (1994), and Korf(1996).

Exercises

14.3.1. Write code for the `compute_f_g_h` function that computes `f` by noting the change in this function caused by moving the tile.

14.3.2. Finding a minimal length solution to the 15 puzzle is much harder than for the 8 puzzle. Try running an IDA* search to solve the minimization form of the 15 puzzle on the following data:

```
nsquared[0]={ 6,2,4,8,14,10,1,3,5,7,11,12,9,13,15,0 }
```

where the goal node is the one we have been using with the empty square in the bottom right position. This example requires 26 moves. In the S^2-1 puzzle, the Manhattan distance between the current position of a tile and its goal position is defined to be the sum of the number of rows plus the number of columns between the current position and the goal position. The sum of these distances over all tiles can be used as an heuristic estimate $h[n]$ in the A* solution to this puzzle. How many nodes does it take to find a solution using this heuristic on the above data? Compare it to the heuristic of Section 14.3.

14.3.3. We have thus far assumed that initial and final positions of the empty square are the same. Consider the general case where they may be different. Write a procedure to determine whether the total number of moves to solve the puzzle will be even or odd. This can be done by counting the number of moves required for the simple problem of getting the empty square from its initial position to its final position. Show that if this number is even (respectively, odd), then all solutions to the original problem require an even (respectively, odd) number of moves.

14.3.4. Using the Manhattan distance as a heuristic, try the more challenging data:

```
nsquared[0] = { 4,5,7,2,9,14,12,13,0,3,6,11,8,1,15,10 };
nsquared[0] = { 14,1,9,6,4,8,12,5,7,2,3,0,10,11,13,15 };
```

where the goal node for these two examples is

```
goal_node = { 0,1,2,3,4,5,6,7,8,9,10,11,12,13,14,15 };
```

The first requires 42 moves, while the second requires 45 moves. Hint: compute the Manhattan distance incrementally.

14.3.5. Order the states added before calling `insert_states`.

14.4 Breadth-first branch-and-bound search

In this section we apply breadth-first search to disjunctive programming applications by means of a technique known as *breadth-first, branch-and-bound search*. The method was pioneered by Land and Doig (1960), Little, Murty, Sweeney and Karel (1963) and Dakin (1965). It is the common ancestor of A* search, the injury method, and the branch-and-bound trick of pruning nodes whose linear relaxation fails to meet the current bound. This technique is also called *best-bound search*.

To keep things simple, we use Model 13.3, New York Supermarkets, to illustrate this method.

Model 14.2 New York Supermarkets Remodeled

Let us approach the problem of Model 13.3 by means of breadth-first search. The basic structure of the model will be much the same as in Chapter 13. This application incorporates a quick-and-dirty solution, the injury method, fuzzy booleans, and branch-and-cut. Here we show how to work these ideas into a breadth-first, branch-and-bound search; this will also be close to A* search and the role of the evaluation function f[n] of the A* code will be played by the optimum of the linear relaxation. New nodes will be generated by searching for the best branching variable while seeking the eureka effect. When a goal node reaches the front of the priority queue, we will know that it is the optimal solution, for it will be the node with the best current evaluation.

The symbolic constants and declarations for both the capacitated warehouse problem and the heuristic reordering will be needed. As in Model 13.3, we have the machinery for modeling the application as a MIP model with fuzzy boolean variables Y[i]:

```
// Definitions and declarations from Model 13.3
#define M 6              // Number of warehouses
#define N 30             // Number of supermarkets
double kap[M],fc[M],d[N],c[M][N];    // Data arrays
continuous X[M][N];      // The amount of j's demand met by i
continuous Y[M];         // Fuzzy booleans for warehouses
continuous Cost;         // The objective function
```

We will also employ a pair of utility variables. A global variable bv of type int will be used for storing the branching variable; a global variable best_so_far of type double will be used to store the value of the best solution found so far.

```
int bv;      // Global utility variable for injury method
double best_so_far;    // Global utility variable for bound
```

14.4 Breadth-first branch-and-bound search

In addition we will need the apparatus from A* search except for g and h:

```
// Definitions and declarations from A* search
#define NIL -99
#define QUEUESIZE 400
int n_lnk[QUEUESIZE], predecessor[QUEUESIZE];
double f[QUEUESIZE];
int ste,end;
```
ij

Finally, we will need to keep track of the feasible region associated with each node in the search. Maintenance of the states is not as simple as in the preceding section. A state here consists of a list of indices of warehouses $i_0, \ldots, i_{k-1}$, the decisions $bit_0, \ldots, bit_{k-1}$ whether or not to open the warehouse, and the integer k itself as well as the evaluation associated with the state.

```
// The next three arrays hold state information
int next_decision[QUEUESIZE];    // The integer k
int decision_list[QUEUESIZE][M]; // bit₀,...,bitₖ₋₁
int permutation[QUEUESIZE][M];   // i₀,...,iₖ₋₁
```

As before, the state also requires the function value f[n]. In this problem the evaluation of the cost associated with a state is simply the optimal value of the linear relaxation of the problem.

In depth-first search one passes from a parent node to a child node, typically by adding a small number of constraints and very often only bounding constraints. In breadth-first search the successive nodes that come to the front of the priority queue can be far removed from one another in terms of their location in the search tree. However, as you jump from partial solution to partial solution, it is likely that the next node has several variable bindings in common with the previous node. Hence let us fix those first. This means that there will be less work required to switch from one node to another. This can be encapsulated in a simple procedure:

```
fix_common_variable_bindings(int ste)
{
int perm;
    and(int k=0;k<next_decision[ste];k++) {
        perm = permutation[ste][k];
        if wp(Y[perm]) == decision_list[ste][k];
        then Y[perm] == decision_list[ste][k];
    }
}
```

The effect of this is to take advantage of the variables Y[k] where the current witness point agrees with the node that is now to be rebuilt. This simplifies the work of the simplex algorithm in restoring the node; this is a programming remark but one that is pertinent to solving MIP problems with breadth-first search.

That done, the rest of the node can be rebuilt:

```
rebuild_node(int ste)
{
    and(int k=0;k<next_decision[ste];k++)
        Y[permutation[ste][k]] == decision_list[ste][k];
    min: Cost;          // Optimize for injury method
}
```

After the node is rebuilt, the next branching variable must be determined in order to generate the children of this node. If the eureka effect kicks in, then this node can be recorded as a goal node and inserted directly in the priority queue. For that the goal_node routine checks if there are no injuries; otherwise, the next branching variable is stored in the global identifier bv.

```
goal_node()
{
    // Set bv = branching variable or fail if eureka effect
    not determine_branching_variable(bv);//As in Model 13.3
}
```

If the linear relaxation of the node has no injuries, the node can be recorded by listing the witness point values of the fuzzy booleans Y[i]:

```
freeze_node(int ste)
{
    next_decision[ste] = M;         // All decisions made
    and(int j=0;j<M;j++) {
        decision_list[ste][j] = wp(Y[j]);
        permutation[ste][j] = j;    // Identity permutation
    }
    best_so_far = f[ste];           // Best known bound
}
```

The analog of bf_try_moves in this situation will build and evaluate the successor nodes by trying first to fix the branching variable at 1 and then at 0:

```
land_and_doig_try_moves(int ste,int & end)
{
    if land_and_doig_add_move(ste,end,1);
    then end = end + 1;
```

14.4 Breadth-first branch-and-bound search

```
    if land_and_doig_add_move(ste,end,0);
        then end = end + 1;
}
```

The subroutine `land_and_doig_add_move` will build the child node; for this a double negation-as-failure will be used to undo the first child so that the second can be built.

```
land_and_doig_add_move(int ste,newstate,bit)
{
    not not {   // Double negation to undo constraints
        Y[bv] == bit;
        min: Cost;                        // Evaluate node
        f[newstate] = wp(Cost);           // Record evaluation
        f[newstate] < best_so_far;        // Check bound
        update_state(ste,newstate,bit);   // Record state
    }
}
update_state(int ste, newstate, bit)
{
int nd;
    nd = next_decision[ste];
    // Copy parent state
    and(int i=0;i<nd;i++) {
        permutation[newstate][i] = permutation[ste][i];
        decision_list[newstate][i] = decision_list[ste][i];
    }
    // Add new elements
    permutation[newstate][nd] = bv;
    decision_list[newstate][nd] = bit;
    next_decision[newstate] = next_decision[ste]+1;
    if next_decision[newstate] == M;   // Goal node case
        then best_so_far = f[newstate];
}
```

Everything is now in place for the main procedure. The code for `21p_main` is similar to that in the previous section. However, here we must enclose the code that builds the new node in a double negation-as-failure so that the variable bindings will be undone after the node is explored. Otherwise, we would not be able to reset the variables with the bindings from the next node that we look at in this breadth-first, branch-and-bound search. To reiterate, the code to rebuild the node that has come to the front of the priority queue will be wrapped in a double negation-as-failure in order to free things up for the next node to be developed.

470   14 Further Search Methods

```
2lp_main()        // Land and Doig Search
{
    init_queue_vars();    // As for A* search

    data();               // As in Model 13.3
    setup();              // Ditto
    quick_and_dirty();    // Ditto
    initialize_bound();   // Set initial bound

    and(int i=end;i<QUEUESIZE-2;i=end){
        if next_decision[ste] == M;
        then {   // Optimal node at front of priority queue
            output();    // New version below
            return;
        }
        not not {
            fix_common_variable_bindings(ste);// Good idea
            rebuild_node(ste); // Node to be developed
            if goal_node();        // No injuries ?
            then freeze_node(ste);    // Eureka effect
            else {  // Generate children
                land_and_doig_try_moves(ste,end);
                insert_states(i,ste,end);// As in A* search
                ste = n_lnk[ste];       // Resume search
            }
        ste != NIL; // Check queue not mined out
        }
    }
    printf("Out of space\n");
}

initialize_bound()
{
    best_so_far = ub(Cost); // Bound from quick_and_dirty
}

output()    // Land and Doig version
{
    printf("ste is %d\n",ste);
    printf("f[ste] is %f\n",f[ste]);
    and(int k=0;k<M;k++)
        if decision_list[ste][k] == 1;
        then printf("Build warehouse %d\n",
                                    permutation[ste][k]);
}
```

14.4 Breadth-first branch-and-bound search

When it comes to branch-and-cut, the breadth-first method is harder to implement than the depth-first method because it must queue the history of the cuts that have been generated on the way to reaching a node. Of course one can still add cuts at the very top level only, which as we know from Chapter 13 can be quite effective. One can also rebuild a node and then add cuts locally. For example, one could emend the routine that rebuilds a state to add the cuts:

```
rebuild_node(int ste)
{
    and(int k=0;k<next_decision[ste];k++)
        Y[permutation[ste][k]] == decision_list[ste][k];
    min: Cost;
    add_cuts();      // From Model 13.3
    // min: Cost; DONE IN add_cuts
}
```

In practice, MIP solvers use a mix of breadth-first and depth-first search. One strategy is to pursue a breadth-first search until the priority queue reaches a certain size. At that point each node taken off the priority queue is developed via a depth-first search. This will account for the entire tree beneath that node. When the size of the queue shrinks sufficiently, breadth-first search can be resumed. From a strategic point of view, the advantage of the breadth-first component of the search is that better global information about the structure of the entire tree can be gathered, and this information can be used locally as each node is developed in either breadth-first or depth-first mode. For example, since the initial breadth-first search explores the top layers of the tree, it might be noted that a certain variable is often chosen as the branching variable. Later at a node where that variable is one of the candidates for branching it can be given priority. Similarly pseudocosts can be compiled across a more diversified portion of the tree. An alternative method that avoids breadth-first search as such is to employ iterative deepening to explore the top layers of the tree and to gather global information before embarking on a full depth-first exploration of the tree.

In MIP models the linear programming relaxations can be computationally demanding in and of themselves. In any search situation where each node requires a significant amount of time, it is critical to minimize the total number of nodes visited. For this reason it is important to have as good initial heuristic solutions as possible so that time is not wasted on underperforming nodes.

The machinery that has been set up in this section can be easily adapted for an important generalization of hill-climbing search known as *beam search*. In an ordinary hill-climbing search, the program makes a dive down the search tree maintaining a single leading edge node. In beam search, the same thing is done but with a multiplicity of nodes. The number of nodes in the beam can be denoted B. The idea is to take the B nodes, to develop all their children, and to retain only the B most promising according to an evaluation function appropriate for the application. A sample problem is given in the exercises.

Having seen breadth-first search applied to a continuous model, let us note some ways of using iterative deepening for disjunctive linear programming. In this situation we can bound the "depth" of the search in two ways: (1) by bounding the depth of the search tree or (2) by weakening the initial bound or bluff for the branch-and-bound search. The first idea will lead to a method for finding a suboptimal solution in an optimization problem. The second will find an optimal solution.

To be more concrete, let us consider the central loop for a standard disjunctive program using the injury method. There are N decisions to be made, bv is the branching variable, and Z is the objective function that we seek to maximize:

```
find_max: Z;
subject_to
    and(;;) {
        max: Z;
        if determine(bv);
        then branch_on(bv);
        else break;
    }
```

To implement method (1), we restrict the number of times the and loop can be re-entered:

```
c_or(int d=1;d<=N;d++)
    find_max: Z;
    subject_to
        and(int i=0;i<=d;i++) {
            if d < N;
            then i < d;
            max: Z;
            if determine(bv);
            then branch_on(bv);
            else break;
        }
```

This method is not guaranteed to find the optimal solution. On the other hand, this method can be used equally well in situations where there is no objective function.

To implement method (2), we start with a high estimate as to the optimal value of Z and work down form there:

```
c_or(int d=HIGH;d>LOW;d=d-DECREMENT)
    find_max: Z;
```

14.4 Breadth-first branch-and-bound search

```
subject_to {
    Z >= d;        // Initial bound
    and(int i=0;i<d;i++) {
        max: Z;
        if determine(bv);
        then branch_on(bv);
        else break;
    }
}
```

Another interesting heuristic is based on the number of injuries remaining after the node is developed. A node with relatively few injuries can be considered "higher" in the tree and developed further. This is a version for continuous models of the *least discrepancy search* of Harvey (1995).

By-and-large search strategies lend themselves beautifully to parallel and distributed computing. Unfortunately this topic is beyond the scope of this book. Yet another important topic that has not been broached is the use of visualization to aid in the development of search strategies.

Exercises

14.4.1. Redo Model 13.1 using breadth-first branch-and-bound search.

14.4.2. Develop the analog of least discrepancy search for continuous models based on an injury count.

14.4.3. This is a job scheduling problem with precedence constraints involving a fixed number of identical machines. Dependencies stipulate that a particular job cannot start until certain others have finished. The goal is to minimize the average completion time of all the jobs. The dependencies are give in Figure 14.5 and the time needed to complete each job is give in Table 14.1. Apply beam search to this problem. Apply beam search to the same problem but with makespan as the objective function. Is beam search necessarily monotonic in the sense that increasing the width of the beam can not worsen the value of the objective function of the best solution found?

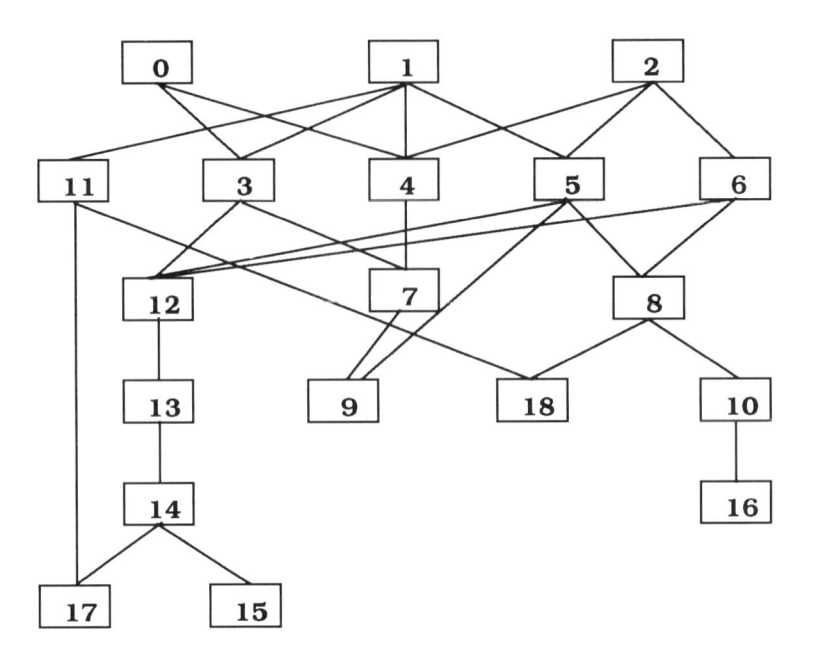

FIGURE 14.5 Job dependencies

TABLE 14.1 Times needed for each job

Job Number	Time	Job Number	Time	Job Number	Time
0	4	7	44	13	16
1	2	8	32	14	13
2	2	9	25	15	33
3	13	10	43	16	22
4	20	11	51	17	11
5	11	12	13	18	44
6	21				

15

Mathematical Underpinnings

In this chapter we present the basic mathematics that underlies the simplex algorithm and give a description of the form of the algorithm known as the *revised simplex method*. The starting point for this mathematical analysis is linear programming duality, which is based on some beautiful insights and theorems that have a long history going back to Fourier (1826).

Familiarity with the basic material of linear algebra is assumed for this chapter.

15.1 Dual multipliers

As we have seen, in many applications the dual relationship between continuous variables and left-hand sides of constraints has a simple intuition behind it, such as the duality between nodes and arcs in network problems. To treat linear programming duality in general, there is no simple universal story, and we must turn to the mathematical structure to express the dual relationship.

Let us recall some basic notions and some mathematical terminology. Typically a set of m constraints on n variables will be given as

$$a_{11}x_1 + \ldots + a_{1n}x_n \leq b_1$$
$$\vdots$$
$$a_{m1}x_1 + \ldots + a_{mn}x_n \leq b_m$$

With this notation, we do not necessarily assume that bounding constraints restricting the feasible region to the first orthant are included. However, nonnegativity requirements will soon reassert themselves and play an important role.

We write this in matrix form as $\mathbf{Ax} \leq \mathbf{b}$. More explicitly,

$$\mathbf{Ax} = \begin{bmatrix} a_{11} & \cdots & a_{1n} \\ & \vdots & \\ a_{m1} & \cdots & a_{mn} \end{bmatrix} \begin{bmatrix} x_1 \\ \vdots \\ x_n \end{bmatrix} \leq \begin{bmatrix} b_1 \\ \vdots \\ b_m \end{bmatrix} = \mathbf{b}$$

As noted in Chapter 4, an objective function

$$c_1 x_1 + \ldots + c_n x_n$$

can be written concisely as $\mathbf{cx}$. The generic linear programming maximization problem can then be expressed as

$$\mathbf{Ax} \leq \mathbf{b}$$
$$\text{max: } \mathbf{cx}$$

When working with vectors and matrices, we will not explicitly state dimensions when they are implied by the notation. By way of example, a matrix inequality $\mathbf{Ax} \leq \mathbf{b}$ stipulates that $\mathbf{b}$ is a column vector and, for all i, b_i is greater than or equal to the scalar product of the i th row of $\mathbf{A}$ with the column vector $\mathbf{x}$. As another example, the equation $\mathbf{yA} = \mathbf{c}$ stipulates that $\mathbf{c}$ is a row vector and, for all j, c_j is the scalar product of the j th column of $\mathbf{A}$ with the row vector $\mathbf{y}$:

$$\mathbf{yA} = \begin{bmatrix} y_1 & \cdots & y_m \end{bmatrix} \begin{bmatrix} a_{11} & \cdots & a_{1n} \\ & \vdots & \\ a_{m1} & \cdots & a_{mn} \end{bmatrix} = \begin{bmatrix} c_1 & \cdots & c_n \end{bmatrix} = \mathbf{c}$$

Let us address the interesting mathematical question of characterizing *all* the constraints that are satisfied by the points in a feasible region. If for all points $\mathbf{x}$ in the feasible region defined by $\mathbf{Ax} \leq \mathbf{b}$ we have $\mathbf{cx} \leq d$, we say that $\mathbf{cx} \leq d$ is a *consequence* of $\mathbf{Ax} \leq \mathbf{b}$, and we denote this relation by

$$\mathbf{Ax} \leq \mathbf{b} \models \mathbf{cx} \leq d$$

In other words, $\mathbf{cx} \leq d$ is a consequence of $\mathbf{Ax} \leq \mathbf{b}$ if the constraint $\mathbf{cx} \leq d$ is redundant when added to $\mathbf{Ax} \leq \mathbf{b}$. The relation $\models$ is read "turnstile" or "double turnstile."

Now we can connect the consequence relation with the maximization problem.

15.1 Dual multipliers

Lemma (First Duality Lemma) *Suppose that* $\mathbf{cx} = z^*$ *is the optimal value of the objective function for the linear program* $\mathbf{Ax} \leq \mathbf{b}$, max: $\mathbf{cx}$. *Then* z^* *is also the minimum z such that* $\mathbf{Ax} \leq \mathbf{b} \models \mathbf{cx} \leq z$.

Proof For all points $\mathbf{x}$ in the feasible region defined by $\mathbf{Ax} \leq \mathbf{b}$, we have $\mathbf{cx} \leq z^*$, and the constraint $\mathbf{cx} \leq z^*$ is a consequence of the constraints $\mathbf{Ax} \leq \mathbf{b}$. Since z^* is the optimal solution to the maximization problem, at some point $\mathbf{p}$ in the feasible region we have $\mathbf{cp} = z^*$. Hence if $z' < z^*$ then $\mathbf{cp} > z'$, and so $\mathbf{cx} \leq z'$ is not a consequence of $\mathbf{Ax} \leq \mathbf{b}$. This completes the proof. ■

The First Duality Lemma establishes the equivalence of a maximization problem and a minimization problem. At first glance the minimization problem of the First Duality Lemma is not a linear program, but an optimization problem of a different species. However, the question arises whether this problem can be formulated as a linear program. We will see that it can be.

In linear algebra a standard device is to generate a new equation from a set of equations by adding some of the given equations together or by multiplying an equation by a constant. These operations are merged by taking sums of multiples of existing equations. This is the basis of Gaussian elimination. The equations that are derived this way are consequences of the original set; they are satisfied by all the solutions to the original set of equations. In fact, a basic result of linear algebra is

Theorem *The following are equivalent*:

1. $\mathbf{Ax} = \mathbf{b} \models \mathbf{cx} = d$
2. *For some* μ, *we have* $\mu\mathbf{A} = \mathbf{c}$ *and* $\mu\mathbf{b} = d$.

This theorem treats the case of equality constraints. Let us see how one can go from inequality constraints $\mathbf{Ax} \leq \mathbf{b}$ to consequence constraints $\mathbf{cx} \leq d$. The simplest way to generate consequences of the set of inequalities $\mathbf{Ax} \leq \mathbf{b}$ is by summing *nonnegative* multiples of them; the restriction to nonnegative multipliers is needed to preserve the inequality $\leq$. In Figure 15.1, the relation between a foursome of constraints and a new constraint derived by adding nonnegative multiples of the original constraints is illustrated. The original constraints are $x_1 + x_2 \leq 5$, $x_1 \leq 4$, $-x_1 \leq 0$, $-x_2 \leq 0$. The new constraint $5x_1 + 3x_2 \leq 23$ is the sum of 3 times the first constraint with 2 times the second and 0 times the other two constraints. We see that the half space defined by the new constraint contains the region defined by the other constraints.

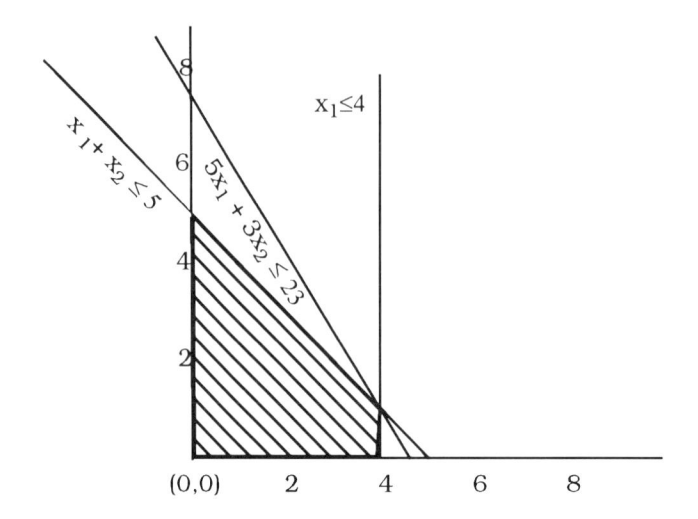

FIGURE 15.1 Nonnegative linear combination of constraints.

The matrix notation for this example is

$$[3\ 2\ 0\ 0]\begin{bmatrix} 1 & 1 \\ 1 & 0 \\ -1 & 0 \\ 0 & -1 \end{bmatrix}\begin{bmatrix} x_1 \\ x_2 \end{bmatrix} \le [3\ 2\ 0\ 0]\begin{bmatrix} 5 \\ 4 \\ 0 \\ 0 \end{bmatrix}$$

Multipliers of constraints are called *Lagrange multipliers*, and the traditional notation for a vector of multipliers such as [3 2 0 0] is $\lambda = [\lambda_1,...,\lambda_m]$. In general, if we start with $\mathbf{Ax} \le \mathbf{b}$ and the row vector λ, the row vector $\lambda\mathbf{A}$ is obtained as the sum of the rows of $\mathbf{A}$ by the Lagrange multipliers $\lambda_1,...,\lambda_m$ and the real number $\lambda\mathbf{b}$ is the scalar product of λ and $\mathbf{b}$. Therefore, if $\lambda \ge \mathbf{0}$, then $\lambda\mathbf{A}$ is a row vector and $(\lambda\mathbf{A})\mathbf{x} \le \lambda\mathbf{b}$ is a constraint, which is a consequence of $\mathbf{Ax} \le \mathbf{b}$. Somewhat more emphatically,

$$\mathbf{Ax} \le \mathbf{b} \models \lambda\mathbf{Ax} \le \lambda\mathbf{b} \qquad \text{for } \lambda \ge \mathbf{0}$$

If $\lambda \ge \mathbf{0}$, we say that $\lambda\mathbf{Ax} \le \lambda\mathbf{b}$ is a *nonnegative linear combination* of the constraints $\mathbf{Ax} \le \mathbf{b}$.

In order to bring the objective function max: $\mathbf{cx}$ and duality back into it, we return to the maximization problem

$$\mathbf{Ax} \le \mathbf{b}.$$
$$\text{max: } \mathbf{cx}$$

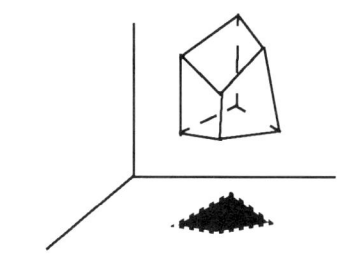

15.1 Dual multipliers

FIGURE 15.2 The projection of a polyhedral set.

Now consider the following linear program in which we have to solve for **y**:

$$\mathbf{y} \geq \mathbf{0}$$
$$\mathbf{yA} = \mathbf{c}$$
$$\min: \mathbf{yb}$$

With the initial maximization problem has been associated a minimization problem, and both are linear programs. The original problem is called the *primal* problem, and the new one is called the *dual* problem.

Theorem (Weak Duality Theorem) *Given a primal/dual pair of linear programs, every solution to the primal maximization problem is majorized by every solution to the dual minimization problem.*

Proof Consider any solution $\lambda \geq 0$, $\lambda \mathbf{A} = \mathbf{c}$ to the dual problem. Then for all **x** in the primal feasible region, we have $\lambda \mathbf{Ax} \leq \lambda \mathbf{b}$, since this is a consequence of $\mathbf{Ax} \leq \mathbf{b}$. Substituting **c** for $\lambda \mathbf{A}$, we find that $\mathbf{cx} \leq \lambda \mathbf{b}$. ∎

Here we see duality emerge in one of its simplest forms, rows and columns of a matrix. In fact, in the dual problem the constraints correspond to the columns of the matrix **A** and the variables **y** to the rows of **A**.

A strength of the Weak Duality Theorem is that it provides a test for optimality. Suppose that a solution to the primal problem has objective function value u^* and that a solution to the dual problem has objective function value u^{**}. If $u^{**} \leq u^*$, then both solutions are optimal and $u^{**} = u^*$.

The next result will require a definition. Let E be a subset of R^{n+1} and let the axes be labeled $(x, x_1, \ldots, x_n)$. The *projection* E_x is defined by the condition

$$(x_1, \ldots, x_n) \in E_x \leftrightarrow \exists x \, (x, x_1, \ldots, x_n) \in E$$

In Figure 15.2 the projection of a polyhedral set in xyz space onto the xy plane is illustrated.

The terms *feasible region* and *polyhedral set* are synonymous. However, in this and the following section, the language is mathematical rather than computational, and we will prefer the term polyhedral set. The next goal is to show that the projection P_x of a polyhedral set P is again a polyhedral set. The strategy for proving this will be to start with the constraints that define P and to generate consequence constraints where the variable x has zero coefficient. If the variable x has a zero coefficient in a constraint, we can consider it as a constraint on the remaining variables, going down a dimension to the space of the $(x_1,...,x_n)$.

The following theorem and proof go back to Fourier (1826). The proof uses a technique known as *Fourier-Motzkin elimination*. Our presentation is adapted from Lassez and Maher (1992). For a fuller exposition, see Kuhn (1956).

Lemma (Quantifier Elimination Lemma) *Let P be a polyhedral subset of R^{n+1} defined by the constraints $\mathbf{Ax} \leq \mathbf{b}$, and let x be a variable. Then the projection P_x is also a polyhedral set. What is more, it is defined by a set of constraints that are nonnegative linear combinations of the constraints $\mathbf{Ax} \leq \mathbf{b}$.*

Proof Rearranging terms, we can put the constraints defining P in the form

$$U_i: \quad a_{i1} x_1 + ... + a_{in} x_n + u_i \leq x, \quad i = 1,...,p$$
$$V_j: \quad x \leq v_j + b_{j1} x_1 + ... + b_{jn} x_n, \quad j = 1,...,q$$
$$W_k: \quad c_{k1} x_1 + ... + c_{kn} x \leq w_k, \quad k = 1,...,r$$

We form a new set of constraints by combining each constraint U_i with each V_j and eliminating x:

$$U_i \times V_j: \quad a_{i1} x_1 + ... + a_{in} x_n + u_i \leq v_j + b_{j1} x_1 + ... + b_{jn} x_n,$$
$$i = 1,..., p, j = 1,...,q$$
$$W_k: \quad c_{k1} x_1 + ... + c_{kn} x_n \leq w_k, \quad k = 1,...,r$$

The constraints W_k are part of the original set of constraints $\mathbf{Ax} \leq \mathbf{b}$. It is easy to see that each of the constraints $U_i \times V_j$ is a nonnegative linear combination of a pair of constraints from the original set. Therefore the constraints $U_i \times V_j$, W_k are all consequences of the original set. Going down a dimension to R^n, these new constraints define a polyhedral set Q. We must have $P_x \subseteq Q$. We have to show that $Q \subseteq P_x$. Let $(x^*_1,...,x^*_n)$ be any point in Q. Let x^* be the *sup* of $a_{i1} x^*_1 + ... + a_{in} x^*_n + u_i, i = 1,...,p$. Then for some i, we have $x^* = a_{i1} x^*_1 + ... + a_{in} x^*_n + u_i$. Since $(x^*_1,...,x^*_n)$ simultaneously satisfies all the constraints $U_i \times V_j$ and W_k, it follows that $(x^*, x^*_1,..., x^*_n) \in P$, and so $Q \subseteq P_x$. ∎

15.1 Dual multipliers

From the lemma it follows that successive projections

$$P_{x_1 x_2 \cdots x_k}$$

are all polyhedral sets defined by constraints that are nonnegative linear combinations of $\mathbf{Ax} \le \mathbf{b}$. It also follows that if the original constraints are inconsistent, then an inequality of the form $a \le b$ with $b < a$ will eventually be generated.

The power of the Quantifier Elimination Lemma lies in the fact that the projection of a polyhedral set is again a set of the same type. Theorems of this kind are called *quantifier elimination theorems* in the mathematical logic literature and *elimination theorems* in the algebra literature. The importance of quantifier elimination was elucidated by Tarski (1951) who established quantifier elimination for real-closed fields. Gaussian elimination is another example of quantifier elimination; in fact, it characterizes the projection of polyhedral sets all of whose defining constraints are equality constraints. In algebraic settings, quantifier elimination results also lead to "transfer theorems" which enable one to conclude a result for an entire family of structures by establishing it for a single structure. For an example of a transfer theorem that uses the above lemma, see Eaves and Rothblum (1992).

The following theorem is fundamental:

Theorem (Duality Theorem) *Suppose that z^* is the optimal value of the objective function in the primal linear programming problem $\mathbf{Ax} \le \mathbf{b}$, max: $\mathbf{cx}$. Then z^* is also the optimal solution to the dual linear program $\mathbf{y} \ge \mathbf{0}$, $\mathbf{yA} = \mathbf{c}$, min: $\mathbf{yb}$.*

Proof Introducing a new variable z for the objective function, we consider the constraints

$$\mathbf{Ax} \le \mathbf{b}$$
$$z \le \mathbf{cx}$$

Applying the Quantifier Elimination Lemma we can eliminate all variables except z; the projection onto the z axis is a one dimensional polyhedral set. Since we are assuming z is bounded from above by z^*, this must be an interval of the form $(-\infty, t]$ or $[a, t]$. Clearly $t \le z^*$. If $t < z^*$, then $\mathbf{Ax} \le \mathbf{b}$, $\mathbf{cx} = z^*$ is infeasible. This is absurd, so $t = z^*$.

Since the constraint $z \le z^*$ must appear in the description of the projection onto the z axis, this constraint can be written as a nonnegative linear combination of the constraints $\mathbf{Ax} \le \mathbf{b}$, $-\mathbf{cx} + z \le 0$. In detail, if $\mathbf{A}$ has m rows, there are $\boldsymbol{\lambda} = [\lambda_1, \ldots, \lambda_m] \ge \mathbf{0}$ and $\lambda_{m+1} \ge 0$ such that the constraint

$$z \le z^*$$

is equal to the sum of the constraints

$$\lambda A x \leq \lambda b$$
$$\lambda_{m+1}(-cx + z) \leq \lambda_{m+1} \, 0$$

But this means that $\lambda_{m+1} = 1$, $\lambda A = c$, and $\lambda b = z^*$. Hence $y = \lambda$ is a solution to the constraints of the dual problem. In other words, we have shown that the optimal value z^* of the primal program is equal to the objective function value of a feasible solution of its dual problem. Therefore the Weak Duality Theorem yields the result. ∎

The components of the solution vector $\lambda \geq 0$ to the dual problem are called the *dual multipliers* of the primal constraints $Ax \leq b$. As we will see, the dual multipliers provide the reduced costs of sensitivity analysis. The duality relation between the primal and dual forms of a linear program is the key to many important applications of linear programming in economics and game theory (see Franklin 1980).

Note that we have shown that the minimization problem of the First Duality Lemma is indeed equivalent to a linear programming problem. In fact, the Duality Theorem provides a solution to the mathematical question of characterizing all the consequences of the constraints $Ax \leq b$.

Theorem (Affine Form of Farkas' Lemma) The following are equivalent:

1. $Ax \leq b \models cx \leq d$.
2. For some $e \leq d$ and some $\lambda \geq 0$, we have $\lambda A = c$ and $\lambda b = e$.

Proof Clearly property (2) implies property (1). For the other direction, suppose that $Ax \leq b \models cx \leq d$. Let e be the optimal solution to the dual problem

$$y \geq 0$$
$$yA = c$$
$$\min: yb$$

The Duality Theorem then yields the result, since $e \leq d$ by virtue of the First Duality Lemma. ∎

This theorem is naturally thought of as a completeness theorem with the dual multipliers λ playing the role of a formal proof.

In this section we have been emphasizing the duality between the rows and columns of a constraint matrix. Another view of linear programming duality is that an element a of R^n can be thought of both as a vector and as a linear functional, namely the map from R^n to R which sends x to the scalar product of a and x. In R^n and in Hilbert space, all linear functionals can be represented this way.

Exercises

15.1.1. The *cone* generated by $\mathbf{x}_1,\ldots,\mathbf{x}_m$ is defined to be the set of all nonnegative linear combinations of $\mathbf{x}_1,\ldots,\mathbf{x}_m$. Prove the following form of Farkas' Lemma: $\mathbf{x}$ is in the cone generated by $\mathbf{x}_1,\ldots,\mathbf{x}_m$ if and only if for all $\mathbf{a}$, $\mathbf{ax}_i \geq 0$ for $i = 1,\ldots,m$ implies $\mathbf{ax} \geq 0$.

15.1.2. Prove that if the dual problem has a feasible solution, then the primal problem has a finite optimum.

15.1.3. Show that for linear programs the dual of the dual is the primal. What is the dual of $\mathbf{Ax} \leq \mathbf{b}$, $\mathbf{x} \geq \mathbf{0}$, max: $\mathbf{cx}$?

15.1.4. Prove Cayley's Theorem: Every group is isomorphic to a group of permutations. Hint: Think "dually."

15.1.5. Suppose the projection P_x is the entire space R^n. How is this detected by Fourier-Motzkin elimination?

15.2 Geometry and algebra

In this section we sketch the basic results that prove that the search space for the optimal solution to a linear programming problem is a finite one. There are three notions of closure that will play an important role: linear closure, affine closure and convex closure. Familiarity with the elements of point set topology is assumed for this section.

Let X be a subset of R^n. The *linear closure* or *linear hull* of X is defined as the smallest subspace containing all points in X. This can also be described as the set of all points that are linear combinations of elements in X, that is, all points of the form

$$\lambda_1 \mathbf{p}_1 + \ldots + \lambda_m \mathbf{p}_m$$
$$-\infty < \lambda_i < +\infty$$
$$\mathbf{p}_1,\ldots, \mathbf{p}_m \in X$$

A linear subspace L of R^n has a finite basis, say, $\mathbf{b}_1,\ldots,\mathbf{b}_k$, with $k \leq n$. If we let $\mathbf{B}$ be the matrix whose columns are the vectors $\mathbf{b}_1,\ldots,\mathbf{b}_k$, the linear space can be written as $L = \{\mathbf{y} : \mathbf{Bx} = \mathbf{y}, \text{for some } \mathbf{x}\}$. Let us note that a linear space can be also defined as an intersection of hyperplanes through the origin. This is a classic example of duality in linear algebra. In fact, let L' be the orthogonal complement of L; that is, $L' = \{\mathbf{a}: \mathbf{ax} = 0 \text{ for all } \mathbf{x} \text{ in } L\}$. Let $\mathbf{a}_1,\ldots,\mathbf{a}_m$ be a basis for L' and let $\mathbf{A}$ be the m by n matrix with rows $\mathbf{a}_1,\ldots,\mathbf{a}_m$. Then L is defined by the equality constraints $\mathbf{Ax} = \mathbf{0}$.

Note the back-and-forth relation between these two descriptions of a linear subspace. In the usual primal description of a subspace as the closure of its basis vectors, the dimension k of the space is equal to the number of columns in the matrix **B** above. In the dual description it is described as the intersection of $n - k$ hyperplanes and $n - k$ is the dimension of the orthogonal complement. When a linear space is described as an intersection of hyperplanes $\mathbf{Ax} = \mathbf{0}$, each hyperplane reduces the dimension of the space by 1 provided that the rows of **A** are linearly independent.

If **p** and **q** are distinct points in R^n, the set of all points of the form

$$(1 - \lambda)\mathbf{p} + \lambda\mathbf{q} = \mathbf{p} + \lambda(\mathbf{q} - \mathbf{p})$$
$$-\infty < \lambda < +\infty$$

is called the *line* through **p** and **q**. A subset A of R^n is said to be an *affine space* if the points $(1 - \lambda)\mathbf{p} + \lambda\mathbf{q} = \mathbf{p} + \lambda(\mathbf{q} - \mathbf{p})$ are in A for all λ whenever **p** and **q** are in A. An affine space is one closed under all straight lines passing through any two points. Straight lines and planes are obvious examples. For this reason, affine spaces are called *flats*. Affine spaces are simply a generalization of linear spaces as the following two results will show:

Theorem *Linear subspaces of R^n are affine spaces that contain the origin, and conversely.*

Proof Clearly every subspace contains the origin and is an affine space. Conversely, an affine space containing the origin is easily verified to be closed under addition and scalar multiplication. ∎

If A is an affine subspace of R^n and if **p** is a vector in R^n, then the *translation* of A by **p** is denoted $A + \mathbf{p}$ and is defined by

$$A + \mathbf{p} = \{\, x + \mathbf{p} : x \in A \,\}.$$

Theorem *Every affine space A is the translation of a unique linear subspace.*

Proof Let **p** be in A and set $L = A + (-\mathbf{p})$. It is straightforward to show that $A = L + \mathbf{p}$ and that if $A = L' + \mathbf{q}$, then $L = L'$. ∎

The *dimension* of an affine space $L + \mathbf{p}$ is defined to be the dimension of L.

Corollary *Every affine space is a finite intersection of hyperplanes and conversely.*

Proof If L is defined by $\mathbf{Ax} = \mathbf{0}$, then $L + \mathbf{p}$ is defined by $\mathbf{Ax} = \mathbf{Ap}$. Conversely, the space defined by $\mathbf{Ax} = \mathbf{b}$ is the translation of the linear space $\mathbf{Ax} = \mathbf{0}$ by any vector **p** such that $\mathbf{Ap} = \mathbf{b}$. ∎

15.2 Geometry and algebra

The case of the one-dimensional affine space will prove important in the Fundamental Theorem of Linear Programming below. This is the situation where the affine space is spanned by a vector **a** situated at a point **p**; geometrically, the affine space is a line through **p**. The primal description of the space is as the set of all $\mathbf{p} + \lambda \mathbf{a}$. The dual description is by means of $n - 1$ hyperplanes $\mathbf{Ax} = \mathbf{Ap}$. By convention, if the affine space reduces to single point, it has dimension 0.

The intersection of all affine spaces that contain a set X is again an affine space and is called the *affine closure* or *affine hull* of X; it is denoted $A\!f\!f(X)$; see Figure 15.3 (a).

A linear combination of points in R^n whose coefficients sum to 1 is called an *affine combination*. The affine closure of X can also be described as the set of all points that are affine combinations of points in X, that is, as all points of the form

$$\lambda_1 \mathbf{p}_1 + \ldots + \lambda_m \mathbf{p}_m$$
$$-\infty < \lambda_i < +\infty$$
$$\lambda_1 + \ldots + \lambda_m = 1$$
$$\mathbf{p}_1,\ldots,\mathbf{p}_m \in X$$

This equivalence requires an argument. For $m = 1$ and $m = 2$, it follows from the definitions; for $m > 2$, it follows by induction. Also note that this sense of the term *affine combination* is different from that of Section 1.4. The contexts are very different, and no confusion should arise.

Recall that a subset X of R^n is *convex* if the line segment joining any two points in X is included in X: For $1 \geq \lambda \geq 0$ the point $\lambda \mathbf{p} + (1 - \lambda) \mathbf{q}$ is in X whenever $\mathbf{p}, \mathbf{q}$ are.

A nonnegative linear combination of points in R^n whose coefficients sum to 1 is called a *convex combination*. Differently put, a convex combination of the points $\mathbf{q}_1,\ldots,\mathbf{q}_m$ has the form

(a) The affine closure   (b) The convex closure

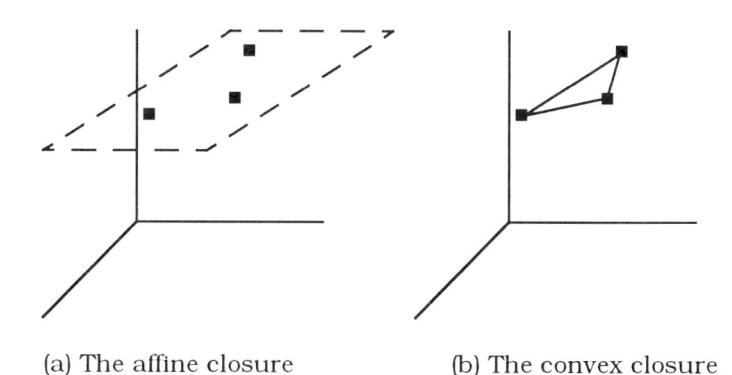

FIGURE 15.3 The affine closure and the convex closure.

$$\lambda_1 \mathbf{q}_1 + \ldots + \lambda_m \mathbf{q}_m$$
$$\lambda_i \geq 0$$
$$\lambda_1 + \ldots + \lambda_m = 1$$

As noted in Chapter 13, each subset X of R^n is contained in a smallest convex set, called its *convex closure* or *convex hull*; see Figure 15.3 (b). The convex closure of X can be described as the intersection of all convex sets containing X. Alternatively, it can be described as the set of points that are convex combinations of points in X. In other words, the convex closure of X consists of all points of the form

$$\lambda_1 \mathbf{p}_1 + \ldots + \lambda_m \mathbf{p}_m$$
$$\lambda_i \geq 0$$
$$\lambda_1 + \ldots + \lambda_m = 1$$
$$\mathbf{p}_1, \ldots, \mathbf{p}_m \in X$$

This equivalence also requires an argument. First note that the set of convex combinations is convex. One then shows by induction on m that a convex set is closed under convex combinations of length m.

We have seen that linear hulls and affine hulls are polyhedral sets. Next we describe the convex hull of a finite set of points in terms of constraints. The result is a corollary to the Quantifier Elimination Lemma.

Corollary *The convex closure of a finite number of points is a polyhedral set.*

Proof Suppose that $\mathbf{p}_1, \ldots, \mathbf{p}_m$ are points in R^n with $\mathbf{p}_i = (p_{i1}, \ldots, p_{in})$. Consider the following constraints in the variables $y_1, \ldots, y_m, x_1, \ldots, x_n$:

$$y_i \geq 0, \quad i = 1, \ldots, m$$
$$y_1 + \ldots + y_m = 1$$
$$p_{1j} y_1 + \ldots + p_{mj} y_m = x_j, \quad j = 1, \ldots, n$$

Eliminating the variables $y_1, \ldots, y_m$ yields a set of constraints in $x_1, \ldots, x_n$ that defines the convex closure of $\mathbf{p}_1, \ldots, \mathbf{p}_m$. ∎

An *extreme point* in a convex set X is one that is not a convex combination of any other points in X. It is easy to see that the convex closure of a finite number of points is the convex closure of a finite number of extreme points: Start with the initial set, drop the first point that is in the convex closure of the other points and continue. Only extreme points are left at the end, and these remain extreme points in their convex closure.

Theorem *If a linear function $\mathbf{cx}$ has a maximum on the convex closure of a finite number of extreme points, then it takes its maximum at one of these points.*

15.2 Geometry and algebra

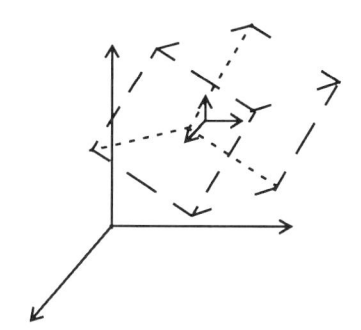

FIGURE 15.4 An open convex set is full dimensional.

Proof Consider the extreme points $\mathbf{p}_1,\ldots,\mathbf{p}_m$ and suppose that the maximum of $\mathbf{cx}$ is at $\mathbf{r}$, where $\mathbf{r} = \lambda_1 \mathbf{p}_1 + \ldots + \lambda_m \mathbf{p}_m$ with $\lambda_i \geq 0$, $\lambda_1 + \ldots + \lambda_m = 1$. Then $\mathbf{cr} = \lambda_1 \mathbf{cp}_1 + \ldots + \lambda_m \mathbf{cp}_m$. If all $\mathbf{cp}_i < \mathbf{cr}$, we have a contradiction. ∎

The *dimension* of a convex set is defined to be that of its affine closure. If the dimension of a convex set in R^n is n, the set is called *full dimensional*.

Let us note two properties of open convex sets:

1. An open convex set is full dimensional.
2. An open convex set has no extreme points.

The first point is illustrated in Figure 15.4. The second point can be illustrated by drawing a sufficiently small line segment with a given point as its midpoint; the point is then a convex combination of the end points of the line segment.

The *interior* of a set X is the union of all open sets included in X. In other words, the interior of X is the largest open subset of X. Polyhedral sets provide a source of closed convex sets, and their interiors provide a source of open convex sets. That is, the points that satisfy the constraints $\mathbf{Ax} \leq \mathbf{b}$ form a closed convex set. On the other hand, the points that satisfy the strict inequalities $\mathbf{Ax} < \mathbf{b}$ form its interior which is an open convex set. If this set is nonempty, it is full dimensional.

The next lemma is a basic topological result that shows that every convex set can be treated as a full-dimensional set by passing to its affine closure.

Lemma *A convex set has nonempty interior in its affine closure.*

Proof Let C be a nonempty convex set. The lemma is immediate if C consists of a single point. So let us suppose that C has more than one element, and without loss of generality, let us assume that the origin is a point in C. Let $\mathbf{p}_0$ denote the origin, and considering points in R^n as vectors, let $\mathbf{p}_1,\ldots,\mathbf{p}_s$ be a maximal set of linearly independent vectors that are elements of C. Then the affine closure of C is the vector space L spanned by $\mathbf{p}_1,\ldots,\mathbf{p}_s$. Let $\mathbf{c}$ be the following point in C, the point known as the *barycenter* of $\mathbf{p}_0,\ldots,\mathbf{p}_s$:

$$\mathbf{c} = \sum_{i=0}^{i=s} \left(\frac{1}{s+1}\right) \mathbf{p}_i$$

Then all points of the form

$$\mathbf{x} = \sum_{i=0}^{i=s} \left(\frac{1}{s+1} + \epsilon_i\right) \mathbf{p}_i$$

where $|\epsilon_i| < 1/(s+1)$ are in C and constitute an open neighborhood of $\mathbf{c}$ in L. ∎

Again, the import of this theorem is that for any convex set, by working in its affine closure, we can treat the set as a full dimensional set. Thus, for example, a convex set that is open in its affine closure has no extreme points.

Intuitively open sets and full dimensional sets are "large." For the sequel we will need a precise notion of "small" set. A set is *nowhere dense* if its only open subset is the empty set. In other words, a set is nowhere dense if it has empty interior. The notion of closed nowhere dense set will serve as a criterion for a set to be "small." As an example of a closed nowhere dense set in R^n, we have any linear or affine subspace of dimension less than n.

Let us note that a finite union of closed nowhere dense sets is again a closed nowhere dense set. For this it suffices to show that the union of a pair of closed nowhere dense sets is again closed and nowhere dense. So let F and G be closed nowhere dense sets. First, the union $F \cup G$ is closed. Next, let U be a nonempty open set. The difference $U - F$ is open, nonempty and disjoint from F; therefore the difference $(U - F) - G$ is open, nonempty, and disjoint from $F \cup G$. Hence U cannot be included in $F \cup G$, which is what we wanted to show.

The following theorem was needed in Chapter 4 to establish the soundness of the algorithm we used to minimize the L_∞ norm of an error term iteratively. Here we generalize it to affine sets rather than just hyperplanes.

15.2 Geometry and algebra

Theorem (Flat Covering Theorem) *Let P be a polyhedral set, and let $A_0,\ldots,A_{n-1}$ be affine spaces such that*

$$P \subseteq \bigcup_{i<n} A_i$$

Then for some $j < n$, we have $P \subseteq A_j$.

Proof Suppose that P is not included in A_j for any j. Then, since $P \subseteq \mathit{Aff}(P)$, the affine closure $\mathit{Aff}(P)$ is not included in A_j for any j. But then $\mathit{Aff}(P) \cap A_j$ is an affine space of dimension strictly less than that of $\mathit{Aff}(P)$. Therefore each $\mathit{Aff}(P) \cap A_j$ is closed and nowhere dense as a subset of $\mathit{Aff}(P)$ and their union is also closed and nowhere dense in $\mathit{Aff}(P)$. It follows that P cannot be covered by the union of the A_j, since P has nonempty interior in $\mathit{Aff}(P)$ and a nonempty open set cannot be covered by a closed nowhere dense set. ∎

This theorem will prove most valuable. For a treatment of it and related theorems in a general context, see Lassez and McAloon (1993).

We introduce some notation, following Schrijver (1986). Given constraints $\mathbf{Ax} \leq \mathbf{b}$ that define the polyhedral set P, let $\mathbf{A}^=\mathbf{x} \leq \mathbf{b}^=$ be the subset consisting of those constraints that are always satisfied as equalities by elements of P and let $\mathbf{A}^+\mathbf{x} \leq \mathbf{b}^+$ be the remaining constraints. Note that it follows from the Flat Covering Theorem that the inequalities $\mathbf{A}^=\mathbf{x} \leq \mathbf{b}^=$, $\mathbf{A}^+\mathbf{x} < \mathbf{b}^+$ are satisfiable provided that P is not empty.

Corollary *The affine closure of the polyhedral set P given by $\mathbf{Ax} \leq \mathbf{b}$ is defined by the constraints $\mathbf{A}^=\mathbf{x} = \mathbf{b}^=$.*

Proof Clearly the affine closure is included in the set defined by the equalities $\mathbf{A}^=\mathbf{x} = \mathbf{b}^=$. For the other direction, the inequalities $\mathbf{A}^=\mathbf{x} = \mathbf{b}^=$, $\mathbf{A}^+\mathbf{x} < \mathbf{b}^+$ are satisfiable by virtue of the Flat Covering Theorem. Let $\mathbf{x}_0$ satisfy $\mathbf{A}^=\mathbf{x} = \mathbf{b}^=$ and let $\mathbf{x}_1$ satisfy $\mathbf{A}^=\mathbf{x} = \mathbf{b}^=$, $\mathbf{A}^+\mathbf{x} < \mathbf{b}^+$. We want to show that $\mathbf{x}_0$ is in the affine closure of P. If $\mathbf{x}_0$ is equal to $\mathbf{x}_1$, we are done. Otherwise, consider the line segment from $\mathbf{x}_1$ to $\mathbf{x}_0$: this segment is the set of all points of the form $\lambda\mathbf{x}_1 + (1-\lambda)\mathbf{x}_0$, $1 \geq \lambda \geq 0$. All points on this segment satisfy $\mathbf{A}^=\mathbf{x} = \mathbf{b}^=$. Let $\lambda' < 1$ be sufficiently close to 1 so that the point $\mathbf{x}_2 = \lambda'\mathbf{x}_1 + (1-\lambda')\mathbf{x}_0$ also satisfies $\mathbf{A}^+\mathbf{x} < \mathbf{b}^+$. Now $\mathbf{x}_0$ is an affine combination of $\mathbf{x}_1$, $\mathbf{x}_2$ and must be in the affine closure of P. ∎

Recall that a set $X \subseteq R^n$ is *bounded* if there are $a_1, b_1, \ldots, a_n, b_n$ such that $\mathbf{x} \in X$ implies $a_1 \leq x_1 \leq b_1, \ldots, a_n \leq x_n \leq b_n$. A bounded polyhedral set is called a *polytope*. The next theorem characterizes polytopes in terms of their extreme points.

Theorem *Every bounded polyhedral set is the convex closure of a finite number of extreme points.*

Proof The proof is by induction on the dimension n of the set. For $n = 0$, the result is immediate. Suppose that the theorem is true for polyhedral sets of dimension less than n, where $n > 0$. Let $\mathbf{Ax} \leq \mathbf{b}$ define a polyhedral set P of dimension n. The constraints $\mathbf{A}^=\mathbf{x} = \mathbf{b}^=$ define the affine closure of P, which is an affine space of dimension n. Let the individual constraints of $\mathbf{A}^+\mathbf{x} \leq \mathbf{b}^+$ be listed as $\mathbf{a}_i\mathbf{x} \leq b_i$. Let H_i be the hyperplane defined by $\mathbf{a}_i\mathbf{x} = b_i$ and let $Y_i = P \cap H_i$. Each Y_i has dimension $< n$; this follows from the fact that this equality constraint is consistent with the constraints $\mathbf{A}^=\mathbf{x} = \mathbf{b}^=$ but is not a consequence of them. It is also the case that not all the Y_i can be empty. By the induction hypothesis, each Y_i is the convex closure of a finite number of points Q_i; let $Q = \cup \, Q_i$. We claim that the points of Q generate P. In fact, let $\mathbf{r} \in P$. If $\mathbf{r}$ is on one of the Y_i, we are done. Otherwise let $\mathbf{p}$ be an extreme point of one of the Y_i. Consider the parametrized straight line $\mathbf{p} + \lambda(\mathbf{r} - \mathbf{p})$ passing through $\mathbf{p}$ and $\mathbf{r}$. Let λ' be the greatest $\lambda > 0$ such that $\mathbf{p} + \lambda(\mathbf{r} - \mathbf{p})$ lies in P. This must exist since P is bounded. Set $\mathbf{q} = \mathbf{p} + \lambda'(\mathbf{r} - \mathbf{p})$. The point $\mathbf{q}$ lies on one of the Y_i because $\mathbf{q}$ must satisfy all the constraints that define P and cannot satisfy the strict form $\mathbf{a}_i\mathbf{x} < b_i$ of all the inequality constraints by virtue of the choice of λ'. But now $\mathbf{q}$ is generated by the points of Q; since $\mathbf{p}$ is an element of Q, it follows that $\mathbf{r}$ is generated by the points of Q. ∎

Recall from Chapter 9 that a *disjunctive set* is a finite union of polyhedral sets.

Corollary *The convex closure of a bounded disjunctive set is a polyhedral set.*

Proof Each of the polyhedral sets in the union is bounded, and so is the closure of its extreme points which are finite in number. The convex closure of the union is the closure of the union of these extreme points and so is a polyhedral set. ∎

For computational purposes in a linear program, we require a witness point at which the optimum for the objective function is attained. Now we give a formal algebraic definition of a vertex on a polyhedral set.

Suppose that a polyhedral set P in R^n is defined by the constraints $\mathbf{Ax} \leq \mathbf{b}$. If the equality forms of a subset of these constraints meet at a single point $\mathbf{v}$ in P, that point is called a *vertex* of P. In other words, a vertex is defined by an n-by-n subsystem $\mathbf{Vx} \leq \mathbf{b}^V$ of the constraints such that the equality form $\mathbf{Vx} = \mathbf{b}^V$ has a unique solution that lies in the polyhedral set; the set of equations is called a *description* of the vertex $\mathbf{v}$. Let us note that if the polyhedral set defined by $\mathbf{Ax} \leq \mathbf{b}$ has a vertex, then we must have $m \geq n$, where m is the number of constraints and n is the number of variables.

Theorem *Every extreme point of the polyhedral set defined by the constraints $\mathbf{Ax} \leq \mathbf{b}$ is a vertex, and conversely.*

15.2 Geometry and algebra

Proof Let $\mathbf{e}$ be an extreme point of the polyhedral set P defined by the constraints $\mathbf{Ax} \leq \mathbf{b}$. Consider the constraints $\mathbf{Ex} \leq \mathbf{b}^E$ that $\mathbf{e}$ satisfies as equalities, $\mathbf{Ee} = \mathbf{b}^E$. These equality constraints define an affine space E. If $\mathbf{e}$ is its only element, we are done. Otherwise, let $\mathbf{Sx} \leq \mathbf{b}^S$ be the remaining constraints; the extreme point $\mathbf{e}$ satisfies the strict form of each of these inequalities. Let U be the open set defined by $\mathbf{Sx} < \mathbf{b}^S$. The intersection $U \cap E$ is an open convex subset of E that is a subset of P and that contains $\mathbf{e}$. Since an open convex set contains no extreme points, it follows that $\mathbf{e}$ is not an extreme point, which is absurd.

For the converse, suppose that $\mathbf{v}$ is a vertex that is the unique solution to the equations $\mathbf{Vx} = \mathbf{b}^V$, where $\mathbf{Vx} \leq \mathbf{b}^V$ is a subsystem of $\mathbf{Ax} \leq \mathbf{b}$. Suppose $\mathbf{v}$ is not an extreme point. Then $\mathbf{v} = \lambda \mathbf{p} + (1 - \lambda) \mathbf{q}$, with $1 > \lambda > 0$ and $\mathbf{p}, \mathbf{q} \in P$. Let $\mathbf{ax} = b$ be an equation among $\mathbf{Vx} = \mathbf{b}^V$ that is not satisfied by $\mathbf{p}$; we can suppose that $\mathbf{ap} < b$. Since $\mathbf{av} = b$, we must have $\mathbf{aq} > b$, which contradicts the fact that $\mathbf{q} \in P$. ∎

Given the algebraic definition of a vertex and given the geometric insight that an optimal solution to a linear programming problem occurs at an extreme point, we have "finitized" the problem of finding the optimal solution to a linear programming problem, when the polyhedral set is bounded. In fact, we have a very naive version of the simplex algorithm: Loop through *all* the definitions of vertices, record the value of the objective function at each vertex, and return the best found. All vertices must be generated if we are to be sure that the optimum has been found. This is a sound and complete algorithm because the number of vertices is finite.

We will show without assuming the polyhedral set to be bounded that this very naive version of the simplex algorithm is sound and complete provided that the polyhedral set has at least one vertex. To that end, we first establish a criterion for a vertex to be optimal that will allow the search for the optimal solution to halt when such a vertex is reached.

Lemma (Optimal Vertex Test) *Consider the linear program $\mathbf{Ax} \leq \mathbf{b}$, max: $\mathbf{cx}$. Let $\mathbf{v}$ be a vertex on the polyhedral set defined by $\mathbf{Ax} \leq \mathbf{b}$ and let $\mathbf{Vx} = \mathbf{b}^V$ be a description of $\mathbf{v}$. Let λ be the solution to the system $\mathbf{y} \mathbf{V} = \mathbf{c}$. Then $\mathbf{v}$ is optimal if $\lambda \geq \mathbf{0}$.*

Proof Let the dimensions of $\mathbf{A}$ be m by n; we have $m \geq n$. Note that $\mathbf{V}$ is an n-by-n nonsingular matrix and λ is an n vector. Extend λ to an m vector $\overline{\lambda}$ by adding 0s for the rows of $\mathbf{A}$ that do not figure in $\mathbf{V}$. Then $\overline{\lambda}$ is a solution to the dual problem. The value of the objective function in the dual problem at $\overline{\lambda}$ is $\overline{\lambda}\mathbf{b} = \lambda\mathbf{b}^V$, and we have $\mathbf{cv} = \lambda\mathbf{Vv} = \lambda\mathbf{b}^V$. Therefore, by the Weak Duality Theorem, the optimal solution of the primal problem is attained at $\mathbf{v}$. ∎

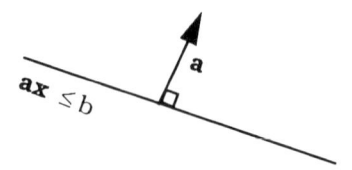

FIGURE 15.5 Exterior normal.

In this situation where $\lambda \geq \mathbf{0}$, we say that the description $\mathbf{Vx} = \mathbf{b}^V$ *certifies optimality* of $\mathbf{v}$ and that $\mathbf{v}$ has *passed* the Optimal Vertex Test. This gives us a test that can serve to improve the naive simplex algorithm: As each vertex is generated, apply the test. If the test is positive, the search can halt. This procedure is sound. In what follows we will show that this procedure is complete.

The geometry that underlies the Optimal Vertex Test is very pretty. For a constraint $\mathbf{ax} \leq b$, the vector $\mathbf{a}$ is perpendicular to the hyperplane $\mathbf{ax} = b$, and it is the normal that points away from the feasible region, as in Figure 15.5. Therefore the Optimal Vertex Test verifies that the vector $\mathbf{c}$ that defines the objective function is a nonnegative linear combination of the normals to the hyperplanes of the constraints that define the vertex. Hence $\mathbf{c}$ is in the cone generated by the normals. In Figure 15.6, we have the feasible region of Figure 15.1. The normal to the level curves of the objective function $c = 15x_1 + 3x_2$ is [15,3]; this vector is captured in the cone of the vectors [1,1] and [1,0], which are the normals to the lines $x_1 + x_2 = 1$ and $x_1 = 4$. Therefore, the optimal value of this objective function is attained at the vertex $(4, 1)$. It is not necessary to look at any other vertices to be assured of this.

A vertex that can be described by more than one set of constraints is said to be *degenerate*. The term degenerate is rather harsh; for exam-

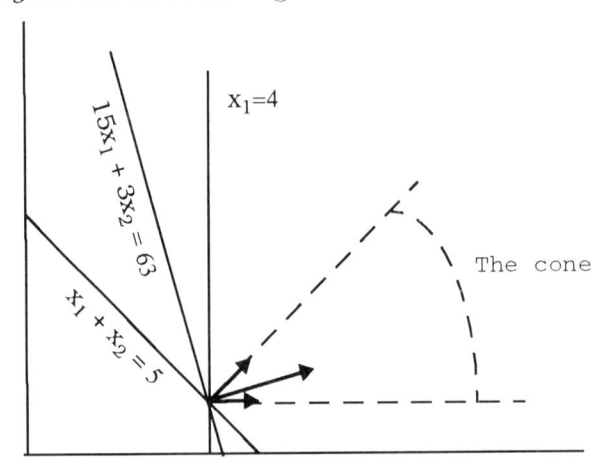

FIGURE 15.6 Capturing the normal to the objective function.

15.2 Geometry and algebra

ple, the apex of the Great Pyramid of Cheops as illustrated in Figure 15.7 is degenerate. This is because only three faces of the pyramid are required to define the apex but four faces meet there, and the apex can be defined four different ways. Another cause of degeneracy is the presence of redundant constraints; an example is the vertex (4,1) of Figure 15.1.

An affine space defined by $n-1$ independent hyperplanes has dimension 1, and so it is a straight line. In what follows, we will be given a line described this way that will be an edge of a polyhedron. Then we will consider movement along this edge.

By way of example, consider the polytope of Figure 15.8. The polytope is defined by the constraints

$$x_1 + x_2 + x_3 \leq 1$$
$$x_1 \geq 0, \ x_2 \geq 0, \ x_3 \geq 0$$

Consider the apex which is described by the equalities

$$x_1 + x_2 + x_3 = 1$$
$$x_1 = 0, \ x_2 = 0$$

As in Figure 15.8 (a), the coordinates of this point are $(0,0,1)$. The edge going from $(0,0,1)$ down to the vertex $(1,0,0)$ is the intersection of the polytope with the line defined by the pair of equality constraints $x_1 + x_2 + x_3 = 1$, $x_2 = 0$. The apex is described by these two equalities plus $x_1 = 0$. Points on the edge down to and through $(0,0,1)$ can be described by these two equalities plus a constraint $x_1 = \eta$, where $0 \leq \eta \leq 1$ as in Figure 15.8 (b). For values of η greater than 1, this point will no longer observe the constraint $x_3 \geq 0$. However, at $\eta = 1$, the constraint $x_1 = \eta$ can be replaced by the equality $x_3 = 0$, and this yields a description of $(1,0,0)$ as a vertex on the polytope as in Figure 15.8 (c).

Suppose that P is a polyhedral set defined by the constraints $\mathbf{Ax} \leq \mathbf{b}$. Let $\mathbf{v}$ be a vertex of P which is described by the constraints $\mathbf{Vx} = \mathbf{b}^V$. A vertex $\mathbf{w}$ is said to be *adjacent* to $\mathbf{v}$ if $\mathbf{w}$ is given by a description which differs from that of $\mathbf{v}$ by one constraint. See Figure 15.8. Note that two

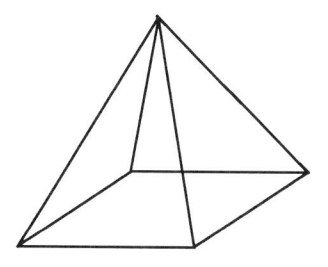

FIGURE 15.7 The Great Pyramid of Cheops.

adjacent degenerate vertices can describe the same extreme point. One more piece of notation will be needed. The vector $\mathbf{e}_i$ denotes the unit vector $[0,\ldots,1,\ldots,0]^T$ with entry 1 in the i th coordinate.

This brings us to the kind of result that no text is ever without.

Theorem (The Fundamental Theorem of Linear Programming) *Let $\mathbf{v}$ be an extreme point on the polyhedral set defined by the constraints of the linear program*

$$\mathbf{Ax} \leq \mathbf{b}$$
$$\max: \mathbf{cx}$$

Then there is a description of $\mathbf{v}$ as a vertex that either (1) certifies optimality or (2) has an adjacent vertex $\mathbf{w}$ such that $\mathbf{cw} > \mathbf{cv}$ or (3) shows that the problem is unbounded.

Proof Let $\mathbf{Vx} = \mathbf{b}^V$ be a description of $\mathbf{v}$ as a vertex. Consider the system of equations $\mathbf{yV} = \mathbf{c}$. Let $\boldsymbol{\lambda}$ be the unique solution. If $\boldsymbol{\lambda} \geq \mathbf{0}$, then the description certifies optimality.

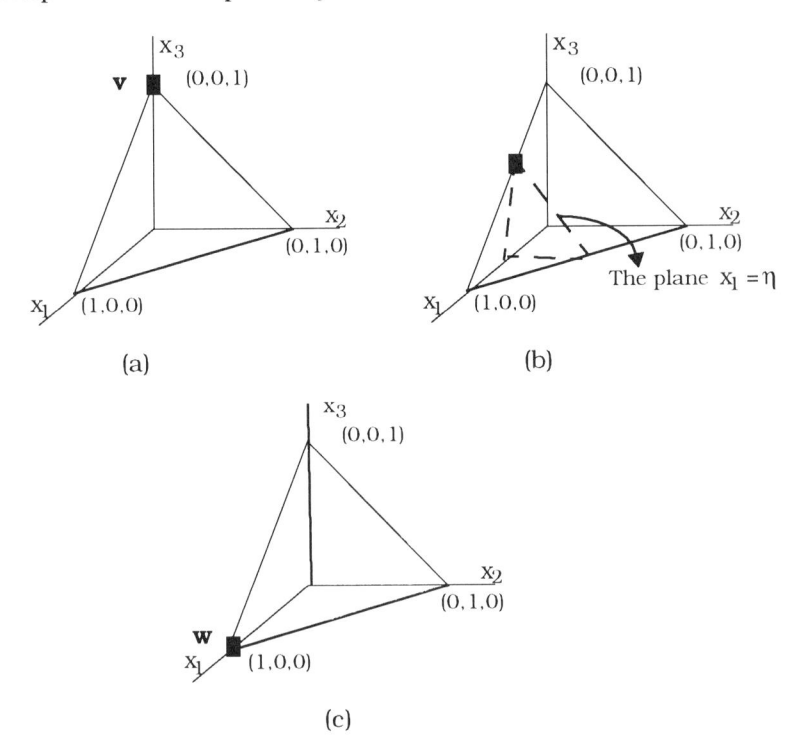

FIGURE 15.8 (a) The vertex $\mathbf{v} = (0,0,1)$; (b) movement along the edge; (c) the vertex $\mathbf{w} = (1,0,0)$.

15.2 Geometry and algebra

For the case where $\lambda_i < 0$ for some i, we first treat the situation where $\mathbf{v}$ is a nondegenerate vertex. Thus we are assuming that $\mathbf{Vx} = \mathbf{b}^V$ is the unique description of $\mathbf{v}$ and that every constraint that is satisfied as an equality by $\mathbf{v}$ is among the $\mathbf{Vx} \leq \mathbf{b}^V$.

Let $\mathbf{V}^i\mathbf{x} = \mathbf{b}_i^V$ be the system of equations obtained by omitting the equality $\mathbf{a}_i\mathbf{x} = b_i$ from $\mathbf{Vx} = \mathbf{b}^V$, and let $\mathbf{Sx} \leq \mathbf{b}^S$ be the constraints that do not figure among the $\mathbf{Vx} \leq \mathbf{b}^V$. The idea will be to move along the line defined by $\mathbf{V}^i\mathbf{x} = \mathbf{b}_i^V$ to witness the fact that $\mathbf{v}$ is not optimal. Since $\mathbf{v}$ is nondegenerate, $\mathbf{v}$ satisfies the strict form of each constraint that does not serve to define it; that is, $\mathbf{v}$ is in the open set U defined by $\mathbf{Sx} < \mathbf{b}^S$. Let $\eta > 0$ be such that the unique solution $\mathbf{x}^*$ to the equations $\mathbf{Vx} = \mathbf{b}^V - \eta\, \mathbf{e}_i$ lies in $U \cap P$. Since $\lambda \mathbf{b}^V = \mathbf{cv}$ and $\lambda = \mathbf{cV}^{-1}$, we have $\mathbf{cx}^* = \mathbf{cV}^{-1}(\mathbf{b}^V - \eta\, \mathbf{e}_i) = \lambda(\mathbf{b}^V - \eta\, \mathbf{e}_i) = \mathbf{cv} - \lambda_i\, \eta > \mathbf{cv}$. Therefore $\mathbf{v}$ is not an optimal solution. Next, consider the set of all such η,

$$E = \{\, \eta > 0 : \text{the solution to } \mathbf{Vx} = \mathbf{b}^V - \eta\, \mathbf{e}_i \text{ lies in } P\, \}$$

If this set is unbounded, the linear programming problem is unbounded. Otherwise, let $\overline{\eta}$ be the largest element in E. Let $\mathbf{w}$ be the point defined by the equations $\mathbf{Vx} = \mathbf{b}^V - \overline{\eta}\, \mathbf{e}_i$, in other words, by the equations $\mathbf{V}^i\mathbf{x} = \mathbf{b}_i^V$, $\mathbf{a}_i\mathbf{x} = b_i - \overline{\eta}$. Clearly $\mathbf{cw} > \mathbf{cv}$. To verify that $\mathbf{w}$ is a vertex, first note that we must have $\mathbf{sw} = b$ for some constraint $\mathbf{sx} \leq b$ from among the $\mathbf{Sx} \leq \mathbf{b}^S$, for otherwise the problem would be unbounded. Next, in order to reach a contradiction, suppose that $\mathbf{s}$ is a linear combination of the rows of $\mathbf{V}^i$. Then if the constraint $\mathbf{sx} = b$ were not a linear combination of the constraints $\mathbf{V}^i\mathbf{x} = \mathbf{b}_i^V$, it would not be consistent with them. So $\mathbf{sx} = b$ is a linear combination of the $\mathbf{V}^i\mathbf{x} = \mathbf{b}_i^V$, and hence $\mathbf{sx} = b$ is satisfied at $\mathbf{v}$. This is absurd. Therefore $\mathbf{w}$ is an adjacent vertex, and $\mathbf{cw} > \mathbf{cv}$.

This completes the proof in the case the vertex is not degenerate. Now to extend to the degenerate case. We can assume that we are in the situation where no description of $\mathbf{v}$ as a vertex certifies optimality. We will need a lemma on closed nowhere dense sets.

Lemma *Let $\{i_1,...,i_n\}$ and $\{j_1,...,j_n\}$ be distinct subsets of $\{1,...,m\}$. Let $\mathbf{A}_1$ be the submatrix of $\mathbf{A}$ consisting of rows $i_1,...,i_n$, and let $\mathbf{A}_2$ be the submatrix of $\mathbf{A}$ consisting of rows $j_1,...,j_n$. Suppose that $\mathbf{A}_1$ and $\mathbf{A}_2$ are nonsingular. For $\mathbf{y} = (y_1,...,y_m)$, let $\mathbf{y}_1 = (y_{i_1},...,y_{i_n})$ and let $\mathbf{y}_2 = (y_{j_1},...,y_{j_n})$. Then the set $E = \{\mathbf{y} : \mathbf{A}_1\,\mathbf{x} = \mathbf{b} + \mathbf{y}_1 \text{ and } \mathbf{A}_2\,\mathbf{x} = \mathbf{b} + \mathbf{y}_2$, for some $\mathbf{x}\}$ is an affine set of dimension strictly less than m and hence is a closed nowhere dense set.*

Proof We will show that $E - \mathbf{b}$ is a linear space of dimension $< m$. Since $\mathbf{A}_1$ and $\mathbf{A}_2$ are invertible, we have $E - \mathbf{b} = \{\mathbf{y} : \mathbf{A}_1\,\mathbf{x} = \mathbf{y}_1, \mathbf{A}_2\,\mathbf{x} = \mathbf{y}_2$, for some $\mathbf{x}\} = \{\mathbf{y} : \mathbf{A}_1^{-1}\,\mathbf{y}_1 = \mathbf{A}_2^{-1}\,\mathbf{y}_2\}$. This set is easily seen to be closed under addition and scalar multiplication. It is a proper subspace since the index sets for $\mathbf{A}_1$ and $\mathbf{A}_2$ are distinct. ∎

Consider the constraints $\mathbf{Vx} \leq \mathbf{b}^V$ that meet at $\mathbf{v}$, and let $\mathbf{Sx} \leq \mathbf{b}^S$ be the remaining constraints. Let $\mathbf{V}_1,\ldots,\mathbf{V}_t$ be the nonsingular submatrices $\mathbf{V}_1,\ldots,\mathbf{V}_t$ of $\mathbf{V}$; each of these provides a description of the vertex $\mathbf{v}$.

For every m vector $\mathbf{y}$, let $\mathbf{y}_i$ denote the n vector obtained by restricting $\mathbf{y}$ to the rows of $\mathbf{V}_i$. For every pair $\mathbf{V}_p, \mathbf{V}_q$ of distinct nonsingular submatrices of $\mathbf{V}$, let $F_{p,q}$ be the affine space $\{\, \mathbf{y} : \mathbf{V}_p \mathbf{x} = \mathbf{b}_p + \mathbf{y}_p,\ \mathbf{V}_q \mathbf{x} = \mathbf{b}_q + \mathbf{y}_q,\ \text{for some } \mathbf{x}\,\}$. By the lemma, each $F_{p,q}$ is a closed nowhere dense set. Since the number of pairs is finite, the union F of the $F_{p,q}$ is a closed nowhere dense set.

Let U be the open set defined by $\mathbf{Sx} < \mathbf{b}^S$. Note that for all sufficiently small $\epsilon > \mathbf{0}$ in $R^m - F$ and for all $\mathbf{V}_i$, all solutions $\mathbf{x}^*$ of

$$\mathbf{V}_i \mathbf{x} = \mathbf{b}^{V_i} + \epsilon^{V_i}$$

lie in U and so satisfy $\mathbf{Sx}^* < \mathbf{b}^S$.

For $\epsilon > \mathbf{0}$ in $R^m - F$, perturb the constraints $\mathbf{Vx} \leq \mathbf{b}^V$ to $\mathbf{Vx} \leq \mathbf{b}^V + \epsilon^V$, leaving the other constraints of the system $\mathbf{Ax} \leq \mathbf{b}$ unchanged. Any vertex described by a subset of the constraints $\mathbf{Vx} \leq \mathbf{b}^V + \epsilon^V$ will be nondegenerate in the perturbed problem; this holds for all ϵ in $R^m - F$.

The idea now is to exploit the fact that there are infinitely many ϵ and only a finite number of descriptions of vertices for the perturbed problems. First let us ensure that at least one of the descriptions of $\mathbf{v}$ continues to describe a vertex in the perturbed problems. In order to guarantee that the point defined by

$$\mathbf{V}_1 \mathbf{x} = \mathbf{b}^{V_1} + \epsilon^{V_1}$$

is a feasible vertex in the perturbed problems under consideration, we will restrict ourselves to an infinite sequence $\epsilon_0, \ldots, \epsilon_k, \ldots$ in $R^m - F$ converging to $\mathbf{0}$. By taking each ϵ_k such that its coordinates corresponding to the rows of $\mathbf{V}_1$ are sufficiently small compared to the entries corresponding to the other rows of $\mathbf{V}$, we can be sure that the vertex described using $\mathbf{V}_1$ satisfies

$$\mathbf{V}_j \mathbf{x} \leq \mathbf{b}^{V_j} + \epsilon_k^{V_j}$$

for $j = 2,\ldots,t$. More precisely, let $\eta_k > \mathbf{0}$ be such that $\eta_k \to \mathbf{0}$ as $k \to \infty$. Let U_k be the non-empty open set consisting of all $\epsilon > \mathbf{0}$ such that $\epsilon < \eta_k$ and

$$\mathbf{V}_1 \mathbf{x} = \mathbf{b}^{V_1} + \epsilon^{V_1} \models \mathbf{V}^{V-V_1} \mathbf{x} < \mathbf{b}^{V-V_1} + \epsilon^{V-V_1}$$

We then choose ϵ_k in $U_k - F$. Therefore, for each ϵ_k we can assume that the point defined by

$$\mathbf{V}_1 \mathbf{x} = \mathbf{b}^{V_1} + \epsilon_k^{V_1}$$

15.2 Geometry and algebra

is a vertex in the perturbed problem. Moreover, for each ϵ_k, let $X(\epsilon_k)$ be the subset of $\{\mathbf{V}_1,...,\mathbf{V}_t\}$ such that j is in $X(\epsilon_k)$ if and only if

$$\mathbf{V}_j \mathbf{x} = \mathbf{b}^{\mathbf{V}_j} + \epsilon_k^{\mathbf{V}_j}$$

describes a feasible vertex in the perturbed problem. By refining the sequence further if necessary, we can assume that the sets $X(\epsilon_k)$ are the same for all k. For simplicity let us suppose that $X(\epsilon_k) = \{\mathbf{V}_1,...,\mathbf{V}_s\}$ for $s \le t$ and let us denote the vertices by $\mathbf{v}_1,...,\mathbf{v}_s$. This brings us to two cases:

Case 1: There is some $\mathbf{v}_j$ and some $\lambda_l < 0$ that witness unboundedness for an infinite subsequence $\epsilon_{r_0},...,\epsilon_{r_k},...$ That is, the point defined by

$$\mathbf{V}_j \mathbf{x} = \mathbf{b}^{\mathbf{V}_j} + \epsilon_{r_k}^{\mathbf{V}_j} - \eta \mathbf{e}_i$$

is feasible for all positive η and all k. Then, letting $k \to \infty$, the point defined by

$$\mathbf{V}_j \mathbf{x} = \mathbf{b}^{\mathbf{V}_j} - \eta \mathbf{e}_i$$

is feasible for all positive η in the original problem which shows that the original problem is unbounded.

Case 2: For some $N \ge 0$ and all $j = 0,...,s$, every vertex $\mathbf{v}_j$ has an adjacent vertex $\mathbf{w}$ such that $\mathbf{cv}_j < \mathbf{cw}$ in the perturbed problem for all ϵ_k, $k \ge N$. There are a finite number of vertices $\mathbf{v}_j$, and with each there is associated a finite number of such adjacent vertices $\mathbf{w}$. So by passing to a subsequence $\epsilon_{r_0},...,\epsilon_{r_k},...$ we can suppose that for each $\mathbf{v}_j$, the same adjacent vertex $\mathbf{w}$ satisfies $\mathbf{cv}_j < \mathbf{cw}$ in the perturbed problem for all $k \ge N$. Denote this map $\mathbf{w} = \varphi(\mathbf{v}_j)$.

We claim that at least one pair $\mathbf{w} = \varphi(\mathbf{v}_j)$ is such that $\mathbf{w}$ has a description that involves a constraint from the set $\mathbf{Sx} \le \mathbf{b}^S$ of those inequalities such that $\mathbf{Sv} < \mathbf{b}^S$. In fact, for each $\mathbf{v}_j$ we have $\mathbf{cv}_j < \mathbf{c}\varphi(\mathbf{v}_j)$ in each perturbed problem. Therefore, starting with any $\mathbf{v}_j$, we have $\mathbf{cv}_j < \mathbf{c}\varphi(\mathbf{v}_j) < \mathbf{c}\varphi(\varphi(\mathbf{v}_j)) < ...$ and so some iteration $\mathbf{w} = \varphi^p(\mathbf{v}_j)$ must be as claimed.

In this final case, let

$$(\mathbf{V}_j)_i \mathbf{x} = b_i^{(\mathbf{V}_j)_i} + \epsilon_{r_k}^{(\mathbf{V}_j)_i}$$
$$\mathbf{sx} = b$$

be the constraints that define the adjacent vertex $\mathbf{w}$, where $\mathbf{sx} = b$ is the new constraint. Recall once again that $\mathbf{sv} < b$. Letting $k \to \infty$, the representation

$$(\mathbf{V}_j)_i \mathbf{x} = b_i^{(\mathbf{V}_j)_i}$$
$$\mathbf{sx} = b$$

of the adjacent vertex certifies the suboptimality of **v** in the original problem. ∎

The Fundamental Theorem has some important corollaries. To start, it enables us to show that the restriction to the first orthant guarantees that a feasible region will always have an extreme point to serve as witness point.

Corollary *Suppose that the constraints* $\mathbf{Ax} \le \mathbf{b}$ *define a polyhedral set P which has a vertex. Let* $\mathbf{cx} \ge d$ *be a new constraint which is consistent with* $\mathbf{Ax} \le \mathbf{b}$; *then the polyhedral set* P_1 *defined by* $\mathbf{Ax} \le \mathbf{b}, \mathbf{cx} \ge d$ *also has a vertex.*

Proof Let **v** be a vertex of P. If **v** satisfies $\mathbf{cx} \ge d$, it is a vertex of P_1 and we are done. Otherwise, **v** is a vertex of the set P_2 defined by $\mathbf{Ax} \le \mathbf{b}, \mathbf{cx} \le d$. By the Fundamental Theorem, the maximum value of the objective function **cx** over P_2 is attained at a vertex **w**. We must have $\mathbf{cw} = d$. Therefore, **w** is a vertex of P_1. ∎

The Fundamental Theorem shows that there is a path from any vertex to an optimal vertex if an optimal solution exists. This path proceeds from vertex to adjacent vertex exchanging one constraint for another each time, a local search *par excellence*. At suboptimal vertices, at least one of the dual multipliers λ_i will be < 0. As the constraint $\mathbf{a}_i\mathbf{x} \le b$ with this multiplier is strengthened to $\mathbf{a}_i\mathbf{x} \le b - \eta$, the objective function increases by $-\eta \lambda_i$.

If the dual multiplier is positive at an optimal vertex, $\lambda_i > 0$, the same analysis shows that if $\mathbf{a}_i\mathbf{x} \le b$ is relaxed to $\mathbf{a}_i\mathbf{x} \le b + \eta$, the increase in the value of the objective function will be $\eta \lambda_i$ so long as $\mathbf{a}_i\mathbf{x} = b + \eta$ is feasible. In the case where the constraint $\mathbf{a}_i\mathbf{x} \le b$ is a bounding constraint on a variable x (a constraint of the form $x \le b$ or $-x \le -b$), the dual multiplier λ_i is called the *reduced cost* of x. Let $\bar{\eta}$ the least upper bound of the set of η such that $x = b + \eta$ (or $-x = -b + \eta$) is feasible, provided this set is bounded; otherwise let $\bar{\eta}$ be $+\infty$. Then $\bar{\eta}$ is the *range* of x. For $\eta < \bar{\eta}$, λ_i is the shadow price of x at b. This is where the functions `rc` and `range` come from.

If the constraint $\mathbf{a}_i\mathbf{x} \le b$ does not serve to describe the optimal vertex, then its dual multiplier is 0. Relaxing the constraint to $\mathbf{a}_i\mathbf{x} \le b + \eta$ will not change the optimal vertex. So in the case where this applies to both the upper and lower bounds on a variable x, the reduced cost of x is 0, and this has infinite range.

Let us turn to the Parametric Analysis Theorem that is used extensively in Chapter 7. A proof can be made using the Duality Theorem and the fact that the optimal solution of the dual problem must occur at a vertex.

15.2 Geometry and algebra

Theorem (Parametric Analysis Theorem) *Suppose that F is a bounded feasible region defined by a consistent set of constraints $\mathbf{A'x} \leq \mathbf{b'}$. Let the variable x and the bound $b_0 \geq 0$ be such that the constraints $\mathbf{A'x} \leq \mathbf{b'}$, $x \leq b_0$ are consistent. For $b \geq b_0$, let $z^*_x(b)$ be the value of the optimal solution to the linear programming problem*

$$\mathbf{A'x} \leq \mathbf{b'}$$
$$x \leq b$$
$$\max: \mathbf{cx}$$

Then z^*_x is a continuous, piecewise linear, concave down, nondecreasing function on the interval $[b_0, \infty)$.

Proof Let $[a, a']$ be the largest closed interval such that for all b in $[a, a']$ the constraints $x = b$, $\mathbf{A'x} \leq \mathbf{b'}$ are consistent. Denote the set of constraints $x \leq b$, $\mathbf{A'x} \leq \mathbf{b'}$ by $\mathbf{Ax} \leq \mathbf{b}$. We have $\mathbf{b} = [b, b_1, ..., b_m]^T$ where $\mathbf{b'} = [b_1, ..., b_m]^T$. For each b in $[a, a']$, $z^*_x(b)$ denotes the maximum value of $\mathbf{cx}$ subject to $\mathbf{Ax} \leq \mathbf{b}$. We want to show that $z^*_x(b)$ is continuous, piecewise linear, and concave down on $[a, a']$. To that end, we consider the dual problems $\mathbf{y} \geq \mathbf{0}$, $\mathbf{yA} = \mathbf{c}$, min: $\mathbf{yb}$. For all b in $[a, a']$, these problems have the same feasible region. By the Corollary to the Fundamental Theorem, this region has at least one vertex. Let $\mathbf{v}_1, ..., \mathbf{v}_s$ be the vertices of the common feasible region of these dual problems. As b varies and $b_1, ..., b_m$ stay fixed, the value of the dual objective function at a vertex $\mathbf{v}_i$ is a linear function of b. Denote these functions $f_1, ..., f_s$. By the Duality Theorem, the optimum function of the primal problem at b is $z^*_x(b) = \min \{ f_1(b), ..., f_s(b) \}$. It is easy to check that the infimum of a finite set of linear functions is piecewise linear, continuous, and concave down. ∎

For a fuller discussion of this theorem and parametric analysis in general, see Franklin (1980). Let us turn to the analysis made in terms of the coefficients in the objective function. As can be seen from the above discussion, the basic Parametric Analysis Theorem holds for the analysis of change in the right-hand-side of multivariable constraints as well as on bounding constraints. In the passage from the primal to the dual linear program, the coefficients in the objective function get mapped to right-handside terms. It is for this reason that the machinery of sensitivity and parametric analysis applies to these objective function coefficients. This yields the Parametric Objective Coefficient Analysis Theorem. From the mathematics one extracts the functions `objcup` and `objcdn`. For a fuller discussion of the theorem from this side, see Chvátal (1983).

Finally, let us discuss the wonderful property of network matrices which guarantees that a vertex solution to a network problem is integral. A matrix is said to be *totally unimodular* if every nonsingular square submatrix has determinant +1 or -1. It is not hard to show that network matrices have this property: the proof proceeds by induction on the size of the submatrices. Extending a network matrix with rows for nonnegativity constraints $-\mathbf{x} \leq \mathbf{0}$ preserves total unimodularity. So, if a vertex in a network problem is defined by the equations $\mathbf{V}\mathbf{x} = \mathbf{b}^V$, with $\mathbf{b}$ integral, the solution must be integral as is easily seen by applying Cramer's Rule.

For further reading, see Rockefellar (1970) and Schrijver (1986).

Exercises

15.2.1. Let f be a function defined on a set X. A point $\mathbf{p}$ is said to be a *global maximum* for f if $f(\mathbf{x}) < f(\mathbf{p})$ for all $\mathbf{x} \in X$. A point $\mathbf{p}$ is said to be a *local maximum* for f if there is an open set U such that $\mathbf{p} \in U$ and such that $f(\mathbf{x}) < f(\mathbf{p})$ for all $\mathbf{x} \in U \cap X$. Assume X is a convex set. Prove the following: If a point $\mathbf{p}$ is a local maximum for the linear objective function $\mathbf{c}\mathbf{x}$, then $\mathbf{p}$ is a global maximum for $\mathbf{c}\mathbf{x}$ on X.

15.2.2. Prove that a network matrix is totally unimodular.

15.2.3. Prove that the cone generated by a finite number of vectors is a polyhedral set.

15.2.4. Prove that if $\mathbf{q}$ is not in the cone generated by $\mathbf{p}_1,\ldots,\mathbf{p}_m$, then there is a hyperplane $\mathbf{a}\mathbf{x} = 0$ such that $\mathbf{a}\mathbf{p}_i > 0$ and $\mathbf{a}\mathbf{q} < 0$.

15.2.5. The proof of the Fundamental Theorem only uses the Weak Duality Theorem. Derive the Duality Theorem from the Fundamental Theorem under the assumption that the region defined by $\mathbf{A}\mathbf{x} \leq \mathbf{b}$ has at least one vertex.

15.2.6. Prove the Baire Category Theorem: A nonempty open subset of R^n cannot be covered by a countable union of closed nowhere dense sets.

15.3 The revised simplex method

Finally, we come to the computational issues of the simplex method. The method in its modern form was pioneered by Dantzig (1951). The revised simplex method dates from Dantzig (1953).

A full-dimensional polytope in R^n with $n + 1$ extreme points is called a *simplex*. Thus in two dimensions, a simplex is a triangle and in three dimensions it is a tetrahedron. A simplex is the basic full-dimensional polytope. This is the starting point for the intuition from which the method got its name; for insight and discussion, see the classic Dantzig (1963).

For computational purposes it is simplest to maintain constraints in *standard form*:

15.3 The revised simplex method

$$Ax = b$$
$$x \geq 0$$
$$\max: cx$$

where A is m-by-n with $m < n$.

In the simplex algorithm the set of inequality constraints

$$a_{11}x_1 + \ldots + a_{1k}x_k \leq b_1$$
$$\vdots$$
$$a_{m1}x_1 + \ldots + a_{mk}x_k \leq b_m$$
$$x_1 \geq 0, \ldots, x_k \geq 0$$

is transformed by adding slack variables $x_{k+1}, \ldots, x_{k+m}$ into

$$a_{11}x_1 + \ldots + a_{1k}x_k + x_{k+1} = b_1$$
$$\vdots$$
$$a_{m1}x_1 + \ldots + a_{mk}x_k + x_{k+m} = b_m$$
$$x_1 \geq 0, \ldots, x_{k+m} \geq 0$$

which is in standard form with $n = m + k$.

With this transformation, the m columns of the matrix A that correspond to the slack variables are linearly independent vectors. Hence there is a set of columns of A that form a basis for R^m. Another way of stating this property is that the rows of the matrix A are linearly independent. These will be our working assumptions: $m < n$ and A has full row rank. However, it will not be necessary to assume that these conditions came about by the introduction of slack variables.

In the standard form, there are m equality constraints which define an affine subspace of R^n of dimension m. The n nonnegativity constraints restrict the feasible region to the first orthant. These constraints together define a polyhedral set in R^n. Vertices are formed as the affine space enters or exits the first orthant. Hence there is a natural class of definitions of vertices to consider: A *basic feasible solution* is defined to be a set of $n - m$ variables $x_{i_1}, \ldots, x_{i_{n-m}}$ such that the system

$$a_{11}x_1 + \ldots + a_{1n}x_n = b_1$$
$$\vdots$$
$$a_{m1}x_1 + \ldots + a_{mn}x_n = b_m$$
$$x_{i_1} = 0, \ldots, x_{i_{n-m}} = 0$$

has a unique solution that lies in the feasible region. Thus a basic feasible solution is a description of an extreme point of the feasible region.

Let us suppose that we have a basic feasible solution defined by $x_{i_1}, \ldots, x_{i_{n-m}}$. These variables are called *nonbasic*, and the remaining variables $x_{j_1}, \ldots, x_{j_m}$ are called *basic*. The columns of the matrix **A** that correspond to basic variables are called *basic columns*, and similarly for the nonbasic case. Note that the square matrix $\mathbf{A}_B$ obtained by restricting **A** to the basic columns is invertible, since the basic feasible solution has a unique solution to its set of equations. Differently put, the basic columns are linearly independent vectors that form a basis for R^m.

It can be shown that every vertex of the polyhedral set is defined by a basic feasible solution. Let us sketch a proof. Given a vertex $\mathbf{e} = (x_1^*, \ldots, x_n^*)$, one first shows that the columns of **A** that correspond to nonzero x^*_i are linearly independent. In fact, let $D = \{ j : x_j^* > 0 \}$, and denote the restriction of **A** to these columns by $\mathbf{A}_D$. The set D has at most m elements. Towards a contradiction, suppose that these columns are not linearly independent. Then the system of equations $\mathbf{A}_D \mathbf{u} = \mathbf{0}$ has a nonzero solution $\overline{\eta}$. For $j \in D$, let η_j be equal to the corresponding entry in $\overline{\eta}$; for $j \notin D$, let $\eta_j = 0$. Let η denote the vector $[\eta_1, \ldots, \eta_n]^T$. We have $\mathbf{A}\eta = \mathbf{0}$. Let τ be a sufficiently small positive number such that $x_j^* - \tau \eta_j > 0$ for all $j \in D$. Consider the distinct points $\mathbf{p} = \mathbf{e} + \tau \eta$ and $\mathbf{q} = \mathbf{e} - \tau \eta$. These points are in the feasible region because $\mathbf{Ap} = \mathbf{Aq} = \mathbf{b}$. Since **e** can be expressed as $.5\mathbf{p} + .5\mathbf{q}$, this contradicts the fact that **e** is an extreme point. We have shown that the columns of D are linearly independent. Now one can extend the columns of D to a set B of m linearly independent columns of **A**, and **e** is described by this basic feasible solution.

Note that, if the extreme point **e** is not degenerate, then its description by means of a basic feasible solution is its only definition. This means that without degeneracy, the notions of vertex and basic feasible solution coincide. As we will see, even in the presence of degeneracy, the class of basic feasible solutions will suffice to run the algorithm. Moreover, a feature of the basic feasible solution representation is that it provides a natural sufficient condition for nondegeneracy: An extreme point is nondegenerate if its representation as a basic feasible solution has no basic variable at 0.

The goal in the simplex algorithm is to find a basic feasible solution that passes the Optimal Vertex Test or, failing that, to detect that the objective function is unbounded on the feasible region. An equality constraint $a_{i1}x_1 + \ldots + a_{in}x_n = b_i$ can be considered as a pair of inequalities, $a_{i1}x_1 + \ldots + a_{in}x_n \leq b_i$ and $-a_{i1}x_1 + \ldots + -a_{in}x_n \leq -b_i$. Therefore, the equality constraints can take positive, negative or zero dual multipliers at an optimal vertex. Only the bounding constraints need have nonnegative dual multipliers. Let us suppose that we are given a basic feasible solution to start with; we will address the issue of getting to this position presently.

If N denotes the set of nonbasic variables, let $\mathbf{A}_N$ denote the restriction of **A** to the nonbasic columns. If B denotes the set of basic variables, we can separate the basic variables from the nonbasic variables and rewrite the constraints $\mathbf{Ax} = \mathbf{b}$ as

15.3 The revised simplex method

$$\mathbf{A_B x_B} + \mathbf{A_N x_N} = \mathbf{b}$$

or

$$\mathbf{A_B x_B} = \mathbf{b} - \mathbf{A_N x_N}$$

Since the matrix $\mathbf{A_B}$ is invertible, we have a matrix equation that describes the basic variables $\mathbf{x_B}$ in terms of the nonbasic variables $\mathbf{x_N}$:

$$\mathbf{x_B} = (\mathbf{A_B}^{-1})\mathbf{b} - (\mathbf{A_B}^{-1})\mathbf{A_N x_N}$$

or

$$\mathbf{x_B} = \mathbf{A_B}^{-1}(\mathbf{b} - \mathbf{A_N x_N})$$

When the problem is rewritten this way, it is called the *reduced problem*, for we have reduced the number of variables by eliminating the basic variables. The objective function can also be expressed algebraically in the reduced problem. If the objective function is $\mathbf{cx} = c_1 x_1 + \ldots + c_n x_n$, we denote the restriction of $\mathbf{c}$ to the basic variables by $\mathbf{c_B}$. Then the objective function value is equal to

$$\mathbf{c_B x_B} + \mathbf{c_N x_N}$$

or

$$\mathbf{c_B}(\mathbf{A_B}^{-1})\mathbf{b} + (\mathbf{c_N} - \mathbf{c_B A_B}^{-1} \mathbf{A_N})\mathbf{x_N}$$

where $\mathbf{c_N}$ denotes the restriction of $\mathbf{c}$ to the nonbasic variables.

To test if a basic feasible solution is the maximal vertex for the objective function $\mathbf{cx} = c_1 x_1 + \ldots + c_n x_n$, we must find dual multipliers that satisfy the Optimal Vertex Test. The bounding constraints all have the form $-x \leq 0$. Those x_j with j in the basis do not serve to define the vertex and must have dual multiplier 0. So one is led to consider the system of equations

$$\mathbf{y A_B} = \mathbf{c_B}$$

The solution to this system is denoted π. At an optimal solution to the linear program the dual multipliers of the bounding constraints for the basic variables are all zero. The components of π must be the dual multipliers of the equality constraints $\mathbf{Ax} = \mathbf{b}$ in order to have multiples of the left hand sides of the constraints sum to $\mathbf{c}$. For each nonbasic variable x_j, the dual multipliers π applied to the x_j column $\mathbf{a}_j$ of the matrix $\mathbf{A}$ will sum to $\pi \mathbf{a}_j$, the scalar product of π with $\mathbf{a}_j$. If x_j is nonbasic, we have the freedom to assign any positive multiplier λ_j to the bounding constraint $-x_j \leq 0$ in order to have $\pi \mathbf{a}_j - \lambda_j = c_j$, where c_j is the coefficient of x_j in the objective function. For this to be possible, it suffices that $\pi \mathbf{a}_j - c_j \geq 0$ hold, for then we can take $\lambda_j = \pi \mathbf{a}_j - c_j$. Therefore, if all the $\pi \mathbf{a}_j - c_j$ are nonnegative, the vertex passes the Optimal Vertex Test. Moreover in this case the values $\pi \mathbf{a}_j - c_j$ are the reduced costs of the nonbasic variables; the basic variables have reduced cost zero, since their bounding constraints do not serve to define the optimal vertex. If the vertex is not optimal, the algorithm must move on to another basic feasible solution.

At a basic feasible solution the $\mathbf{x}_N$ are all 0, and so the values of the basic variables at this vertex are equal to $\mathbf{A}_B^{-1}\mathbf{b}$ and the value of the objective function is $\mathbf{c}_B(\mathbf{A}_B^{-1})\mathbf{b}$. If the basic feasible solution passes the Optimal Vertex Test, the optimal solution can, therefore, be read off easily.

When the current vertex is not optimal, the simplex algorithm moves to a vertex that differs from the current one by exactly one bounding constraint. Geometrically this corresponds to moving to an adjacent vertex. To do this, a nonbasic variable and a basic variable will change places.

This brings us to the subject of *pricing*. First we must select a new variable to come into the basis. The simplest criterion is to select the one with highest value $\pi \mathbf{a}_j - c_j$. This is known as *reduced cost pricing*. The term comes from the fact that we are working with the reduced problem and from its interpretation in economic models. Having selected a column, say j, to enter the basis, we must find a basic column for it to replace. Recall that the values of the basic variables are determined by the values of the nonbasic variables, each of which is set to 0. In order to improve the objective function, we want the entering variable to take a nonzero value. When the variable to enter the basis "leaves" its current value of 0, this will cause a change in all the basic variables. Now the situation comes down to a collection of single linear equations in one variable: Keeping all the nonbasic variables fixed at 0 except for the entering variable parametrizes each of the basic variables as a function of the entering variable via the matrix equation

$$\mathbf{x}_B = \mathbf{A}_B^{-1}(\mathbf{b} - \mathbf{A}_N \mathbf{x}_N)$$

This matrix equation describes a system of linear equations for the "dependent" variables $\mathbf{x}_B$ in terms of the "independent variables" $\mathbf{x}_N$. To reiterate, fixing all the $\mathbf{x}_N$ at 0 except for the incoming variable x_j defines each basic variable as a linear function of this one independent variable.

Let us write these functions as $y_k = a_t x_j + b_k$. To start, $x_j = 0$ and $y_k \geq 0$. As x_j increases, if y_k increases, x_j cannot replace y_k in the basis; if y_k decreases to 0, let τ_k be the solution to $0 = ax_j + b_k$. *If no variable decreases, the objective function is unbounded.* Otherwise, the variable to be replaced with the incoming variable is one with minimal τ_k. Let us denote its index by kk. This switch will preserve feasibility, and the new basic columns will indeed form a basis. The first point is clear. For the second, note that for each k, τ_k is the coefficient of $\mathbf{a}_k$ when the incoming vector $\mathbf{a}_j$ is written as a linear combination of the current basis vectors;

15.3 The revised simplex method

since τ_{kk} must be nonzero, $\mathbf{a}_j$ and $\mathbf{a}_{kk}$ can be exchanged. (This is an application of the Exchange Property; see the exercises.)

When the switch is decided upon, some bookkeeping must be done. The lists of nonbasic and basic columns have to be updated, the exiting variable must be set to 0 and all the basic variables must be reset to their new values. This is straightforward.

We have ourselves a search problem. We have to go through basic feasible solutions, passing from one to an adjacent one until an optimal vertex is found or unboundedness is detected. For this search various hill-climbing techniques are used that exploit different pricing strategies. Reduced cost pricing uses the heuristic that for equal increases in the value of the incoming variable, this choice will increase the objective function value the most. However, another choice might actually increase the value of the objective function more if it can enter the basis with a higher value. For more, much more, on pricing strategies, we refer the reader to Nazareth (1987) and to Forrest and Goldfarb (1992).

A step in the simplex algorithm going from one basic feasible solution to the next is called an *iteration* or a *pivot*. All the usual pricing strategies produce an iteration that does not decrease the value of the objective function. If the value of the objective function is strictly increased at each iteration, then we know that the search must terminate since there are a finite number of basic feasible solutions. So the only possible gap is that the hill climb can fall into a cycle of iterations at a degenerate vertex without increasing the value of the objective function. In this situation, geometrically the extreme point is not changing, but its description as a vertex in terms of the constraints of the system is changing. The simplex algorithm works at the latter level.

Cycling is rare, but it can and does happen. There are both theoretical and computational responses to this kind of problem (e.g., see Nazareth 1987). At the computational level, sometimes the accumulated roundoff error of floating point arithmetic will perturb or randomize things enough to cause the search to break out of the cycle. Prudent code will detect that the pricing strategy is not improving the objective function after a certain number of iterations and the code will take appropriate action until the cycle is exited.

Deliberately perturbing things as we did in the proof of the Fundamental Theorem is an idea for avoiding cycling that goes back to Charnes (1952). In the context of vertices represented as basic feasible solutions, degeneracy can manifest itself only if the right hand side vector $\mathbf{b}$ is a linear combination of fewer than m columns of $\mathbf{A}$. Each subspace generated by fewer than m columns is closed and nowhere dense in R^m. Hence arbitrarily small ϵ can be found such that $\mathbf{b} + \epsilon$ lies outside of all these subspaces where $\epsilon = (\epsilon_1, \epsilon_2, \ldots, \epsilon_m)$. If slacks are used to get to standard form, $\mathbf{Ax} = \mathbf{b} + \epsilon$ will be feasible. Algorithmically one does not introduce infinitesimal values for the ϵ_k but simply treats them as formal variables. This makes for much less work than the proof of the

Fundamental Theorem would seem to call for. First note that the formula to decide the exiting variable in the perturbed problem is based on

$$\mathbf{x}_B = (\mathbf{A}_B^{-1})(\mathbf{b} + \boldsymbol{\epsilon}) - (\mathbf{A}_B^{-1})\mathbf{A}_N\mathbf{x}_N$$

or

$$\mathbf{x}_B = (\mathbf{A}_B^{-1})\boldsymbol{\epsilon} + (\mathbf{A}_B^{-1})\mathbf{b} + (\mathbf{A}_B^{-1})\mathbf{A}_N\mathbf{x}_N$$

The expressions to determine the exiting variable will have the form $a + a_1\epsilon_1 + \ldots + a_m\epsilon_m$ and can be identified with the vector $(a, a_1, \ldots, a_m)$. By applying the lexicographic order to these vectors, one can decide the exiting variable. Similarly, when the pivot is made, the new value of the objective function can be written as $\bar{a} + \bar{a}_1\epsilon_1 + \ldots + \bar{a}_m\epsilon_m$ or as the vector $(\bar{a}, \bar{a}_1, \ldots, \bar{a}_m)$. The critical observation is that with each iteration the value of the vector $(\bar{a}, \bar{a}_1, \ldots, \bar{a}_m)$ increases lexicographically. At some point we must have an optimal vertex or a vertex that breaks out of the cycle. This is the basis of the *lexicographic rule* to prevent cycling of Dantzig, Orden and Wolfe (1955). For a full discussion, see Chvátal (1983) and Saigal (1995).

The simplest strategy to describe for avoiding cycling is *Bland's Rule*: Always choose the variable of least index to enter the basis and always choose the variable of least index to exit the basis. This anticycling rule is a product of the remarkable combinatorics of the simplex method, which blends algebra and geometry in a very special way; these combinatorics are captured abstractly in the study of oriented matroids (See Bland 1977a,b).

This completes the description of the basic version of the revised simplex method. Nothing significant needs to be changed in the above discussion if the lower bounds on the variables are assumed to be other than 0, as supporting other lower bounds is straightforward. In practice, variables also have finite upper bounds, and these constraints are not maintained as part of the matrix **A** but rather are treated the same as the lower bound constraints with some small differences: An entering variable that is at its upper bound will decrease in value to enter the basis, and an exiting variable can increase to its (finite) upper bound, as well as decrease to its lower bound. The pricing mechanism must take also these changes into account. In particular, if a nonbasic variable is at its upper bound, it is a candidate to enter the basis if and only if its reduced cost is positive.

We launched into this discussion *in medias res*. Now let us address the question of starting things off. If the origin satisfies the constraints, then slack variables can form the basis and all the other variables can be initialized to 0. This will happen if all the constraints are resource constraints, for example. In more complex situations, one classic technique is to use *artificial variables*; this method is described in all texts on linear programming; e.g. see Chvátal (1983). Another is to proceed

15.3 The revised simplex method

incrementally, one constraint at a time. Since the incremental approach does not require any further definitions, let us apply Occam's Razor and choose to sketch how this technique works.

Given

$$a_{11}x_1 + \ldots + a_{1n}x_n + x_n = b_1$$
$$\vdots$$
$$a_{m\text{-}1}x_1 + \ldots + a_{m\text{-}1n}x_n + x_n = b_{m\text{-}1}$$
$$x_1 \geq 0, \ldots, x_n \geq 0$$

with basis B, suppose that

$$a_{m1}x_1 + \ldots + a_{mn}x_n + x_{n+1} = b_m$$

is a new constraint with slack x_{n+1}. We can form a new basis by adding x_{n+1} to B. If the matrix equation $\mathbf{x}_B = \mathbf{A}_B^{-1}(\mathbf{b} - \mathbf{A}_N\mathbf{x}_N)$ gives x_{n+1} a nonnegative value, we have a basic feasible solution. If x_{n+1} is given a negative value, then change its lower bound to that value and make its upper bound be 0; we now have a basic feasible solution for this modified set of constraints. To get back to the original constraints, maximize x_{n+1}. If x_{n+1} reaches 0, the constraints are consistent and the lower bound on x_{n+1} can be restored to 0 and the upper bound freed; if x_{n+1} fails to reach 0, the new constraint is not feasible. This is a straightforward and simple approach; however, it is not always the approach of choice. Finding a "good" initial vertex can be challenging and important (see Bixby 1992).

Let us run through a small example. Consider the maximization problem

$$x \geq 0, \ y \geq 0$$
$$x \leq 2, \ y \leq 3$$
$$x + y \geq 1$$
$$.5\,x + y \leq 3$$
$$\max: 6x + y$$

To start, the lower bounds of x and y are both 0, and one sets x and y at 0. The upper bounds are 2 and 3. Both variables are nonbasic. To address the multivariable constraint $x + y \geq 1$, one introduces a slack variable s and writes the constraint as

$$-x + -y + s = -1, \ s \geq 0$$

Then s is placed in the basis with value -1. This yields a point (0,0,-1) that is not feasible. So the lower bound of s is changed temporarily to -1 and its upper bound to 0 in order to try to move s up to 0. The matrix $\mathbf{A}$ is $\begin{bmatrix}-1 & -1 & 1\end{bmatrix}$, the column $\mathbf{b}$ is $\begin{bmatrix}-1\end{bmatrix}$. We seek to maximize s. The matrix $\mathbf{A}_B$ is simply $\begin{bmatrix}1\end{bmatrix}$, and so the inverse $\mathbf{A}_B^{-1}$ is also $\begin{bmatrix}1\end{bmatrix}$.

Since we are maximizing s, we have $\mathbf{c}_B = \begin{bmatrix} 1 \end{bmatrix}$. Solving $\pi = \mathbf{c}_B \mathbf{A}_B^{-1}$, we have $\pi = \begin{bmatrix} 1 \end{bmatrix} \begin{bmatrix} 1 \end{bmatrix} = \begin{bmatrix} 1 \end{bmatrix}$. The reduced costs of x and y are now computed. For x, the reduced cost is $\begin{bmatrix} 1 \end{bmatrix} \begin{bmatrix} -1 \end{bmatrix} - 0 = -1$. For y, it is the same. So both are candidates to enter the basis and both have the same reduced cost. Let us try y. The equation from the system $\mathbf{x}_B = \mathbf{A}_B^{-1}(\mathbf{b} - \mathbf{A}_N \mathbf{x}_N)$ computing s as a function of x and y is

$$[s] = [1]\left([-1] - [-1\ -1]\begin{bmatrix} x \\ y \end{bmatrix}\right)$$

which reduces, as one would expect, to

$$s = -1 + x + y$$

Then writing s as a linear function of y with x at 0, one has

$$s = -1 + y$$

and we solve for the (temporary) upper bound of s

$$0 = -1 + y$$

to find $y = 1$. So y enters the basis at 1 and s exits at 0. The lower bound of s is reset to 0 and the upper bound is freed.

Now one treats the constraint $.5\,x + y \le 3$ by introducing a slack variable t to form $.5\,x + y + t = 3$ and one places t along with y in the basis. The value of t is $3 - .5*0 - 1*1 = 2$. This is nonnegative, so we have a basic feasible solution.

We want to maximize $6x + y$. The matrix $\mathbf{A}_B$ and its inverse $\mathbf{A}_B^{-1}$ are

$$\mathbf{A}_B = \begin{bmatrix} -1 & 0 \\ 1 & 1 \end{bmatrix};\quad \mathbf{A}_B^{-1} = \begin{bmatrix} -1 & 0 \\ 1 & 1 \end{bmatrix}$$

The vector $\mathbf{c}_B$ is $[1\ 0]$. Solving $\pi = \mathbf{c}_B \mathbf{A}_B^{-1}$, we find $\pi = [-1\ 0]$. Computing the reduced costs $\pi \mathbf{a}_j - c_j$ of the nonbasic variables we find that x is

$$[-1\ 0]\begin{bmatrix} -1 \\ 0.5 \end{bmatrix} - 6 = -5$$

and that s is

$$[-1\ 0]\begin{bmatrix} 1 \\ 0 \end{bmatrix} = -1$$

15.3 The revised simplex method

So with reduced cost pricing, we take x as our candidate to enter the basis. Expressing the basis variables as functions of x and s via the matrix equation $\mathbf{x}_B = \mathbf{A}_B^{-1}(\mathbf{b} - \mathbf{A}_N \mathbf{x}_N)$, we have

$$\begin{bmatrix} y \\ t \end{bmatrix} = \begin{bmatrix} -1 & 0 \\ 1 & 1 \end{bmatrix} \left(\begin{bmatrix} -1 \\ 3 \end{bmatrix} - \begin{bmatrix} -1 & 1 \\ 0.5 & 0 \end{bmatrix} \begin{bmatrix} x \\ s \end{bmatrix} \right) = \begin{bmatrix} -x + s + 1 \\ 0.5x - s + 2 \end{bmatrix}$$

Setting s to its value 0, we find that

$$y = -x + 1$$
$$t = .5x + 2$$

As x increases y decreases, and so we set y to its lower bound and solve $0 = -x + 1$ to find $x = 1$. As x increases, t also increases. The slack variable t has no upper bound. Thus x can increase to its upper bound of 2 without violating the upper bound on t, but this would violate the lower bound on y. Therefore x can enter the basis at 1 and y can exit at 0. The nonbasic variables are now $y = 0$, $s = 0$. We recompute the values of the basis variables as $x = 1$ and $t = 2.5$.

At this point the matrix $\mathbf{A}_B$ and its inverse $\mathbf{A}_B^{-1}$ are

$$\mathbf{A}_B = \begin{bmatrix} -1 & 0 \\ 0.5 & 1 \end{bmatrix}; \quad \mathbf{A}_B^{-1} = \begin{bmatrix} -1 & 0 \\ 0.5 & 1 \end{bmatrix}$$

The vector $\mathbf{c}_B$ is [6 0]. Solving $\pi = \mathbf{c}_B \mathbf{A}_B^{-1}$, we find that $\pi = [-6\ 0]$. Computing the reduced costs $\pi \mathbf{a}_j - c_j$ of the nonbasic variables, we find that the reduced cost of y is

$$\begin{bmatrix} -6 & 0 \end{bmatrix} \begin{bmatrix} -1 \\ 1 \end{bmatrix} - 1 = 5$$

and for s we have

$$\begin{bmatrix} -6 & 0 \end{bmatrix} \begin{bmatrix} 1 \\ 0 \end{bmatrix} - 0 = -6$$

So we have s as our candidate to enter the basis. Expressing the basis variables as functions of y and s via $\mathbf{x}_B = \mathbf{A}_B^{-1}(\mathbf{b} - \mathbf{A}_N \mathbf{x}_N)$,

$$\begin{bmatrix} x \\ t \end{bmatrix} = \begin{bmatrix} -1 & 0 \\ 0.5 & 1 \end{bmatrix} \left(\begin{bmatrix} -1 \\ 3 \end{bmatrix} - \begin{bmatrix} -1 & 1 \\ 1 & 0 \end{bmatrix} \begin{bmatrix} y \\ s \end{bmatrix} \right) = \begin{bmatrix} -y + s + 1 \\ (-0.5)y - 0.5s + 2.5 \end{bmatrix}$$

Setting y to its value 0, we find that

$$x = s + 1$$
$$t = -.5s + 2.5$$

As s increases, x increases. Thus we set x to its upper bound and solve $2 = s + 1$ to find $s = 1$. As s increases, t decreases. Setting the slack variable t to 0, we solve $0 = -.5s + 2.5$. So s can increase to 5 without violating the lower bound on t. The upper bound on x is reached earlier at $s = 1$. Therefore s can enter the basis at 1, and x can exit at 2. The nonbasic variables are now $x = 2$, $y = 0$. We recompute the values of the basic variables as $s = 1$ and $t = 2$.

The matrix $\mathbf{A}_B$ and its inverse $\mathbf{A}_B^{-1}$ are now

$$\mathbf{A}_B = \begin{bmatrix} 1 & 0 \\ 0 & 1 \end{bmatrix}; \quad \mathbf{A}_B^{-1} = \begin{bmatrix} 1 & 0 \\ 0 & 1 \end{bmatrix}$$

The vector $\mathbf{c}_B$ is [0 0]. Solving $\pi = \mathbf{c}_B \mathbf{A}_B^{-1}$, we find that $\pi = [0\ 0]$. Computing the reduced costs $\pi \mathbf{a}_j - c_j$ of the nonbasic variables, we find that the reduced cost of x is

$$\begin{bmatrix} 0 & 0 \end{bmatrix} \begin{bmatrix} 1 \\ 0 \end{bmatrix} - 6 = -6$$

and for y we have

$$\begin{bmatrix} 0 & 0 \end{bmatrix} \begin{bmatrix} 0 \\ 1 \end{bmatrix} - 1 = -1$$

Since x is at its upper bound, it is not a candidate to enter the basis. We must select y. Expressing the basis variables as functions of x and y by means of the matrix equation $\mathbf{x}_B = (\mathbf{A}_B^{-1})\mathbf{b} - (\mathbf{A}_B^{-1})\mathbf{A}_N\mathbf{x}_N$, we find

$$\begin{bmatrix} s \\ t \end{bmatrix} = \begin{bmatrix} 1 & 0 \\ 0 & 1 \end{bmatrix} \left(\begin{bmatrix} -1 \\ 3 \end{bmatrix} - \begin{bmatrix} -1 & -1 \\ 0.5 & 1 \end{bmatrix} \begin{bmatrix} x \\ y \end{bmatrix} \right) = \begin{bmatrix} x + y - 1 \\ (-0.5)x - y + 3 \end{bmatrix}$$

Setting x to its value 2, we have

$$s = y + 1$$
$$t = -y + 2$$

As y increases, s increases, and so s must stay in the basis. As y increases, t decreases. Setting the slack variable t to 0, we solve $0 = -y +$

15.3 The revised simplex method

2; so y can increase to 2 without violating the lower bound on t. Therefore, y can enter the basis at 2 and t can exit at 0. The nonbasic variables are now $x = 2$, $t = 0$. We recompute the values of the basic variables as $y = 2$ and $s = 3$.

The matrix $\mathbf{A}_B$ and its inverse $\mathbf{A}_B^{-1}$ are now

$$\mathbf{A}_B = \begin{bmatrix} -1 & 1 \\ 1 & 0 \end{bmatrix}; \quad \mathbf{A}_B^{-1} = \begin{bmatrix} 0 & 1 \\ 1 & 1 \end{bmatrix}$$

The vector $\mathbf{c}_B$ is [1 0]. Solving $\pi = \mathbf{c}_B \mathbf{A}_B^{-1}$, we find $\pi = $ [0 1]. We compute the reduced costs $\pi \mathbf{a}_j - c_j$ of the nonbasic variables x and t. For x we find

$$\begin{bmatrix} 0 & 1 \end{bmatrix} \begin{bmatrix} -1 \\ 0.5 \end{bmatrix} - 6 = -5.5$$

and for t we have

$$\begin{bmatrix} 0 & 1 \end{bmatrix} \begin{bmatrix} 0 \\ 1 \end{bmatrix} - 0 = 1$$

Neither variable can enter the basis. Therefore we have reached a basic feasible solution that satisfies the Optimal Vertex Test. At $x = 2$, $y = 3$ the optimal value 15 of the objective function is attained.

There are many important computational aspects to implementing the simplex algorithm. We have mentioned that pricing is important; it is also computationally expensive. Another such aspect is the computation of $\mathbf{A}_B^{-1}$. This requires some numerical linear algebra. First, the role of the inverse is played by a factorization of the matrix $\mathbf{A}_B$ into upper and lower triangular matrices, typically by means of a technique known as the LU decomposition. Second, when an iteration is made, the matrix $\mathbf{A}_B$ is only changed by one column. This is taken into account and the LU decomposition of the new matrix is computed incrementally from the last one. Third, an effort is made to keep the LU factorization as sparse as possible. In LP problems, typically the matrices $\mathbf{A}$ and $\mathbf{A}_B$ are very sparse, but the inverse of a sparse matrix need not be sparse. Furthermore, to cope with numerical instability that arises in this kind of computation, various scaling and equilibrating strategies are employed. The reader is referred to Chvátal (1983).

At this point we have reduced the task of maintaining linear constraints to a procedural hill-climbing algorithm. All is discrete, and the continuous variable has exited the stage. It remains, however, in the theorems of the mathematicians and the imaginations of the programmers.

Exercises

15.3.1. Give an example of a linear program in standard form to show that in the degenerate case, an extreme point can be described as a vertex in a way that is not a basic feasible solution and that certifies optimality for an objective function. Give an example of a non-degenerate vertex whose description as a basic feasible solution has a non-basic variable at 0. Hint: For these examples, poke around the origin.

15.3.2. Suppose that a problem with a network matrix and integral right-hand-side is put into standard form for the simplex algorithm by adding slack and surplus variables. Show that the algorithm will find an integral solution.

15.3.3. Prove the following form of the *Exchange Property*: Suppose that $\mathbf{p}_1,\ldots,\mathbf{p}_m$ are linearly independent and suppose that $\mathbf{r} = \lambda_1 \mathbf{p}_1 + \ldots + \lambda_m \mathbf{p}_m$ with $\lambda_1 \neq 0$; then $\mathbf{r},\mathbf{p}_2,\ldots,\mathbf{p}_m$ are linearly independent and $\mathbf{p}_1$ is in the linear space spanned by $\mathbf{r},\mathbf{p}_2,\ldots,\mathbf{p}_m$.

Getting Started

The accompanying diskette contains the 2LP system. This is an Integrated Development Environment (IDE) complete with editor and windowing system to assist in working with 2LP models. The system on the diskette is for PCs running a version of the Windows operating system. For additional information and for other versions of the system, one can access the home page:

   http://www.brooklyn.cuny.edu/lbslab

In particular, Unix versions of 2LP for several platforms and a Linux version for PCs are available at this location.

To install the accompanying software on a PC running Windows 95 or Windows NT, insert the diskette in Drive A (or B) and run `a:\install` (or `b:\install`). Then follow the simple instructions.

In order to run this version of 2LP under Windows 3.1 or Windows 3.11, the Win32s support system has to be installed. If you run `a:\install`, it will inform you whether or not this software is already available on your system; if it is, it will proceed as with Windows 95. The Win32s software is available free of charge on the internet from the Microsoft Corporation at

   ftp://ftp.ncsa.uiuc.edu/Mosaic/Windows/Win31x/Win32s

This material includes a file `readme.txt` which provides instructions for installation. Win32s also comes with several C/C++ compilers.

The installation will set up a collection of icons in a group called 2LP:

   2LP   2LP Help   2LP Dialog Box Help   Going Out to C   2LP Uninstall

Clicking on *2LP* will bring up the IDE, clicking on *2LP Help* will provide you with information on the 2LP language, clicking on *2LP Dialog Box Help* will provide you with guidance to the features of the environment. Clicking on *Going Out to C* will open a text file with instructions for linking to external libraries and *2LP Uninstall* will undo the installation.

The 2LP installation will place the system in a directory, whose default name is TWOLP. In this directory, one will find the executables for the 2LP system. There is a subdirectory MODELS which contains code for models in this text. The subdirectory EXTDLL contains code and samples for linking 2LP with external C code, as described in *Going Out to C*. For installing 2LP on a PC that is on a network, consult the readme.txt file on the installation diskette.

For convenience, we give a list of the 2LP keywords below.

```
2lp_main                integral
absgap                  lb
and                     max
boolean                 min
break                   nint
c_either                not
c_or                    or
ceil                    objcup
continue                objcdn
continuous              printf
double                  random
else                    range
either                  rc
exit                    return
extern                  seed
fabs                    sigma
find_all                string
find_max                subject_to
find_min                then
floor                   ub
if                      void
int                     wp
```

The keywords extern, boolean and void are for use with external functions and are not discussed in this text. For usage, consult *2LP Help* and *Going Out to C*.

References

Aho, A. V., Hopcroft, J. E., and Ullman, J. D. (1974) *The Design and Analysis of Computer Algorithms*, Addison-Wesley.

Ait-Kaci, H. (1991) *Warren's Abstract Machine: A Tutorial Reconstruction*, MIT Press.

Andersen, K. A., and Hooker, J. N. (1996) A linear programming framework for logics of uncertainty, *Decision Support Systems* **16**, pp. 39-53.

Applegate, D., and Cook, W. (1991) A computational study of the job-shop scheduling problem, *ORSA Journal on Computing* **3**, pp. 149-156.

Anderson, S., and Patny, S. (1994) Keeping the Big Apple fiscally fit, *OR/MS Today*, August 1994, pp. 46-49.

Balas, E. (1967) Discrete programming by the filter method, *Operations Research* **15**, pp. 915-957.

Balas, E. (1971) Intersection cuts - a new type of cutting planes for integer programming, *Operations Research* **19**, pp. 19-39.

Balas, E. (1985) Disjunctive programming and a hierarchy of relaxations for discrete optimization problems, *SIAM Journal Algebraic and Discrete Methods* **6**, pp. 149-156.

Balas, E., Ceria, S., and Cornuéjols, G. (1993) A lift and project cutting plane algorithm for mixed 0-1 programs. *Mathematical Programming* **58**, pp. 295-324.

Barth, P. (1996) *Logic Based 0-1 Constraint Programming*, Kluwer.

Bartholdi, J. J., Orlin, J. B., and Ratliff, H. D. (1980) Cycle scheduling via integer programs with circular ones, *Operations Research* **28** (5), Sept.-Oct., pp. 1074-1085.

Beale, E. M. L. (1955) On minimizing a convex function subject to linear inequalities, *Journal of the Royal Statistical Society* **17**, pp. 173-184.

Berge, C. (1991) *Graphs*, North-Holland.

Bertsekas, D. (1991) *Linear Network Optimization: Algorithms and Codes*, MIT Press.

Beaumont, N. (1990) An algorithm for disjunctive programming, *European Journal of Operational Research* **48**, pp. 362-371.

Birkhoff, G., and von Neumann, J. (1936) The logic of quantum mechanics, *Annals of Mathematics* **37**, pp. 823-843.

Bixby, R. (1992) Implementing the simplex method: The initial basis, *ORSA Journal on Computing* **4**, pp. 267-284.

Bixby, R., Kennedy, K., and Kremer, U. (1994) Automatic data layout using 0-1 integer programming, *Proceedings of the International Con-*

ference on *Parallel Architectures and Compilation Techniques* (PACT94), pp. 111-122.

Bland, R. G. (1977a) A combinatorial abstraction of linear programming, *Journal of Combinatorial Theory (B)* **23**, pp. 33-57.

Bland, R. G. (1977b) New finite pivoting rules for the simplex method, *Mathematics of Operations Research* **2**, pp. 103-107.

Bosch, R. (1993) A Big Mac attack, *OR/MS Today*, August, pp. 30-31.

Brearley, A., Mitra, G., and Williams, H. P. (1975) An analysis of linear programming problems prior to applying the simplex method, *Mathematical Programming* **8**, pp. 54-83.

Brooke, A., Kendrick, D., and Meeraus, A. (1988) *GAMS: A User's Guide*, The Scientific Press.

Carlier, J., and Pinson, E. (1989) An algorithm for solving the job shop problem, *Management Science* **35**, pp. 164-176.

Carlier, J., and Pinson, E. (1990) A practical use of Jackson's preemptive schedule for solving the job-shop problem, *Annals of Operations Research* **26**, pp. 269-287.

Censor, Y. and S. Zenios (to appear) *Parallel Optimization*, Oxford University Press.

Chvátal, V. (1973) Edmonds polytopes and a hierarchy of combinatorial problems, *Discrete Mathematics* **4**, pp. 305-337.

Chvátal, V. (1983) *Linear Programming*, Freeman.

Charnes, A. (1952) Optimality and degeneracy in linear programming, *Econometrica* **20**, pp. 160-170.

Church, A. (1956) *Introduction to Mathematical Logic*, Princeton University Press.

Colmerauer, A. (1987) Opening the Prolog III universe, BYTE Magazine, August.

Colmerauer, A. (1990) A introduction to Prolog III, *Communications of the ACM* **33**, pp. 69-90.

Cook, S. A. (1971) The complexity of theorem-proving procedures, *Proceedings of the Third Annual ACM Symposium on the Theory of Computing*, ACM, New York, pp. 151-158.

Cook, W., Coullard, C. R. and Turán, G. (1987) On the complexity of cutting plane proofs, *Discrete Applied Mathematics* **18**, pp. 25-38.

Cook, W., Rutherford, T., Scarf, H., and Shallcross, D. (1993) An implementation of the generalized basis reduction algorithm for integer programming, *ORSA Journal on Computing* **5**, pp. 206-212.

Cox, J., McAloon, K., and Tretkoff, C. (1992) Computational complexity and constraint logic programming, *Annals of Mathematics and Artificial Intelligence* **5**, pp. 163-190.

Coxeter, H. S. M., and Moser, W. O. J. (1980) *Generators and Relations for Discrete Groups*, Fourth Edition, Springer-Verlag, New York.

Cplex (1995) *The Cplex Callable Library, Manual*, Cplex Optimization.

Crowder, H. P., and Padberg, M. W. (1980) Solving large-scale symmetric traveling salesman problems to optimality, *Management Science* **26**, pp. 495-509.

References

Crowder, H. P., Johnson, E., and Padberg, M. W. (1983) Solving large-scale zero-one linear programming problems, *Operations Research* **31**, pp. 803-834.

Culberson, J. C., and Schaeffer, J. (1994) Efficiently searching the 15-puzzle, Technical Report TR 94-08, University of Alberta, Edmonton, Alberta, Canada.

Dakin, R. J. (1965) A tree-search algorithm for mixed integer programming problems, *The Computer Journal* **8**, pp. 250-255.

Dantzig, G. (1951) Maximization of a linear function of variables subject to linear inequalities, *Activity Analysis of Production and Allocation*, ed. by T. C. Koopmans, Wiley, pp. 339-347.

Dantzig, G. (1953) Computational algorithm of the revised simplex method, Report RM 1266, The Rand Corporation, Santa Monica, CA.

Dantzig, G. (1955) Linear programming under uncertainty, *Management Science* **1**, pp. 197-206.

Dantzig, G. (1963) *Linear Programming and Extensions*, Princeton University Press.

Dantzig, G., Fulkerson, D., and Johnson, S. (1954) Solution of a large scale traveling salesman problem, *Operations Research* **2** (4), pp. 393-410.

Dantzig, G., Fulkerson, D., and Johnson, S. (1959) On a linear programming, combinatorial approach to the traveling salesman problem, *Operations Research* **7** (1), pp. 58-66.

Dantzig, G., Orden, A., and Wolfe, P. (1955) The generalized simplex method for minimizing a linear form under linear inequality restraints, *Pacific Journal of Mathematics* **5**, pp. 183-195.

Dantzig, G., and Van Slyke, R. M. (1967) Generalized upper bounding techniques, *Journal of Computer and System Sciences* **1**, pp. 213-226.

Davis, M., and Putnam, H. (1960) A computing procedure for quantification theory, *Journal of the ACM* **1**, pp. 201-215.

Dennett, D. C. (1995) *Darwin's Dangerous Idea, Evolution and the Meanings of Life*, Simon and Schuster.

Dincbas, M., van Hentenryck, P., Simonis, H., and Aggoun, A. (1988) The constraint logic programming language CHIP, in *Proceedings of the 2nd International Conference on Fifth Generation Computer Systems*, pp. 249-264.

Dobkin, D., Lipton, R. J., and Reiss, S. (1979) Linear programming is log-space hard for P, *Information Processing Letters* **8**, pp. 96-97.

Dorfman, R., Samuelson, P., and Solow, R. (1958) *Linear Programming and Economic Analysis*, McGraw-Hill.

Driebeek, N. J. (1966) An algorithm for the solution of mixed-integer programming problems, *Management Science* **12**, pp. 576-587.

Eaves, B. C., and Rothblum, U. (1992) Dines-Fourier-Motzkin quantifier elimination and an application of corresponding transfer principles over ordered fields, *Mathematical Programming* **53**, pp. 307-321.

Erkut, E. (1994) Big Mac attack revisited, *OR/MS Today*, June, pp. 50-52.

Falkowski, B.-J. and Schmitz, L. (1986) A note on the queens' problem, *Information Processing Letters* **23**, pp. 39-46.

Fang, S.-C., and Puthenpura, S. (1993) *Linear Optimization and Extensions: Theory and Algorithms*, Prentice-Hall.

Ferris, M., and Mangasarian, O. (1995) Breast cancer diagnosis via linear programming, *Computational Science and Engineering* **2**, pp. 70-71.

Fischer, H., and Thompson, G. L. (1963) Probabilistic learning combinations of local job-shop scheduling rules, in *Industrial Scheduling*, ed. J. F. Muth and G. L. Thompson, Prentice-Hall, pp. 225-251.

Fitting, M. (1983) *Proof Methods for Modal and Intuitionistic Logics*, Reidel (now Kluwer), second edition forthcoming.

Floyd, R. (1967) Assigning meaning to programs, *Mathematical Aspects of Computer Science* **XIX**, American Mathematical Society, pp. 19-32.

Fourier, J.-B. J. (1826) Solution d'une question particulière du calcul des inégalités, *Nouveau Bulletin des Sciences par la Société philmathique de Paris*, pp. 99-100 (reprinted, G. Olms, Hildesheim, 1970, pp. 317-319).

Forrest, J., and Goldfarb, D. (1992) Steepest edge simplex algorithms for linear programming, *Mathematical Programming* **57**, pp. 341-374.

Fourer, R., Gay, D., and Kernighan, B. (1993) *AMPL: A Modeling Language for Mathematical Programming*, Duxbury Press.

Franklin, J. (1980) *Methods of Mathematical Economics*, Springer-Verlag.

Garey, M. R., and Johnson, D. S. (1979) *Computers and Intractability: a Guide to the Theory of NP-completeness*, Freeman.

Gasser, R. U. (1995) Harnessing Computational Resources for Efficient Exhaustive Search, Ph.D. Thesis, Swiss Federal Institute of Technology.

Geoffrion, A. (1969) An improved implicit enumeration approach to integer programming, *Operations Research* **17**, pp. 178-190.

Ginsberg, M. (1993) *Essentials of Artificial Intelligence*, Morgan-Kaufman.

Girard, J.-Y. (1993) On the unity of logic, *Annals of Pure and Applied Logic* **59**, pp. 201-217.

Glover, F. (1965) A bound escalation method for the solution of integer linear programs, Graduate School of Industrial Administration Reprint No. 175, Carnegie-Mellon University.

Glover, F. (1986) Future paths for integer programming and links to artificial intelligence, *Computers and Operations Research* **13**, pp. 533-549.

Glover, F., Taillard, E., and de Werra, D. (1993) A user's guide to tabu search, *Annals of Operations Research* **41**, pp. 3-28.

Gödel, K. (1930), Die Vollständigkeit der Axiome des logischen Functionenkalküls, *Monatschefte für Mathematik und Physik* **38**, pp. 173-

198. English translation: The completeness of the axioms of the functional calculus of logic, in J. van Heijenoort, *From Frege to Gödel, A Source Book in Mathematical Logic*, Harvard University Press 1967, pp. 582-591.

Goldberg, D. E. (1989) *Genetic Algorithms in Search, Optimization and Machine Learning*, Addison-Wesley.

Gomory, R. E. (1958) Outline of an algorithm for integer solutions to linear programs, *Bulletin of the American Mathematical Society* **64**, pp. 275-278.

Gomory, R. E. (1963) An algorithm for integer solutions to linear programs, *Recent Advances in Mathematical Programming*, ed. by R. L. Graves and P. Wolfe, McGraw-Hill, pp. 269-302.

Gondran, M. and Minoux, M. (1984) *Graphs and Algorithms*, Wiley.

Greenberg, H. (1968) A branch-and-bound solution to the general scheduling problem, *Operations Research* **8**, pp. 353-361.

Greenberg, H. J. (1993a) *A Computer-Assisted Analysis System for Mathematical Programming Models and Solutions*, Kluwer.

Greenberg, H. J. (1993 b) *Modeling by Object-Driven Linear Elemental Relations*, Kluwer.

Gribbin, J. (1984) *In Search of Schrödinger's Cat*, Bantam Books.

Gries, D. (1981) *The Science of Programming*, Springer-Verlag.

Griffin, T. (1990) A formulas-as-type notion of control, *Proceedings of the 17th ACM Symposium on Principles of Programming Languages*, pp. 47-58.

Haken, A. (1985) The intractability of resolution, *Theoretical Computer Science* **39**, pp. 297-308.

Halldórsson, M. M. (1993) A still better performance guarantee for approximate graph coloring, *Information Processing Letters* **45**, pp. 19-23.

Hailperin, T. (1976) *Boole's Logic and Probability*, Studies in Logic and the Foundations of Mathematics **85**, North-Holland.

Hansen, P. (1986) The steepest ascent mildest descent heuristic for combinatorial programming, *Congress on Numerical Methods in Combinatorial Optimization*, Capri.

Hanson, W., and Martin, R. (1990) Optimal bundle pricing, *Management Science* **36** (2), pp. 155-174.

Harche, F., Hooker J. N., and Thompson, G. L. (1994) A computational study of satisfiability algorithms for propositional logic, *ORSA Journal of Computing* **45** (4), pp. 423-435.

Haralick, R., and Elliot, G. (1980) Increasing tree search efficiency for constraint satisfaction problems, *Artificial Intelligence* **14**, pp. 263-313.

Harvey, W. (1995) Nonsystematic Backtracking Search, Thesis, Department of Computer Science, Stanford University.

Hoffman, K., and Padberg, M. (1991) Improving LP representation of zero-one linear programs for branch and cut, *ORSA Journal on Computing* **3**, pp. 121-134.

Holland, J. (1992) *Adaptation in Natural and Artificial Systems*, MIT Press. First edition: University of Michigan Press, 1975.

Hooker, J. (1992) Generalized resolution for 0-1 linear inequalities, *Annals of Mathematics and Artificial Intelligence* **6**, pp. 271-286.

Hooker, J., and Yan, H. (1994) Verifying logic circuits by Bender's decomposition, in *Principles and Practice of Constraint Programming*, ed. V. Saraswat and P. van Hentenryck, MIT Press, pp. 269-290.

Hughes, R. I. G. (1989) *The Structure and Interpretation of Quantum Mechanics*, Harvard University Press.

Ignizio, J., and Cavalier, T. (1994) *Linear Programming*, Prentice-Hall.

Jackson, J. R. (1955) Scheduling a production line to minimize maximum tardiness, Recent Report No. 43, Management Science Research Project, University of California, Los Angeles.

Jaffar, J., and Lassez, J.-L. (1987) Constraint logic programming, *Proceedings of the 14th ACM Symposium on Principles of Programming Languages*, pp. 111-119.

Jaffar, J., and Maher, M. (1994) Constraint logic programming: a survey, *Journal of Logic Programming* **19/20**, pp. 503-581.

Jaffar, J., and Michaylov, S. (1987) Methodology and implementation of a CLP System, *Proceedings of the 1987 International Logic Programming Conference*, ed. J.-L. Lassez, MIT Press.

Jaumard, B., Hansen, P., and Poggi de Aragaö, M. (1991) Column generation methods for probabilistic logic, *ORSA Journal on Computing* **3**, pp. 135-148.

Jeroslow, R. (1980) A cutting plane game for facial disjunctive programs, *SIAM J. Control and Optimization* **18** (3) pp. 264-280.

Jeroslow, R. (1989) *Logic-Based Decision Support, Mixed Integer Model Formulation*, Annals of Discrete Mathematics Monograph, North-Holland.

Jeroslow, R., and Lowe, J. (1984) Modeling with integer variables, *Mathematical Programming Study* **22**, pp. 167-184.

Johnson, W. W. (1879) Notes on the "15" puzzle, I, *American Journal of Mathematics* **2**, pp. 397-399.

Jones, N. D., and Laaser, W. T. (1976) Complete problems for deterministic polynomial time, *Theoretical Computer Science* **3** (1) pp. 105-118.

Jourdan, J. (1995) Concurrence et Coopération de Modèles Multiples dans les Langages de Contraintes, Thesis, University of Paris.

Kall, P., and Wallace, S. (1994) *Stochastic Programming*, Wiley.

Kamath, A., Karmarker, N., Ramakrishnan, K., and Resende, M. (1991) A continuous approach for inductive inference, Bell Laboratory Technical Report.

Karmarkar, N. (1984) A new polynomial time algorithm for linear programming, *Combinatorica* **4**, pp. 373-395.

Karger D., Motwani, R., and Sudan, M. (1994) Approximate graph coloring by semidefinite programming, *Proc. 35th IEEE Symposium on the Foundations of Computer Science*, pp. 2-13.

Karp, R. (1972) Reducibility among combinatorial problems, in *Complexity of Computer Computations*, ed. J. W. Thatcher and R. E. Miller, Plenum Press, pp. 85-103.

Khachian, L. G. (1979) A polynomial algorithm for linear programming, *Doklady Akad. Nauk USSR* **244** (5) pp. 1093-96. Translated in *Soviet Math. Doklady* **20**, pp. 191-194.

Kirkpatrick, S., Gelatt, C., and Vecchi, M. (1983) Optimization by simulated annealing, *Science* **220**, pp. 671-680.

Knuth, D. (1969) *The Art of Computer Programming, Volume 2, Seminumerical Algorithms*, Addison-Wesley.

Korf, R. E. (1985) Depth-first iterative-deepening: An optimal admissible tree search, *Artificial Intelligence* **27**, pp. 97-109.

Korf, R. E. and Taylor, L. (1996) Finding optimal solutions to the twenty-four puzzle, preprint.

Kozen, D. C. (1992) *The Design and Analysis of Algorithms*, Springer-Verlag.

Kuhn, H. W. (1956) Solvability and consistency for linear equations and inequalities, *American Mathematical Monthly* **63**, pp. 217-232.

Krivine, J.-L. (1994) Classical logic, storage operators and second-order lambda-calculus, *Annals of Pure and Applied Logic* **68**, pp. 53-78.

Land, A. H., and Doig, A. G. (1960) An automatic method of solving discrete programming problems, *Econometrica* **28,** pp. 497-520.

Land, A. H., and Powell, S. (1979) Computer codes for problems of integer programming, *Annals of Discrete Mathematics* **5**, pp. 221-269.

Lassez, J.-L., and Maher, M. (1992) On Fourier's algorithm for linear arithmetic constraints, *Journal of Automated Reasoning* **9**, pp. 373-379.

Laurière, J.-L. (1978) A language and a program for stating and solving combinatorial problems, *Artificial Intelligence* **10**, pp. 29-127.

Lassez, J.-L., and McAloon, K. (1993) A constraint sequent calculus, *Constraint Logic Programming, Selected Research*, ed. F. Benhamou and A. Colmerauer, MIT Press, pp. 33-44.

Lenstra, H. W., Jr. (1983) Integer programming with a fixed number of variables, *Mathematics of Operations Research* **8**, pp. 538-548.

LePape, C. (1994) Scheduling as intelligent control of decision-making and constraint propagation, in *Intelligent Scheduling*, ed. M. Zweben and M. Fox, Morgan Kaufmann, pp. 67-98.

Levin, L. (1973) Universal sorting problems, in Russian in *Problemy Peredaci Informacii* **9**, pp. 115-166; English translation in *Problems of Information Transmission* **9**, pp. 265-266.

Lewis, H. R., and Papadimitriou, C. H (1981) *Elements of the Theory of Computation*, Prentice-Hall.

Lin, S. (1965) Computer solutions of the traveling salesman problem, *Bell System Technical Journal* **44**, pp. 2245-2269.

Lin, S., and Kernighan, B. W. (1973) An effective heuristic algorithm for the traveling salesman problem, *Operations Research* **21**, pp. 498-516.

Little, J. D. C., Murty, K. G., Sweeney, D. W. and Karel, C. (1963) An algorithm for the traveling salesman problem, *Operations Research* **11**, pp. 972-989.

Lovász, L., and Schrijver, A. (1991) Cones of matrices and set functions and 0-1 optimization, *SIAM J. of Optimization* **1** (2) pp. 166-190.

Lucas, E. (1882) *Récréations Mathématiques*, Volumes 1-4, Gauthiers-Villars.

Maier, D., and Warren, D. S. (1988) *Computing with Logic*, Benjamin/Cummings.

Mackworth, A. K., and Freuder, E. C. (1993) The complexity of constraint satisfaction revisited, *Artificial Intelligence* **25**, pp. 65-74.

Mangasarian, O. L. (1993) Mathematical programming in neural networks, *ORSA Journal on Computing* **5**, pp. 349-380.

McAloon, K., and Tretkoff, C. (1990) Subrecursive constraint logic programming, *Proceedings of the NACLP 1990 Workshop on Logic Programming Architectures and Implementation*, ed. J. Mills.

McAloon, K., and Tretkoff, C. (1995) 2LP: Linear programming and logic programming, in *Principles and Practice of Constraint Programming*, ed. V. Saraswat and P. van Hentenryck, MIT Press, pp. 101-116.

McAloon, K., and Tretkoff, C. (1996) Logic, modeling, and programming, *Annals of Operations Research*, to appear.

Minsky, M. L., and Papert, S. (1988) *Perceptrons, An Introduction to Computational Geometry*, expanded edition, MIT Press.

Mitchell, M. (1996) *An Introduction to Genetic Algorithms*, MIT Press

Minton, S., Johnston, M., Philips, A., and Laird, P. (1990) Solving large-scale constraint satisfaction and scheduling problems using a heuristic repair method, *Proceedings of the Eight National Conference on Artificial Intelligence*, pp. 17-24.

Muth, J. F., and Thompson, G. L. (1963) *Industrial Scheduling*, Prentice-Hall.

Nazareth, J. L. (1987) *Computer Solutions of Linear Programs*, Oxford University Press.

Nemhauser, G. L., and Wolsey, L. A. (1988) *Integer and Combinatorial Optimization*, Wiley.

Nemhauser, G., Savelsbergh, M., and Sigismondi G. (1994) MINTO, A Mixed INTeger Optimizer, *Operations Research Letters* **15**, pp. 47-58.

Newell, A., and Simon, H. (1972) *Human Problem Solving*, Prentice-Hall.

Osman, I. and Kelly, J. (1996) *Meta-Heuristics: Theory and Applications*, Kluwer.

Papadimitriou, C. H. (1994) *Computational Complexity*, Addison-Wesley.

Papadimitriou, C. H., and Steiglitz, K. (1982) *Combinatorial Optimization, Algorithms and Complexity*, Prentice-Hall.

Percus, O. and Kalos, M. (1989) Random number generators for MIMD parallel processors, *Journal of Parallel and Distributed Computing* **6**, pp. 477-497.

References

Pudlák, P. (1996) Lower bounds for resolution and cutting planes proofs and monotone computations, to appear in the *Journal of Symbolic Logic*.

Puget, J.-F. (1994) A C++ implementation of CLP, ILOG Technical Report.

Ratner, D., and Warmuth, M. (1990) The (n^2-1)-puzzle and related relocation problems, *Journal of Symbolic Computation* **10**, pp. 111-137.

Rao, V. N., Kumar V., and Ramesh, K. (1987) A parallel implementation of iterative deepening A*, *Proceedings of the Fifth National Conference on Artificial Intelligence*, pp. 878-882.

Reingold, E., Nievergelt, J., and Deo, N. (1977) *Combinatorial Algorithms, Theory and Practice*, Prentice-Hall.

Robinson, J. A. (1965) A machine oriented logic based on the resolution principle, *Journal of the ACM* **12**, pp. 23-41.

Rockafellar, R. T. (1970) *Convex Analysis*, Princeton University Press.

Rosenblatt F. (1962) *Principles of Neurodynamics*, Spartan Books, New York.

Roy, A., and Mukhopadhyay, S. (1991) Pattern classification using linear programming, *ORSA Journal on Computing* **3**, pp. 66-80.

Rudd, A., and Schroeder, M. (1982) The calculation of minimum margin, *Management Science* **28** (12), pp. 1368-1379.

Rumelhart, D. E., and McClelland, J. L. (1987) *Parallel Distributed Processing*, 2 volumes, MIT Press.

Saigal, R. (1995) *Linear Programming, A Modern Integrated Analysis*, Kluwer.

Salkin, H. M., and Mathur, K. (1989) *Foundations of Integer Programming*, North-Holland.

Sankaran, J. (1989) Bidding Systems for Certain Nonmarket Allocations of Indivisible Items, Thesis, University of Chicago.

Savelsbergh, M. (1994) Preprocessing and probing techniques for mixed integer programming problems, *ORSA Journal on Computing* **6**, pp. 445-454.

Schrage, L. (1991) LINDO, *An Optimization Modeling System*, Scientific Press.

Schrijver, A. (1986) *Theory of Linear and Integer Programming*, Wiley.

Selman, B., Levesque, H., and Mitchell, D. (1992) A new method for solving hard satisfiability problems, *Proceedings of the Tenth National Conference on Artificial Intelligence*, pp. 440-446.

Seymour, P. (1980) Decomposition of regular matroids, *Journal of Combinatorial Theory (B)* **28**, pp. 305-359.

Shortliffe, E. (1976) *Computer Based Medical Consultations: MYCIN*, American Elsevier.

Slate, D. J., and Atkin, L. R. (1977) *CHESS 4.5 – The Northwestern University Chess Program*, Springer Verlag.

Stigler, G. J. (1945) The cost of subsistence, *Journal of Farm Economics* **27**, pp. 303-314.

Story, W. E. (1879) Notes on the "15" puzzle II, *American Journal of Mathematics* **2**, pp. 399-404.

Tarski, A. (1951) *A Decision Method for Elementary Algebra and Geometry*, revised edition, University of California Press, Berkeley.

Tijdeman, R. (1974) Hilbert's seventh problem: on the Gel'fond-Baker method and its applications, *AMS Symposia in Pure Mathematics* **XXVIII**, ed. F. Browder, pp. 241-268.

Tomlin, J. A. (1971) An improved branch-and-bound method for integer programming, *Operations Research* **19**, pp. 1070-075.

Troelstra, A. S. (1977) Aspects of constructive mathematics, *The Handbook of Mathematical Logic*, ed. J. Barwise, North-Holland, pp. 973-1053.

van Hentenryck, P. (1989) *Constraint Satisfaction in Logic Programming*, MIT Press.

van Laarhoven, P. J. M., Aarts, E. H. L. and Lenstra, J. K. (1992) Job shop scheduling by simulated annealing, *Operations Research* **40**, pp. 113-125.

Warren, D. H. D. (1983) An abstract Prolog instruction set, Technical Note 309, AI Center, SRI International, Menlo Park CA

Williams, H. P. (1994) *Model Building in Mathematical Programming*, Wiley.

Zweben, M. and Fox, M. (1994) *Intelligent Scheduling*, Morgan Kaufmann.

List of Models

Model 1.1	The Wildlife Conservation Park	13
Model 1.2	The Library Fund	18
Model 1.3	The Fortune-teller	28
Model 1.4	Inside vs. Outside	33
Model 2.1	The Fixed-Rate Mortgage	40
Model 2.2	The Chicken or the Egg	45
Model 2.3	Call 911	48
Model 2.4	The Linear Knapsack	56
Model 2.5	The Paris Bourse	58
Model 3.1	The Pacific Rim	65
Model 3.2	British Cooking	71
Model 3.3	Motorcycles Inc.	81
Model 3.4	The Committee	84
Model 3.5	The Library Fund Reprised	90
Model 4.1	Risk vs Reward	99
Model 4.2	The Fixed-Rate Mortgage Reviewed	107
Model 4.3	Batch Scheduling	118
Model 4.4	The Grid	123
Model 4.5	Fairer Co-op Tax	131
Model 4.6	Extreme Vertices	136
Model 5.1	Parallel Sessions	144
Model 5.2	Cell Discrimination	158
Model 6.1	Double Data	166
Model 6.2	The Committee Realigned	171
Model 6.3	The Body Shop	178
Model 6.4	We Open in Venice	183
Model 6.5	The Workstations on the Assembly Line	191
Model 7.1	The Price of Resource X	204
Model 7.2	The Grid Analyzed	220
Model 8.1	The Pacific Rim Revisited	230
Model 8.2	The 0-1 Solution I	234
Model 8.3	Salt and Mustard	238
Model 8.4	The 0-1 Solution, II	243
Model 8.5	Who Has the Zebra?	246
Model 9.1	Motorcycles Inc. Reconsidered	264

Model 9.2	Silicon Logic	268
Model 9.3	The Body Shop Recast	273
Model 10.1	Corporate Paper	292
Model 10.2	The N-Queens	304
Model 10.3	The Twelve Coins	315
Model 10.4	Expert Automotive Repair	324
Model 11.1	All the N-Queens Solutions	332
Model 11.2	Fast Food	349
Model 11.3	Mountaineering	355
Model 11.4	Mathematicians and Physicists	363
Model 12.1	Military History	380
Model 12.2	More Fast Food	388
Model 12.3	Yet Another Crew Compilation	392
Model 13.1	The Capitol University Budget	401
Model 13.2	The Jobber's Fee	418
Model 13.3	New York Supermarkets	426
Model 13.4	British Cooking Refined	434
Model 14.1	The S^2-1 Puzzle	447
Model 14.2	New York Supermarkets Remodeled	466

Name Index

A
Aarts, E. 189
Aho, A. 276
Ait-Kaci, H. 256
Andersen, K. 30
Anderson, S. 48
Applegate, D. 391
Aristotle 175
Atkin, L. 452

B
Balas, E. 275, 425, 432
Barth, P. 272, 417
Bartholdi, J. 52, 118
Beaumont, N. 275
Bertsekas, D. 117
Birkhoff, G. 259
Bixby, R. 143, 507
Bland, R. 506
Boole, G. 28, 30
Bosch, R. 349
Brearley, A. 411
Brooke, A. 408
Brouwer, L. E. J. 253

C
Carlier, J. 121, 182, 432
Cavalier, T. 164
Censor, Y. 95
Ceria, S. 425
Charnes, A. 505
Chebychev, P.L. 135
Cheops 493
Church, A. 276, 287
Chvátal, V. 135, 417, 499, 506, 511
Colmerauer, A. 256
Cook, S. 278
Cook, W. 391, 417, 425
Cornuéjols, G. 425
Coullard, C. 417
Cox, J. 256
CPLEX 387
Crowder, H. 432
Culberson, J. 464

D
Dakin, R. 390, 466
Dantzig, G. 2, 45, 95, 395, 432, 500, 506
Davis, M. 323
de Werra, D. 189
Dempster, A. 30
Dennett, D. 175
Deo, N. 333
Dobkin, D. 328
Doig, A. 371, 466
Dorfman, R. 203
Driebeek, N. 441

E
Eaves, B. 481
Elliot, G. 310
Erkut, E. 349

F
Falkowski, B.-J. 304
Fang, S.-C. 277
Farkas, J. 482
Fermat, P. 175
Ferris, M. 164
Fibonacci, L. 301
Fischer, H. 177, 182
Fitting, M. 165
Floyd, R. 283
Forrest, J. 505
Fourer, R. 408
Fourier, J.-B. 2, 475, 480
Franklin, J. 482, 499
Freuder, E. 310
Friedman, H. 260
Fulkerson, D. 432

G
Garey, M. 276, 281
Gasser, R. 464
Gauss, C. F. 304
Gay, D. 408
Gelatt, C. 189
Geoffrion, A. 432
Ginsberg, M. 259
Girard, J.-Y. 255
Glover, F. 189, 432
Gödel, K. 276, 287
Goldberg, D 196
Goldfarb, D. 505
Gomory, R. 424, 432

Greenberg, H. 275
Greenberg, H.J. 411
Gries, D. 140, 283
Griffin, T. 255

H
Hailperin, T. 28
Hall, P. 95
Hansen, P. 30, 189
Hanson, W. 424
Haralick, R. 310
Harvey, W. 473
Hilbert, D. 253
Hoffman, K. 400
Holland, J. 196
Hooker, J. 30, 424
Hopcroft, J. 276
Hughes, R. 259

I
Ignizio, J. 164

J
Jackson, J. 121
Jaffar, J. 256
Jaumard, B. 30
Jeroslow, R. 323, 424, 425
Johnson, D. 276, 281
Johnson, E. 432
Johnson, S. 432
Johnston, M. 310
Jones, N. 328
Jourdan, J. 310

K
Kalos. M. 175
Kamath, A. 300
Kamesh, K. 452
Karel, C. 466
Karger, D. 149, 302
Karmarkar, N. 2, 277, 300
Kelly, J. 189
Kemeny, G. 45
Kendrick, D. 408
Kennedy, K. 143
Kernighan, B. 188, 408
Khachian, L. 2, 277
Kirkpatrick, S. 189
Knuth, D. 175
König, D. 287
Korf, R. 452, 464

Kozen, D. 58
Kremer, U. 143
Krivine, J.-L. 255
Kuhn, H. 480
Kumar, V. 452

L
Laaser, W. 328
Laird, P. 310
Land, A. 371, 386, 466
Lassez, J.-L. 256, 480, 489
Lenstra, J.K. 189
Levesque, H. 300
Levin, L. 278
Lewis, H. 322
Lin, S. 183, 185, 188, 281
Lipton, R. 328
Little, J.D.C. 466
Lovász, L. 425, 522
Lowe, J. 424
Lucas, E. 236, 303

M
Mackworth, A. 310
Maher, M. 256, 480
Maier, D. 313
Mangasarian, O. 164
Martin, R. 424
Mathur, K. 378, 441
McAloon, K. 242, 256, 489
McClelland, J. 164
Meerhaus, A. 408
Michaylov, S. 256
Minsky, M. 156
Minton, S. 310
Mitchell, D. 300
Mitchell, M. 196
Mitra, G. 411
Motwani, R. 302
Motzkin, T. 2, 480
Mukhopadhyay, S. 164
Murty, K.G. 466
Muth, J. 182

N
Nazareth, J. 505
Nemhauser, G. 432
Newell, A. 408
Nievergeld, J. 333
Nilsson, N. 30

O

Occam, William of 507
Orden, A. 506
Orlin, J. 52, 118
Osman, I. 189

P

P. Erdös 243
Padberg, M. 400, 432
Papadimitriou, C. 276, 277, 278, 279, 322
Papert, S. 156
Pascal, B. 175
Patny, S. 48
Percus, O. 175
Philips, A. 310
Pinson, E. 121, 182, 432
Poggi de Aragaö, M. 30
Powell, S. 386
Pudlak, P. 417
Puget, J.-F. 250
Puthenpura, S. 277
Putnam, H. 323

R

Ramakrishnan, K. 300
Rao, V. 452
Ratliff, H. 52, 118
Ratner, D. 448
Reingold, E. 333
Reiss, S. 328
Resende, M. 300
Robinson, J.A. 322
Rockefellar, R.T. 500
Rosenblatt, F. 156
Rothblum, U. 481
Roy, A. 164
Rudd, A. 118
Rumelhardt, D. 164
Russell, B. 253
Rutherford, T. 425

S

Saigal, R. 277, 506
Salkin, H. 378, 441
Samuelson, P. 203
Sankaran, J. 131
Savelsbergh, M. 432
Scarf, H. 425
Schaeffer, J. 464
Schmitz, L. 304
Schrijver, A. 135, 425, 432, 489, 500
Schroeder, M. 118
Selman, B. 300
Seymour, P. 122
Shafer, G. 30
Shallcross, D. 425
Shortliffe, E. 327
Shrage, L. 131
Siegel, C.L. 260
Sigismondi, G. 432
Simon, H. 408
Slate, D. 452
Solow, R. 203
Steiglitz, K. 277, 278, 279
Stuckey, P. 256
Sudan, M. 302
Sweeney, D.W. 466

T

Taillard, E. 189
Tarski, A. 481
Thompson, G. 177, 182
Tomlin, J. 441
Tretkoff, C. 242, 256
Troelstra, A. 139
Turán, G. 417
Turing, A. 276

U

Ullmann, J. 276

V

van Hentenryck, P. 256
van Laarhoven, P. 189
Van Slyke, R. 395
Vecchi, M. 189
von Neumann, J. 259

W

W. Schur 242
Wallace, S. 95
Warmuth, M. 448
Warren, D. S. 313
Warrren, D. H. D. 256
Williams, H. P. 71, 411, 434
Wolfe, P. 506
Wolsey, L. 432

Y

Yap, R. 256

Z
Zenios, S. 95

Subject Index

Symbols
%, 50
|=, 476

Numerics
0-1 integer programming, 233, 278
0-1 knapsack problem, 279
2-SAT, 282
3-SAT, 280

A
A* search, 446, 459, 466
absgap, 387
abstract data type, 8
adjacent vertex, 493
admissible heuristic, 345, 459
affine closure, 485
affine combination, 18, 485
affine expression, 25
Affine Form of Farkas' Lemma, 482
affine hull, 485
affine set, 481
affine space, 484
airline crew scheduling, 392
ALICE, 250
AMPL, 408
and loop, 39, 40
annealing schedule, 189
antecedent, 103
arc, 22, 112, 116, 117, 118, 122
arc consistency, 310
arithmetic operator, 18
array, 17, 32
artificial intelligence, 168, 233
artificial variables, 506
aspiration criteria, 189
assignment operator, 3
assignment problem, 121

B
back-propagation, 164
backtracking, 189, 228, 255, 256, 279
barycenter, 488
basic column, 502
basic feasible solution, 501
basic variable, 502
basis, 502
beam search, 471
best-bound search, 466
best-child-first search, 297, 353
big M constraint, 407
binary search, 396
Bland's Rule, 506
blending problem, 16, 71
block, 42
bluff, 353, 472
bound, 337, 339
bounded set, 54, 489
bounding constraint, 53
braces, 42
branch-and-bound search, 337, 341, 348, 349, 353, 363, 364, 386, 391, 466
branch-and-cut, 425
breadth-first search, 386, 443, 466
break, 108, 375
bubble sort, 301, 454

C
C, 3, 5, 103, 140, 255, 411, 455
C++, 3, 67, 103, 140, 140, 455
c_either, 140
c_or loop, 143
capacitated warehouse location problem, 425
capital budgeting problem, 401
Cartesian space, 3
ceil, 24
certainty calculus, 327
certainty factor, 102, 327
Chebychev approximation, 135
child node, 229, 289
CHIP, 250
choice point, 228, 255
classical disjunction, 139, 253
classical logic, 227
clause, 268
clone, 190
CLP(R), 256
CNF, 268, 416
coherent, 288, 289, 299, 309

531

column generation, 30
combinatorial explosion, 236, 263
comment, 7
commutativity, 97, 98
complete enumeration, 235
complete proof system, 320, 323
completeness, 283, 300
completeness theorem, 287, 322, 482
conditional disjunction, 140, 256, 443, 446
conditional or, 140
conditional statement, 104
cone, 483, 492, 500
confidence factor, 102
conjunction, 97
Conjunctive Normal Form, 268
connectionist, 164
co-NP, 280
co-NP-Complete, 280
co-NP-Hard, 281
consequence, 476
consequent, 103
consistent set of constraints, 6
constrain-and-generate, 144, 243
constraint, 1, 2, 5, 18, 26
constraint logic programming (CLP), 256
constraint propagation, 409
context-free language, 26
`continue`, 440
`continuous`, 3, 8, 17, 25, 64
convex closure, 414, 486
convex combination, 485
convex hull, 414, 486
convex set, 413, 485
correctness, 283
co-SAT, 280
crossover, 190, 196
cut, 413
cutting plane, 413
cutting planes proof, 417
cycling, 505

D

Davis-Putnam method, 323
deadheading, 392
decision problem, 277
declarative, 1, 3
degenerate, 492
demand constraint, 48, 87, 283

dense data, 79, 115
depth-first search, 229, 331, 443
derivative, right sided, 203
deviation variable, 153
diet problem, 13, 349, 388
differentiable, 203
dimension, 484
discriminating function, 155
disjunct, 139
disjunction, 97, 139
disjunctive constraint, 261
disjunctive linear programming, 253
Disjunctive Normal Form, 273
disjunctive programming, 261, 278
disjunctive requirement, 261
disjunctive set, 261, 490
distributed computing, 452, 473
divide-and-conquer, 395
DNF, 273
`double`, 17, 25
downside risk, 95
Drop/Add heuristic, 142
dual multiplier, 482
dual problem, 479
duality, 30, 67, 122, 203, 391
duality relation, 30
Duality Theorem, 481, 482
dynamic programming principle, 457

E

elimination theorem, 481
ellipsoid method, 2, 277
`else`, 104
empty clause, 321
enumeration, 189, 235
enumerative, 288, 289, 291, 296, 300, 302, 309
Euclidean distance, 136
Euclidean norm, 135
eureka effect, 372, 379, 415
evaluation function, 170
even permutation, 453
Exchange Property, 505, 512
exclusive-or, 395
`exit`, 222
expected value, 94
extreme point, 8, 128, 486

Subject Index

F

fabs, 25
facial disjunctive program, 425
failure, 6, 9
fair, 289, 299, 302
fathomed, 373
feasible region, 6, 480
feasible set of constraints, 6
Fibonacci sequence, 301
find_all, 336
find_max, 349
find_min, 348
First Duality Lemma, 477, 482
first fail strategy, 310, 394
first order logic, 287, 327
fitness, 190
fixed charge problem, 418
flat, 130, 484
Flat Covering Theorem, 130, 131, 135, 152, 489
floor, 24
flow, 112
flow shop problem, 190
formal language, 31, 276
formal proof, 320
formal theorem, 320
FORTRAN, 455
Fourier-Motzkin elimination, 2, 277, 480
Fundamental Theorem of Linear Programming, 485
fuzzy boolean, 404, 421, 424, 427
fuzzy logic, 249, 394, 404, 413

G

GAMS, 408
gap, 387
Gaussian elimination, 2, 277, 477, 481
generalized network, 122
generalized upper bound, 395
generate-and-test, 233
genetic algorithm, 189, 191, 196
global variable, 58, 64
goal constraint, 150
goal node, 229
goal programming, 150
goal state, 311
graph coloring, 144, 281, 301
Great Pyramid of Cheops, 493
greedy algorithm, 57, 201
greedy strategy, 110
gub, 395, 398, 400

H

half space, 3, 135, 477
heap, 463
heuristic, 48, 170
hill-climbing, 174, 178, 185, 188, 291, 353, 505, 511
horizontal condition, 306
Horn clause, 323
hyperplane, 3, 130, 483

I

IDA* search, 463
idempotence, 97, 103
if, 104
ILOG Solver, 250
implication, 97, 103
implicit enumeration, 245
incumbent solution, 343
initial bound, 353
injury method, 289, 311, 371, 378, 379, 391, 398, 466
int, 17, 26
integer knapsack problem, 279, 337
integer programming (IP), 278
integral, 173
integrality requirement, 264, 268
interior, 487
interior point method, 2, 277
interpolated function, 211
intuitionistic disjunction, 139
intuitionistic negation, 165
iterative deepening A* search, 463
iterative deepening search, 369, 444

J

Jackson's preemptive schedule, 121, 346
job shop problem, 177

K

k-consistency, 310
knapsack problem, 56, 99, 100, 109, 110, 201

L

L_∞ norm, 135

L_1 norm, 135
L_2 norm, 135
Lagrange multiplier, 478
lazy evaluation, 103, 140
lb, 216
leaf node, 229
least discrepancy search, 473
left-hand side, 30
lexicographic order, 288
lexicographic rule, 506
lift-and-project method, 425
Lin's algorithm, 188
line, 3, 484
linear algebra, 2
linear closure, 483
linear combination, 18
linear functional, 482
linear hull, 483
linear programming (LP), 2, 277
linear relaxation, 48, 401, 412, 413, 417
LISP, 103, 140, 255, 455
literal, 268
local declaration, 63
local optimum, 185
local search, 52, 142, 165, 175, 188, 281, 443
local variable, 58, 64
log space computable, 328
logical consequence, 329
logical requirement, 261, 378
lookahead, 174, 245, 344, 360
loop body, 40
loop condition, 40, 43
loop control variable, 40, 68
loop header, 39
loop invariant, 283, 284
lower bound, 53
LU decomposition, 511

M

machine learning, 164
makespan, 178, 192, 275
Manhattan distance, 460, 465
marking, 307, 310
mathematical logic, 276, 371
matroid, 506
max, 11, 24, 27
metaheuristic, 189
min, 11, 24, 27
minimum cover problem, 88, 281

mixed integer program (MIP), 278
mixed integer programming (MIP), 278, 282, 401, 408, 411
modus ponens, 320
monotonic, 288, 289, 299, 309
multiperiod problem, 18, 58, 71, 89
multivariable constraint, 54
mutation, 190, 196

N

natural selection, 190
negation, 97, 165
negation-as-failure, 165
net present value, 402
network, 22, 112
network matrix, 116, 499
network problem, 116, 122
network with gains, 122
neural net, 164
nint, 24
node, 22, 112, 118, 122
nonbasic column, 502
nonbasic variable, 502
nondeterminism, 279, 280
nondeterministic polynomial time (NP), 277
nonnegative linear combination, 478
norm, 134
not, 165
nowhere dense set, 488
NP ≠ co-NP, 280
NP-Complete, 279
NP-Hard, 281
n-ply lookahead, 170
N-queens, 304, 312, 332, 333, 363

O

objcdn, 209, 499
objcup, 208, 499
object, 8
objective function, 11
Occam's Razor, 507
offspring, 190
operations research, 2, 233, 276, 282
Optimal Vertex Test, 491, 492
optimization problem, 10
optimum function, 200
or, 140, 227

Subject Index

or loop, 237
orthant, 5
outlier, 159
outsourcing, 33

P

P, 277
P ≠ NP, 279
parallel computing, 452, 473
parameter passed by reference, 67
parameter passed by value, 67
parametric analysis, 169
Parametric Analysis Theorem, 202, 498
Parametric Objective Coefficient Analysis Theorem, 224, 499
parent node, 229
parsing, 5
partial correctness, 283
Pascal, 103, 140, 455
pattern recognition, 155
P-Complete, 328
penalty method, 378, 433
perceptron, 156
Perceptron Learning Theorem, 157
persistent disjunction, 227, 256, 443, 446
perturbation, 199, 496, 505
pigeonhole principle, 322, 415
pivot, 505
point set topology, 483
polygon, 6, 12
polyhedral set, 6, 257, 261, 414, 480
polyhedron, 6, 8, 9, 11, 379, 493
polynomial time, 276
polynomial time hierarchy, 281
polytope, 54, 489
postcondition, 312
precedence, 26
precondition, 312
preemption, 119
pre-solve, 411
pricing, 504
primal problem, 479
printf, 5
priority queue, 459
probabilistic algorithm, 177

probabilistic logic, 28, 30, 327
procedure, 63
product mix problem, 64, 67, 81, 200, 208
progressive roundoff, 51, 142, 171
projection, 479
Prolog, 165, 455
Prolog III, 256
pruning, 229
pseudocost, 386, 407
pseudorandom sequence, 175

Q

quadrant, 5
quadratic assignment problem, 196, 378
quadratic programming, 282
quantifier elimination, 481
queue, 463
quick-and-dirty solution, 174, 406, 424, 428

R

ragged array, 163, 393
random, 175
randomization, 175, 391, 407
range, 204, 498
range constraint, 55
ratio constraint, 76, 199
rc, 204, 498
recursion, 256, 312, 314, 316, 317, 319, 443, 446
recursion theory, 276, 371
reduced cost, 204, 498
reduced cost pricing, 504
reduced problem, 503
reduction, 278
relational operator, 3, 18
reproduction, 190
resolution, 320
resolvent, 321
resource allocation problem, 64
resource constraint, 64
return, 168
revised simplex method, 256, 475, 500
right-hand side, 30
robust choice point, 238, 240, 255

S

satisfiability problem (SAT), 278

satisfying assignment, 268
scalar, 67
scenario, 89
Scheme, 255
search space, 229
seed, 175
segregated storage problem, 433
semantic proof method, 320
semi-definite programming, 302
sensitivity analysis, 199, 349
set covering problem, 84, 171, 391
set partitioning problem, 391
shadow price, 203, 204, 219
shuffle strategy, 391, 407
sibling node, 229, 289, 309
sidetracking, 256
sigma loop, 39
simplex, 500
simplex method, 2, 16, 256, 277, 475, 500
simulated annealing, 188
singular point, 202
slack variable, 152, 208
slice (subarray), 70
smooth, 203
solution set, 261
solvable in polynomial time, 277
sound proof system, 320, 323
soundness, 283, 300
sparse data, 79, 111, 115, 123, 125
specially ordered set of type 1, 400
specially ordered set of type 2, 400
stack, 302, 306, 314, 396, 437
stage one decision, 89
standard form, 500
state, 289, 364, 443
state space search, 311, 363, 443
statement, 42, 64
step assignment, 40, 43
step function, 203
stochastic model, 93
storeback parameter, 171
storeback variable, 171
strict evaluation, 103, 140
strict relational operator, 18, 273
string, 22, 64
structural change, 202
subject_to, 349
success, 9

surplus variable, 152, 208
symbolic constant, 32

T
tabu search, 188, 386
temporal requirement, 264, 273
terminal node, 229
test, 26
then, 104
theoretical computer science, 276
threshold requirement, 264
totally unimodular, 499
trail stack, 310
trailing, 307, 309, 396, 437
training set, 160
transfer theorem, 481
transportation problem, 118, 426
transshipment problem, 112, 118
traveling salesperson problem (TSP), 183, 278, 279
tree search, 229, 289
Turing machine, 276, 328
type, 25

U
ub, 216
uncertainty, 89
uncertainty factor, 102
unit clause, 321
unit resolution, 271, 322
unit vector, 135
upper bound, 53
UR, 323

V
vector (one-dimensional array), 70
vertex, 8, 128, 490
vertical condition, 306
visualization, 473

W
Warren Abstract Machine (WAM), 256
Weak Duality Theorem, 479, 482
what-if, 98, 199
witness point, 8, 24
wp, 24
wrt, 212

WILEY-INTERSCIENCE SERIES IN DISCRETE MATHEMATICS AND OPTIMIZATION

ADVISORY EDITORS

RONALD L. GRAHAM
AT & T Bell Laboratories, Murray Hill, New Jersey, U.S.A.

JAN KAREL LENSTRA
Centre for Mathematics and Computer Science, Amsterdam, The Netherlands
Erasmus University, Rotterdam, The Netherlands

ROBERT E. TARJAN
Princeton University, New Jersey, and
NEC Research Institute, Princeton, New Jersey, U.S.A.

AARTS AND KORST • Simulated Annealing and Boltzmann Machines: A Stochastic Approach to Combinatorial Optimization and Neural Computing
ALON, SPENCER, AND ERDÖS • The Probabilistic Method
ANDERSON AND NASH • Linear Programming in Infinite-Dimensional Spaces: Theory and Application
BARTHÉLEMY AND GUÉNOCHE • Trees and Proximity Representations
BAZARRA, JARVIS, AND SHERALI • Linear Programming and Network Flows
CHONG AND ZAK • An Introduction to Optimization
COFFMAN AND LUEKER • Probabilistic Analysis of Packing and Partitioning Algorithms
DASKIN • Network and Discrete Location: Modes, Algorithms and Applications
DINITZ AND STINSON • Contemporary Design Theory: A Collection of Surveys
GLOVER, KLINGHAM, AND PHILLIPS • Network Models in Optimization and Their Practical Problems
GOLSHTEIN AND TRETYAKOV • Modified Lagrangians and Monotone Maps in Optimization
GONDRAN AND MINOUX • Graphs and Algorithms *(Translated by S. Vajdā)*
GRAHAM, ROTHSCHILD, AND SPENCER • Ramsey Theory, Second Edition
GROSS AND TUCKER • Topological Graph Theory
HALL • Combinatorial Theory, Second Edition
JENSEN AND TOFT • Graph Coloring Problems
LAWLER, LENSTRA, RINNOOY KAN, AND SHMOYS, EDITORS • The Traveling Salesman Problem: A Guided Tour of Combinatorial Optimization
LEVITIN • Perturbation Theory in Mathematical Programming Applications
MAHMOUD • Evolution of Random Search Trees
MARTELLO AND TOTH • Knapsack Problems: Algorithms and Computer Implementations
McALOON AND TRETKOFF • Optimization and Computational Logic
MINC • Nonnegative Matrices
MINOUX • Mathematical Programming: Theory and Algorithms *(Translated by S. Vajdā)*
MIRCHANDANI AND FRANCIS, EDITORS • Discrete Location Theory
NEMHAUSER AND WOLSEY • Integer and Combinatorial Optimization
NEMIROVSKY AND YUDIN • Problem Complexity and Method Efficiency in Optimization *(Translated by E. R. Dawson)*
PACH AND AGARWAL • Combinatorial Geometry
PLESS • Introduction to the Theory of Error-Correcting Codes, Second Edition
SCHRIJVER • Theory of Linear and Integer Programming
TOMESCU • Problems in Combinatorics and Graph Theory *(Translated by R. A. Melter)*
TUCKER • Applied Combinatorics, Second Edition

CUSTOMER NOTE: IF THIS BOOK IS ACCOMPANIED BY SOFTWARE, PLEASE READ THE FOLLOWING BEFORE OPENING THE PACKAGE.

This software contains files to help you utilize the models described in the accompanying book. By opening the package, you are agreeing to be bound by the following agreement:

This software product is protected by copyright and all rights are reserved by the author and John Wiley & Sons, Inc. You are licensed to use this software on a single computer. Copying the software to another medium or format for use on a single computer does not violate the U. S. Copyright Law. Copying the software for any other purpose is a violation of the U. S. Copyright Law.

This software product is sold as is without warranty of any kind, either express or implied, including but not limited to the implied warranty of merchantability and fitness for a particular purpose. Neither Wiley nor its dealers or distributors assumes any liability of any alleged or actual damages arising from the use of or the inability to use this software. (Some states do not allow the exclusion of implied warranties, so the exclusion may not apply to you.)

WILEY